I0055273

Handbook of Bioenergy

Handbook of Bioenergy

Editor: Brian Shelton

CALLISTO REFERENCE

www.callistoreference.com

Callisto Reference,
118-35 Queens Blvd., Suite 400,
Forest Hills, NY 11375, USA

Visit us on the World Wide Web at:
www.callistoreference.com

© Callisto Reference, 2019

This book contains information obtained from authentic and highly regarded sources. Copyright for all individual chapters remain with the respective authors as indicated. All chapters are published with permission under the Creative Commons Attribution License or equivalent. A wide variety of references are listed. Permission and sources are indicated; for detailed attributions, please refer to the permissions page and list of contributors. Reasonable efforts have been made to publish reliable data and information, but the authors, editors and publisher cannot assume any responsibility for the validity of all materials or the consequences of their use.

ISBN: 978-1-64116-173-2 (Hardback)

Trademark Notice: Registered trademark of products or corporate names are used only for explanation and identification without intent to infringe.

Cataloging-in-Publication Data

Handbook of bioenergy / edited by Brian Shelton.
    p. cm.
Includes bibliographical references and index.
ISBN 978-1-64116-173-2
1. Biomass energy. I. Shelton, Brian.
TP339 .H36 2019
662.88--dc23

Table of Contents

Permissions

List of Contributors

Index

Preface

I am honored to present to you this unique book which encompasses the most up-to-date data in the field. I was extremely pleased to get this opportunity of editing the work of experts from across the globe. I have also written papers in this field and researched the various aspects revolving around the progress of the discipline. I have tried to unify my knowledge along with that of stalwarts from every corner of the world, to produce a text which not only benefits the readers but also facilitates the growth of the field.

Bioenergy refers to the renewable energy produced from materials that are derived from biological sources. An organic material that stores sunlight in the form of chemical energy is called biomass. Wood, straw, sugarcane, manure, etc. are some of the most commonly used organic materials for the production of bioenergy. Biomass fuels are often by-products, waste products or residues of animal husbandry, farming and forestry. Corn, soybeans, willow, switchgrass, etc. are some of the plants that are grown for the production of biomass. Microbial fuels cells can be used to attain bioenergy by the conversion of chemical energy stored in wastewater into electrical energy. The conversion is aided by the metabolic processes of electrogenic microorganisms. This book brings forth some of the most innovative concepts and elucidates the unexplored aspects of bioenergy. The various studies that are constantly contributing towards advancing technologies and evolution of this field are examined in detail. Students, researchers, experts and all associated with bioenergy will benefit alike from this book.

Finally, I would like to thank all the contributing authors for their valuable time and contributions. This book would not have been possible without their efforts. I would also like to thank my friends and family for their constant support.

Editor

Transport biofuels in global energy–economy modelling – a review of comprehensive energy systems assessment approaches

ERIK O. AHLGREN [iD], MARTIN BÖRJESSON HAGBERG[a] and MARIA GRAHN

Department of Energy and Environment, Chalmers University of Technology, Göteborg, Sweden

Abstract

The high oil dependence and the growth of energy use in the transport sector have increased the interest in alternative nonfossil fuels as a measure to mitigate climate change and improve energy security. More ambitious energy and environmental targets and larger use of nonfossil energy in the transport sector increase energy–transport interactions and system effects over sector boundaries. While the stationary energy sector (e.g., electricity and heat generation) and the transport sector earlier to large degree could be considered as separate systems with limited interaction, integrated analysis approaches and assessments of energy–transport interactions now grow in importance. In recent years, the scientific literature has presented an increasing number of global energy–economy future studies based on systems modelling treating the transport sector as an integral part of the overall energy system and/or economy. Many of these studies provide important insights regarding transport biofuels. To clarify similarities and differences in approaches and results, the present work reviews studies on transport biofuels in global energy–economy modelling and investigates what future role comprehensive global energy–economy modelling studies portray for transport biofuels in terms of their potential and competitiveness. The results vary widely between the studies, but the resulting transport biofuel market shares are mainly below 40% during the entire time periods analysed. Some of the reviewed studies show higher transport biofuel market shares in the medium (15–30 years) than in the long term (above 30 years), and, in the long-term models, at the end of the modelling horizon, transport biofuels are often substituted by electric and hydrogen cars.

Keywords: comprehensive energy systems assessment approaches, energy–transport interactions, futures, global energy–economy modelling, transport biofuel market shares, transport biofuels, transport sector

Introduction

The high oil dependence and the growth of energy use in the transport sector have increased interest in alternative transport fuels as a measure to mitigate climate change and improve energy security. Local air pollution is also a driver for finding alternatives to conventional petrol and diesel based on crude oil. Alternatives to conventional diesel and petrol include biofuels, hydrogen, electricity or synthetic fuels from, for example coal or natural gas.

Biofuels (in this study 'biofuel' is used to denote biobased transport fuels) currently only contribute to a small share of the energy supply to the transport sector;

while the total global final fuel use in the sector is about 100 EJ (OECD/IEA, 2012), the use of biofuels is only about 2.5 EJ. However, several governments and intergovernmental organizations have policy targets aiming at a future increase in biofuel use; for example, in the EU, the share of fuels from renewable sources in the transport sector should amount to at least 10% of the total transport fuel use by 2020 (EC, 2009, 2015).

While the stationary energy sector (e.g., electricity and heat generation) and the transport sector previously to a large extent could be considered as separate systems with limited interaction, more ambitious energy and environmental targets and an increased utilization of alternative energy carriers in the transport sector can be expected to have system effects over sector boundaries due to several reasons; competition for biomass resources, which can be used both for biofuel production and/or heat/power production (ultimately due to land scarcity); system interactions due to plants

[a]Present address: IVL Swedish Environmental Research Institute, PO Box 530 21, Göteborg, SE 400 14, Sweden

Correspondence: Erik O. Ahlgren
e-mail: erik.ahlgren@chalmers.se

coproducing several outputs, such as biofuels, heat and electricity; and electric cars and hydrogen production based on electrolysis affecting the electricity generation system by increasing demand and, possibly, by evening out the load curve and allowing more intermittent generation. Environmental and climate concerns also stress interaction over sector boundaries as both the stationary energy sector and the transport sector give rise to greenhouse gas (GHG) emissions and fill up the common (politically and/or environmentally set) emission quota. As economical resources are limited, a system-wide allocation strategy is imperative.

Methodological approaches in which the parts of the energy and transport system are investigated separately have been, and still are, common in environmental and energy systems planning and future studies. However, as the importance of dynamic interactions over sector boundaries increases, an expanded systems view in which the coevolution of an integrated energy and transport system is analysed increases in importance. In recent years, a growing number of energy–economy future studies based on systems modelling treating the transport sector as an integrated part of the energy system and/or economy have emerged in the scientific literature.

Global energy–economy systems modelling can be an important tool in future studies on how to achieve a more environmentally friendly transport and energy system. With regards to biofuels, the modelling can give significant insights on feasible future market penetration levels.

Thus, important insights regarding the potential future role of biofuels, with potential system-wide effects taken into account, can be provided. The interpretation and implications of the model results presented in the literature can, however, be complex.

There are several modelling studies applying a system-wide perspective on the future role of biofuels, but synthesis studies in this field are rare. Girod *et al.* (2013) present a modelling analysis on the climate impact of transportation but the presentation of biofuel results is rather limited. In their study, which is not a review but rather a modelling comparison, five global models are run with common global income and population assumptions.

To clarify similarities and differences in approaches and results of modelling studies providing insights on biofuel futures, the present work seeks to review and synthesize studies carried out within this field. Thus, the aim of the study was to determine what future role do comprehensive global energy–economy modelling studies portray for biofuels in terms of their future potential and competitiveness. The specific questions guiding the study are:

- What future utilization levels for biofuels do the studies depict as likely/cost-effective?
- What factors influence differences in results?
- What overall insights can be reached based on the aggregate results of the studies?

This review is based on a systematic selection of studies. The selection criteria are rather restrictive in order to increase chances of drawing valid conclusions based on the selected material. The selection criteria limit the review to scientifically published (in peer-reviewed journals) modelling studies with a global energy system coverage. Only more recent publications (publication after year 2000 and until year 2015) are included. The included studies should also have a comprehensive systems approach treating the transport sector as an integral part of the overall energy system and/or economy. In addition, included studies should be applying a medium-term (15–30 years) to long-term (above 30 years) time horizon. They should further preferably focus on the transport sector or, otherwise, be of relevance from a biofuel perspective (implying that they present biofuel-specific results). These selection criteria have resulted in seventeen studies to be covered by this review (including one, IEA (2008), which not entirely fulfil the selection criteria but which is added since one of the selected studies, Fulton *et al.* (2009), is building upon it and it adds essential material), a sufficiently large number of studies to enable the formation of justifiable general insights.

The studies

The bulk of recently published modelling studies utilizing a global approach and analysing questions related to future use of biofuels are based on bottom-up, optimization energy system modelling. In the models used in these studies, fossil energy resources are generally represented by an, over the studied time period, accumulated available resource base and related extraction costs. Renewable options such as biomass are also limited, but their availabilities are generally linked to a model year, that is a maximum potential use of biomass per year is assumed. The models are to different degrees regionalized; while some models see the world as one global region with, for example unlimited possibilities of trade and allocation of emission reductions between countries and continents, others are disaggregated into different geographical world regions. In the latter case, this allows for the inclusion of model features such as restrictions in trade between regions, regional caps for CO_2 emissions and regional targets for biofuel use. In global models, energy prices are to large degree decided endogenously as a function of the final demand for a certain resource, although the studies also at times include sensitivity analyses of different energy price developments. The studies are briefly presented below:

Takeshita & Yamaji (2008) examine the potential role of FT synfuels in competition with other fuel options, and Takeshita (2012) assesses cobenefits of CO_2 reduction and reduction air pollutants from road vehicles. Both are using the REDGEM70 model.

Turton (2006) describes a sustainable automobile transport scenario using the model ECLIPSE. In the study, multiple sustainable development objectives are taken into account, including continued economic growth with reduced income disparities between different world regions, climate change mitigation and security of energy supply.

Azar et al. (2003), Grahn et al. (2009a,b) and Hedenus et al. (2010) use the GET model to study the cost-effectiveness of optimal fuel choices in the transport sector under various assumptions of future developments of carbon policy, carbon capture and storage, and electricity generation technologies.

Gielen et al. (2002, 2003) study the optimal use of biomass for GHG emissions reductions using the BEAP model.

Gül et al. (2009) utilize a global MARKAL model, denoted the Global Multi-regional MARKAL model (GMM), to analyse long-term prospects of alternative fuels in personal transport, focusing on biofuels and hydrogen. In this study, the bottom-up energy system model is linked to the climate change model MAGICC (in a similar manner as Turton, 2006).

Fulton et al. (2009) present transport-related results and modelling from the IEA study 'Energy Technology Perspectives' (IEA, 2008) in which a combination of the MARKAL-based IEA-ETP model and the IEA Mobility Model (MoMo) is utilized.

Anandarajah et al. (2013) give special focus to the road transport sector (using a version of the TIAM model referred to as TIAM-UCL) and investigate the role of hydrogen and electricity for transport sector decarbonization.

Akashi & Hanaoka (2012) examine the technological feasibility of large cuts in GHG emissions using the AIM/Enduse [Global] model.

Van Ruijven and van Vuuren (2009) explore the energy system impacts of different future hydrocarbon prices, using the global energy model TIMER.

Kitous et al. (2010) present a long-term assessment of the worldwide energy system in scenarios ranging from a baseline to a very low GHG stabilization using the POLES model.

Kyle & Kim (2011) assess global light-duty vehicle (LDV) transport and the implications of vehicle technology advancement and fuel switching on GHG emissions and primary energy demands by using the GCAM model and by simulating five different technology scenarios.

Table 1 summarizes the seventeen selected global modelling studies and some of their respective model features. In section 2.1, the models utilized in the selected studies are presented in more detail, and section 2.2 presents the scenarios applied.

Models utilized

The REDGEM70 model (Takeshita & Yamaji, 2008; Takeshita, 2012) is a bottom-up, global energy systems linear optimization model regionally disaggregated into 70 regions. The model has a long-term time horizon from 2000 to 2100. It considers a number of energy conversion technologies as well as carbon capture and storage (CCS) in power generation, oil refinery and production of synthetic fuels. The model includes several technologies for production of alternative transport fuels, for example hydrogen (H_2), methanol (MeOH), dimethyl ether (DME), Fischer Tropsch (FT)-diesel; bioethanol (EtOH) and biodiesel. The comparably high regional disaggregation level enables capturing of trade flows between world regions and associated distribution and infrastructural costs.

The integrated assessment model ECLIPSE incorporates the energy systems model ERIS with macroeconomic and passenger transport demand models and is further linked to the climate model MAGICC (Turton, 2006). The ERIS model is a bottom-up optimization model for studies of the global energy system. It has been developed to include non-CO_2 GHG emissions, forest sinks and CCS. Furthermore, endogenous technology learning is applied for a number of technologies, meaning that the cost of a technology in the model depends on the level of its deployment.

GET is a bottom-up energy system model based on linear optimization of system cost for the study of long-term development of the global energy system under carbon constraints (Azar et al., 2003; Grahn et al., 2009a,b; Hedenus et al., 2010). It is driven by exogenously given energy demands in four different stationary end-use sectors as well as transportation demands divided into different transport modes. Many published studies using GET focus on cost-effective fuel choices in the transport sector and system-wide effects associated with this. In later applications, the model has been regionalized and the model's heat sector representation has been improved.

The BEAP model (Gielen et al., 2002, 2003) is a further example of a bottom-up optimization (of system cost) global energy systems model. It is based on mixed integer programming, in which the development of the

Table 1 Selected global modelling studies and their related model features. Optimization refers to system cost optimization

Reference	Model – Regionalization	Model characteristics	End-year
Takeshita & Yamaji (2008); Takeshita (2012)	REDGEM70 – 70 regions	Optimization, Partial Equilibrium, Perfect Foresight, Bottom-Up	2100
Turton (2006)	ECLIPSE – 11 regions	Optimization, General Equilibrium, Perfect Foresight, Hybrid, Endogenous Technology Learning, Elastic Demand	2100
Azar et al. (2003); Grahn et al. (2009a,b); Hedenus et al. (2010)	GET – 1; 6/10; 1 region(s)*	Optimization, Partial Equilibrium, Perfect Foresight, Bottom-Up	2100
Gielen et al. (2002, 2003)	BEAP – 12 regions	Optimization, Partial Equilibrium, Perfect Foresight, Bottom-Up, Elastic demand	2040
Gül et al. (2009)	GMM (MARKAL) – 6 regions	Optimization, Partial Equilibrium, Perfect Foresight, Bottom-Up, Endogenous Technology Learning	2100
Fulton et al. (2009), IEA (2008)	ETP (MARKAL) + MoMo (model-linking) – 22 regions (MoMo)	Optimization (ETP)/Simulation (MoMo), Partial Equilibrium, Perfect Foresight, Bottom-Up, Endogenous Technology Learning, Elastic Demand	2050
Anandarajah et al. (2013)	TIAM-UCL (TIMES) – 16 regions	Optimization, Partial Equilibrium, Perfect Foresight, Bottom-Up, Endogenous Technology Learning, Elastic demand	2100
Akashi & Hanaoka (2012)	AIM/Enduse [Global] – 32 regions	Optimization, Partial Equilibrium, Dynamic recursive, Bottom-Up	2050
van Ruijven & van Vuuren (2009)	TIMER – 26 regions	Simulation, System Dynamics, Bottom-Up, Endogenous Technology Learning	2050
Kitous et al. (2010)	POLES – 12 regions	Simulation, Partial Equilibrium, Recursive, Bottom-up, Endogenous Technology Learning, Elastic Demand	2100
Kyle & Kim (2011)	GCAM – 14 regions	Simulation, Partial equilibrium, Dynamic recursive (myopic), Elastic Demand	2095

*The four different studies apply GET model versions with various regionalizations: 1, 6, 10 and 1 regions, respectively.

system is decided through maximization of the sum of the consumers' and producers' surplus. Focusing on biomass systems, the BEAP model covers the global energy, food and materials system and divides the world in 12 regions. The regions are characterized by natural resource availability, labour costs and technology availability. Trade of resources, energy carriers, food products and materials between the regions are possible but result in increased transportation causing additional emissions and costs.

MARKAL is a well-established energy system model framework, which can be combined with different databases and, in such way, form different model applications. MARKAL models are of bottom-up optimization (of system cost) type and generally based on linear programming. The Global Multi-regional MARKAL model (GMM) is a global 6-world region MARKAL model (Gül et al., 2009). GMM has a detailed representation of alternative fuel chains. In terms of biofuels, it includes biodiesel, FT-diesel, ethanol, methanol, DME and synthetic natural gas (SNG) derived from biomass. Several hydrogen production routes are represented, including routes based on biomass gasification.

MoMo is a spreadsheet model aimed at estimating and projecting travel indicators, energy consumption, pollutant emissions and GHGs generated for worldwide mobility (Fulton et al., 2009). In this context, the MoMo model is used to generate transport energy demand projections that are then fed into the IEA-ETP optimization model framework.

The ETSAP-TIAM model is a TIMES-based model representing the global energy system (Anandarajah et al., 2013). TIMES (an acronym for The Integrated MARKAL-EFOM System) is an update of the MARKAL modelling framework. The basics of the two modelling frameworks are the same; that is, also TIMES models can be described as bottom-up energy systems models based on system cost optimization. Compared to MARKAL, TIMES includes several enhanced features, for example a more flexible seasonal and diurnal time division.

The AIM/Enduse model framework (Akashi & Hanaoka, 2012), in a similar manner as MARKAL and TIMES, has been utilized combined with different databases and in different studies to analyse national energy systems as well as the global energy system. The global version of AIM/Enduse model, AIM/Enduse [Global], splits the world into 32 regions over a time horizon from 2005 to 2050. In contrast to earlier mentioned global models, the AIM/Enduse [Global] does not apply perfect foresight but is a dynamic recursive model indicating that technology and fuel selection occur one model year at a time, influenced by previous model years (installed capacities, etc.) but uninformed of future developments regarding energy prices and technology costs.

While the above-described models rely largely on optimization in the choice of future fuel and technologies, three of the selected studies apply models of a more simulatory approach and also seek to incorporate other aspects in the technology choices made. These models are presented below.

The TIMER model, which is part of the integrated assessment model IMAGE, describes the long-term dynamics of the production and consumption of energy carriers in 26 global regions (van Ruijven & van Vuuren, 2009). Here, costs combined with preferences are used in sectoral multinomial logit models in the selection of technologies. The multinomial logit model allocates most of the investments for the technologies with the lowest costs, but if there are other only slightly more costly technologies, a small share of the investment is made into these also (this is in contrast to strict linear programming optimization in which the lowest cost option takes it all if no other constraints apply).

The POLES model can also be described as utilizing a simulating approach. It is a recursive simulation model of the global energy system and has been used in various studies at both national and international levels (Kitous et al., 2010). Integrating a detailed regional, sectorial and technological specification, the POLES model allows assessments of GHG mitigation policies. Explicit technological description is used for secondary fuel production as well as on the demand side for buildings and vehicles. Econometric functions allow evolving consumption patterns to be taken into account. These functions include both behavioural changes and investment decisions.

The GCAM model (previously known as MiniCAM) is a long-term, global, technologically detailed, partial-equilibrium integrated assessment model that includes representations of energy, agriculture, land use and climate systems (Kyle & Kim, 2011). The model calculates an equilibrium for energy goods and services, agricultural goods, land and GHG emissions.

Scenarios applied

Many of the global modelling studies apply climate policies with exogenously determined targets for future atmospheric CO_2 concentration levels. The use of biofuels in the transport sector is contrasted to fossil transport fuels and often also to other potential low-carbon transport options, which generally are based on either hydrogen or electricity. Table 2 summarizes the model input data related to transport sector technology representation and scenario assumptions.

While many of the studies present a number of model scenarios with different input data and assumptions, here we focus on scenarios with stringent climate polices. Most of the studies apply a stabilization target for atmospheric CO_2 concentration, but some studies instead apply an exogenous CO_2 penalty cost. In the latter case, the resulting emissions or CO_2 stabilization level is an output of the model (for comparison purposes, this output has been included in Table 2 within parentheses). The scenarios include climate ambitions from medium (such as 550 ppm CO_2 concentration) to high levels (such as 400 ppm). The assumed biomass potential, that is the maximum amount of biomass that can be used for energy purposes per year in the models, also varies between the studies.

The representation of fuels and technologies in the transport sector is of importance for the outcome of the models and also for how the outcome should be interpreted. Many of the studies treat biofuels in an aggregate way and thus only include a single generic bio-based fuel option: denoted biomass to liquid (BtL), synthetic fuel, methanol or simply 'biofuel'. Other studies include a range of biofuel options. The representation of non-biofuel low-carbon transport fuels as well as vehicle technologies varies between the studies.

Model results

Biofuel utilization

In the presentation of results, summarized in Table 3 and visualized in Figs 1 and 2, four of the 17 studies are excluded: Grahn et al. (2009a), Gielen et al. (2002), Anandarajah et al. (2013) and Kyle & Kim (2011). Gielen et al. (2002) was excluded since the model utilized is the same and scenarios similar to Gielen et al. (2003), and the biofuels presentation is considerably more extensive in the latter. In Anandarajah et al. (2013), it is not possible to identify the biofuel share. Grahn et al. (2009a) as well as Kyle & Kim (2011) present clear biofuel results but only for the light-duty vehicle segment and, thus, their results are not directly comparable with the rest. Further, Fulton et al. (2009) builds upon IEA (2008) and, thus, only the results from Fulton et al. (2009) are presented in Figs 1 and 2.

The resulting biofuel utilization and market shares vary in a wide range. For most model–scenario

Table 2 Climate ambition, biomass potential and fuel and technology representation in road transport for the selected studies. Blanks indicate that info was unclear or could not be obtained

Reference	Climate policy or target	Max biomass per year	Technology representation road transport		
			Biofuels	Other low-carbon options	Vehicle technologies
Takeshita & Yamaji (2008)	550 ppm	300 EJ (2050); 250 EJ (2100)*	Biodiesel, EtOH, biogas, FT- liq., DME, MeOH, H_2	H_2	ICEV, HEV, FCV
Takeshita (2012)	400 ppm	300 EJ (2050); 250 EJ (2100)*	Biodiesel, EtOH, biogas, FT- liq., DME, MeOH, H_2	H_2, Electricity	ICEV, HEV, EV, PHEV, FCV
Turton (2006)	550 ppm	235 EJ (2050); 320 EJ (2100)	H_2, alcohol, FT- liq.	H_2	ICEV, HEV, FCV
Azar et al. (2003)	400 ppm	200 EJ	MeOH, H_2	H_2	ICEV, FCV
Grahn et al. (2009a)	450 ppm	205 EJ	BtL, H2	H2, Electricity	ICEV, HEV, EV, PHEV, FCV
Grahn et al. (2009b)	450 ppm	205 EJ	BtL, H_2	H_2	ICEV, FCV
Hedenus et al. (2010)	400 ppm	200 EJ	Synthetic fuel, H_2	H_2, Electricity	ICEV, HEV, EV, PHEV, FCV
Gielen et al. (2003)	80 \$/tCO_2 cost (75% GHG red. compared to the 1995 level)	Depends on land prices and on costs for intensification of agriculture calculated by the model	MeOH, FT-gasoline, EtOH	No	ICEV
Gül et al.(2009)	450 ppm	195 EJ	Biodiesel, FT-diesel, EtOH, MeOH, DME, bio-SNG, H_2	H_2, Electricity	ICEV, HEV, EV, PHEV, FCV
Fulton et al. (2009) IEA (2008)	450 ppm	Not clear (results = 150 EJ)	Biodiesel, EtOH, BtL (BtL, biodiesel, LC ethanol)	H_2 Electricity	ICEV, HEV, EV, PHEV, FCV
Anandarajah et al. (2013)	Global mean temp. not rise more than 2 °C	Probably about 100 –150 EJ†	Biodiesel, EtOH, H_2	H_2 Electricity	ICEV, HEV, EV, PHEV, FCV
Akashi & Hanaoka (2012)	Cost incr. from 0 to 600 \$/tCO_2 in 2000–2050 (50% GHG red. compared to the 1990 level)	364 EJ	'Biofuel'	H_2, Electricity	ICEV, HEV, EV, PHEV, FCV
van Ruijven & van Vuuren (2009)	100 \$/tCO_2 cost (10–45% CO_2 red. compared to the 1990 level)		'Biofuel'	H_2	ICEV
Kitous et al. (2010)	400 ppm	200 EJ	'Biofuel', H_2	H_2 Electricity	ICEV, HEV, EV, PHEV, FCV
Kyle & Kim (2011)	Cost incr. from 10 to 400 \$/tCO_2 in 2020–2095 (450 ppm)		BtL, biomass-based gas	H_2, Electricity	ICEV, HEV, EV, PHEV, FCV

ICEV, internal combustion engine; HEV, hybrid electric vehicle; FCV, fuel cell vehicle; EV, Electric vehicle (battery-powered); PHEV, plug-in hybrid electric vehicle.
*Supporting info from Takeshita (2009).
†Supporting info from Erb et al. (2009).

combinations, the biofuel share stays below 40% and some of the studies show very low levels (0–10%). Studies showing biofuel market shares above 40% rely not only on 'regular' biofuels but also on hydrogen based on bio-energy with carbon capture and storage (BECCS). Even though market shares for biofuels in

Table 3 Biofuel-related results of global climate policy scenarios*

Reference	Transport and biofuel results for climate policy scenario	Comments and sensitivity
Takeshita & Yamaji (2008)	The utilization of FT products in the transport sector amounts to 21 EJ in 2050 and 78 EJ in 2100. About half of this is FT-kerosene used in aviation. FT production is combined with BECCS after 2070. Petroleum products continue to have a dominating position in the transport sector throughout the century. *Biofuel share 2050: 10%; (transport)* *Biofuel share 2100: 23% (transport)*	High biopotential; medium CO_2 reduction. In the stationary sector, H_2 produced from biomass accounts for a significant part of the energy use. Likewise to the FT synfuel production, H_2 production is combined with BECCS after 2070. High total final transport energy demand (340 EJ in 2100) lowers biofuel share, although the biofuel use in absolute terms is high.
Takeshita (2012)	Electricity and biomass-derived FT products gain market shares starting from 2040. In 2050, use of FT products from biomass in road transport is about 2 EJ and, in 2100, 13 EJ. At the end of the century, remaining parts are petroleum products (68 EJ), electricity (39 EJ) and a small amount of H_2 (1 EJ). *Biofuel share 2050: 2% (road transport)* *Biofuel share 2100: 11% (road transport)*	High biopotential; high CO_2 reduction. The share of plug-in hybrids in light-duty vehicles reaches 90% in 2100. CCS and fuel switching are mentioned as important CO_2 reduction measures in the stationary sectors.
Turton (2006)	Oil and gas dominate transport fuel supply in first half of the century, but then a large increase in biofuels is seen. In 2100, biomass to alcohol accounts for about 55 EJ, or 26%, of transport sector final energy use; biomass to H_2 accounts for about 86 EJ, or 41% of transport sector final energy use. H_2 is produced primarily with BECCS. *Biofuel share 2050: 6% (transport)* *Biofuel share 2100: 67% (transport)*	High biopotential; medium CO_2 reduction. A large increase in nuclear is allowed in the scenario. This makes nuclear dominate the electricity system (nuclear electricity generation amounts to 220 EJ in 2100). Direct thermal needs are supplied mainly by a combination of gas, H_2 and electricity (rather than biomass or coal). Electric vehicles are unavailable in the model.
Azar et al. (2003)	Oil remains the only fuel in transport (excluding trains) until 2040–2050 when a transition to H_2 begins. In 2100, H_2 is the only fuel used in transport. H_2 is produced from fossil fuels with CCS and from solar energy. No biofuels enters the scenario. *Biofuel share 2050: 0% (transport)* *Biofuel share 2100: 0% (transport)*	Low biopotential; high CO_2 reduction. Higher H_2-related costs, larger biomass potential or restrictions for bio-industrial heat give a transient period with biofuels. Nuclear is restricted to current levels and a conservative potential for CCS is assumed. Electric vehicles are unavailable in the model.
Grahn et al. (2009b)	With regional CO_2 emission caps (RC), the biofuel utilization peaks at 2050 with 15 EJ and goes down to 8 EJ in 2100. Total transport fuel use adds up to 223 EJ in 2100. Of this, 56% is non-biomass-based H_2 and remaining parts are primarily natural gas and petroleum products. A global CO_2 cap gives lower biofuel utilization (3 EJ in 2100). *Biofuel market share 2050: 9% (transport) - RC* *Biofuel market share 2100: 4% (transport) - RC*	Low biopotential; high CO_2 reduction. Sensitivity analysis shows that biofuel usage peak at medium CO_2 reduction targets and that higher biomass supply potential increases biofuel use in results. If HEVs, PHEVs and BEVs are included, biofuel use decreases. In the study, nuclear is restricted to current levels and a conservative potential for CCS is assumed.
Hedenus et al. (2010)	Around 2040 biofuel PHEVs are introduced in LDV transport and dominate this sector after 2070. For heavy vehicles, a shift from diesel ICE to H_2 FCVs occurs around 2050. In 2100, 27 EJ of biofuel is used. Total final energy use in transport is 194 EJ. H_2 accounts for about half of the supply and electricity about 20%. Natural gas and petroleum products account for the remaining part. Solar thermal energy dominates both the electricity sector and H_2 production. *Biofuel share 2050: 10% (transport)* *Biofuel share 2100: 14% (transport)*	Low biopotential; high CO_2 reduction. Nuclear and CCS are unavailable in the base scenario. Alternative scenarios in which nuclear and CCS dominate the electricity sector, the biofuel utilization in 2100 is 52 EJ (26%) and 81 EJ (35%), respectively. Compared with other GET model versions (Azar et al., 2003) and Grahn et al. (2009b), the use of biomass for high temperature industrial heat is restricted.

(continued)

Table 3 (continued)

Reference	Transport and biofuel results for climate policy scenario	Comments and sensitivity
Gielen *et al.* (2003)†	Use of biofuels (ethanol, methanol and synthetic diesel/gasoline) and natural gas-based methanol increase over time. In 2020, approximately 50 EJ gasoline/diesel, 39 EJ biofuels and 22 EJ methanol (based on natural gas) are used in the transport sector. *Biofuel share 2020: 35% (road transport)* *(Biofuel share 2050: 70% (road transport))*	High biopotential; high CO_2 reduction. Majority of the biomass used is allocated for the production of transport fuels. Less stringent CO_2 reduction scenarios reduce biofuel utilization. The model lacks low-carbon options in the transport sector other than biofuels (such as electricity or H_2).
Gül *et al.* (2009)‡	Biofuel production (for all sectors, but primarily transport) peaks at 31 EJ around 2075 and then decreases to 14 EJ in 2100. H_2 becomes the main transport fuel and FCVs dominate the personal transport sector. Favoured H_2 production technology is coal-based production with CCS, but also H_2 production from nuclear and wind power via electrolysis are major sources. *Biofuel share 2050: 25% (of vehicle km in personal road transport)* *Biofuel share 2100: 7% (of vehicle km in personal road transport)*	Low biopotential; high CO_2 reduction. With medium CO_2 reduction (550 ppm), no dip in biofuel production is seen at the end of the century. Biofuel production is 34 EJ in 2100. High total energy demand; primary energy demand is close to 1700 EJ in 2100. Nuclear accounts for 400 EJ of this (about 150 EJ electricity) and (non-bio) renewables 400 EJ.
Fulton *et al.* (2009), IEA (2008)	For the so-called BLUE map scenario, about 29 EJ biofuel is used in transport. Further, 13 EJ H_2, 12 EJ electricity and about 57 EJ petroleum products are used. For the next 10–15 years, cane ethanol from Brazil is mentioned as a low-cost biofuel option, while over time, lingo-cellulosic ethanol and FT fuels are highlighted. *Biofuel share 2050: 26% (transport)*	Low biopotential; medium/High CO_2 reduction. In 2050, around 25% substitution of liquid fossil fuels by biofuels is seen in several different climate policy scenarios. CCS and nuclear account for about half of the electricity generation in 2050. Other important sources are solar, wind and hydro.
Anandarajah *et al.* (2013)	Biofuels play a minor role. H_2 accounts in 2050 for around 20% of transport energy consumption. Electricity plays a major role and is used in both plug-in hybrid vehicles and battery electric vehicles. H_2 is mainly produced from centralized large coal plants with CCS in the medium term while in the longer term, electrolysis plays a key role. *Biofuel share: not clear (but low)*	Low biopotential; high CO_2 reduction. Bioenergy is prioritized for use in the power generation and industry, often in combination with CCS. With more biomass available, deployment of bio-CCS is increased. If CCS is not an available option, use of biomass as heating fuel and biomass use in industry increase (rather than biofuel production).
Akashi & Hanaoka (2012)	HEV passenger cars are introduced on a large scale after 2015 and reach more than 60% of the market by 2035 (share of pkm). FCVs are rapidly deployed after 2035. In 2050, the transport biofuel use (excluding H_2) is about 50 EJ. H_2 produced from biomass with BECCS amounts to 13 EJ. The remaining part, 75 EJ, is mainly petroleum products (although small amounts of natural gas and electricity are also seen). *Biofuel share 2050: 45% (transport)*	High biopotential; medium/high CO_2 reduction. Wind, solar, biomass and hydro together account for about 75% of the total power generation in 2050. Increase in nuclear capacity is restricted (an increase of about 150% from 2005 is allowed). In the results, a major shift from coal to gas occurs in industry (no biomass).
van Ruijven & van Vuuren (2009)	Exogenously forced low, medium and high fossil fuel price scenarios are tested. In the high price scenario with climate policy, the biofuel use is 50 EJ in 2030 but decreases as more fuel efficient vehicles and H_2, produced from coal with CCS, are introduced. In 2050, the use of biofuels is about 27 EJ (23%), and the remaining part is primarily H_2. Lower fossil fuel prices give somewhat higher use of biofuels, significantly less use of H_2 and higher use of petroleum products. *Biofuel share 2050: 23–27% (transport)*	Medium/high CO_2 reduction. Exogenous prices imply that there will be no response in oil prices due to less oil demand. The authors point out that this is only likely if the high oil prices are caused by depletion. If not, the analysis represents an initial effect which will be partly cancelled out by price decreases in the longer run. Nuclear and CCS are allowed large shares in electricity generation.

Table 3 (continued)

Reference	Transport and biofuel results for climate policy scenario	Comments and sensitivity
Kitous *et al.* (2010)	About 10% of the total biomass use is used for production of biofuel and H_2 throughout the studied time horizon (should correspond to about 10–12 EJ at the end of the century). In 2050–2100, electric and plug-in vehicles account for almost 60% of the total light vehicle stock and, in 2100, H_2-fuelled cars (both ICE and FC) have a 35% market share. H_2 production is primarily based on nuclear. *Biofuel share: not clear (but low)*	Low biopotential; high CO_2 reduction. Biomass (with CCS) and other renewables account for around 65% of the electricity generation. Remaining part is primarily based on nuclear and natural gas. About 80% of the total biomass use is in electricity generation at the end of the century.
Kyle & Kim (2011)	In scenarios dominated by liquid hydrocarbons in LDV transport, biomass accounts for about 10% of LDV primary energy supply, or about 7 EJ in 2050 and 10 EJ in 2095. Other primary energy carriers to the sector include crude oil, unconventional oil, coal and natural gas. CCS is applied. *Biofuel share 2050: <10% (LDV transport)* *Biofuel share 2095: <10% (LDV transport)*	High CO_2 reduction. Study focuses on primary energy supply rather than final energy use. Input data or results for stationary energy system are not explicit.

Biofuel shares are shown in italics.

*High, medium and low biomass potentials refer to >300, 250–300 and <250 EJ annually, respectively. High, medium and low CO_2 reduction refer to CO_2 atmospheric concentration stabilization levels of 450 ppm or less, 500–550 ppm and above 550 ppm, respectively.

†Supporting info from Grahn *et al.* (2007).

‡Supporting info from Gül *et al.* (2009).

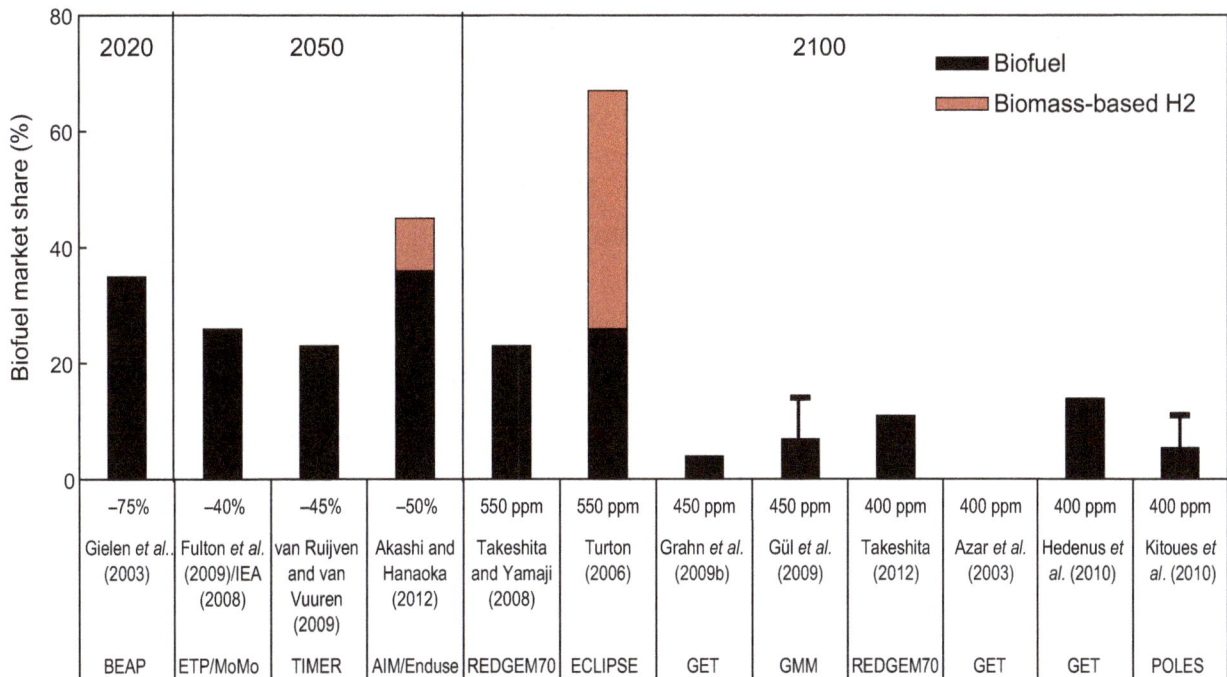

Fig. 1 Biofuel market shares at the model end-year for the thirteen comparable reviewed global studies. The presented shares for Gül *et al.* (2009) and Kitous *et al.* (2010) have been calculated assuming a total final transport energy use of 200 EJ, and the uncertainty bars correspond to 100–300 EJ of total final energy (due to lack of directly comparable data in the articles).

most of the scenarios stay at low–medium levels (0–40%), many of the scenarios show a significant increase in biofuel use in absolute terms compared with today's

level of 2.5 EJ (out of the total final transport sector fuel use of about 100 EJ; OECD/IEA, 2012). Thus, the results suggest an increase in biofuel use compared with

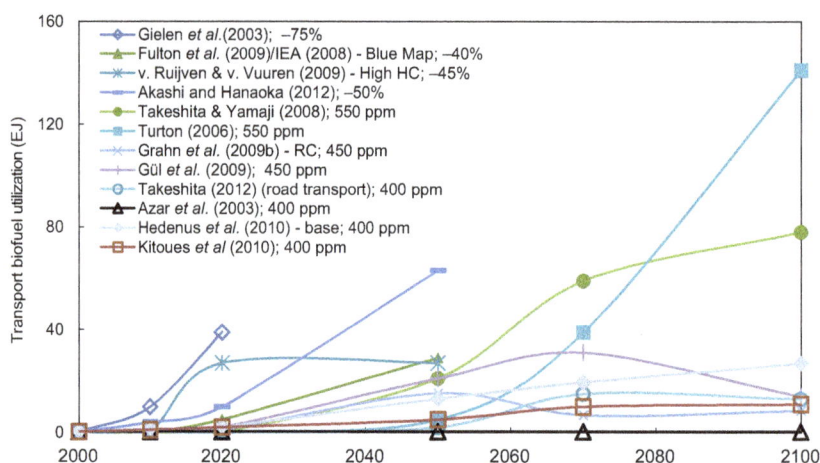

Fig. 2 Development of biofuel utilization over the modelled time horizons for each of the thirteen studies. After the reference, the end-year CO2 atmospheric concentration constraint is provided.

today's level but show, at the same time, that biofuels tend to not dominate the future transport sector.

Many of the studies only include a single aggregate biofuel option and, thus, provide no insights in regard to which biofuel type is preferable. Among the studies that do point out specific biofuel options, Takeshita & Yamaji (2008) and Takeshita (2012) highlight FT liquids (synthetic diesel, gasoline and kerosene) as an advantageous alternative, partly due to its potential to fuel the aviation sector. Akashi & Hanaoka (2012) and Turton (2006) point out bio-hydrogen combined with BECCS and Turton (2006) also favour bio-alcohol over FT liquids. Fulton et al. (2009) mention ethanol as well as FT liquids.

Factors influencing the biofuel utilization

From comparing the scenario results of the different studies, factors of importance for the global biofuel utilization can be identified. These include the assumed biomass potential, the assumed climate ambition and the model technology representation for the transport sector as well as for the stationary energy system.

The future potential availability of biomass for energy purposes depends on competition for land and water including land use and biodiversity issues, food demand as well as agricultural productivity, which all are linked to large uncertainties. The reviewed global modelling studies show significant differences in regard to assumed biomass potentials. For example, Akashi & Hanaoka (2012) and Turton (2006), at the end of their modelled time horizons, assume biomass potentials of 364 EJ and 320 EJ, respectively, while Grahn et al. (2009b) and Kitous et al. (2010) assume levels around 200 EJ. This could be one reason explaining that the former present a widespread use of biofuels in their

results, while the latter show significantly lower shares of biofuels.

Several of the studies also highlight biomass availability as a central constraint for the utilization of biofuels. Gül et al. (2009) conclude that the key limiting factor for a further deployment of biofuels is the availability of biomass and that biomass is more cost-effectively utilized in electricity and heat production in a carbon-constrained world. Sensitivity analyses testing robustness show that an increased biomass supply potential generally also increases the deployment of biofuels under stringent climate scenarios (e.g., Azar et al., 2003; Grahn et al., 2009a,b although there are exceptions (Anandarajah et al., 2013).

In regard to technology representation in the transport sector, the availability of low-carbon options in addition to biofuels is of significance for the competitiveness of biofuels and, in particular, optimism with regard to the development of hydrogen FCVs and/or electric vehicles does reduce the competitiveness of biofuels. As the models generally apply a long time horizon and often assume decreasing costs for new technologies over time, this is particularly true towards the end of the studied time horizons.

Turton (2006) and Akashi & Hanaoka (2012) are among the studies obtaining the highest biofuel utilization (together with Takeshita & Yamaji (2008)). As shown in Fig. 1, this is a result of utilization of both 'conventional' biofuels and a considerable share of biomass-based hydrogen production in combination with BECCS. Several studies exclude the latter alternative (hydrogen production with BECCS) in their models. Whether this option is included or not is of relevance for the competitiveness of biomass-based hydrogen production compared with non-biomass-based options.

Not only is the representation of technology options in the transport sector of significance for the resulting biofuel utilization, but also the technology representation of the stationary energy system. The availability of future low-cost, non-biomass-based low-carbon electricity generation can be significantly contributing to a high biofuel use, as this lowers the demand for biomass in the stationary energy system. In particular, this can be seen in scenarios allowing a high use of nuclear power generation and/or electricity generation based on CCS (the two low-carbon electricity generation options with a high potential and lowest cost in the reviewed studies). Assumptions regarding these technologies and their future deployment differ widely partly due to political and public acceptance issues.

Another aspect of technology representation in the stationary energy system of importance for the resulting biofuel utilization is to what degree biomass can supply industrial process heat demands. When Hedenus *et al.* (2010) increase the level of detail in regard to representation of process heat demand and introduce limitations for the amount of biomass allowed in the GET model, this results in higher biofuel utilization than in other GET modelling studies. Similar limitations may be of significance also in other models.

The impact of the assumed climate objectives on the biofuel utilization is not entirely straightforward. Generally, no-policy scenarios show a low use of bioenergy in general and biofuels in particular due to the availability of cheaper energy sources, such as coal.

With increasing climate ambitions and thus higher CO_2 emission penalties, bioenergy increases in competitiveness compared with fossil fuel options. For 'medium' climate ambitions (e.g., 550 ppm), a certain amount of biofuels is also cost-effective in many of the reviewed studies. However, for very stringent climate targets, results are more diverse. Grahn *et al.* (2009b) and Gül *et al.* (2009) suggest that the cost-effective biofuel usage tends to peak at medium CO_2 reduction targets. While fossil-based transport fuels are likely to dominate at less ambitious reduction targets, more stringent targets increase the cost-effective biofuel usage, but with CO_2 reduction targets in line with a 450-ppm atmospheric CO_2 concentration stabilization or 2-degree maximum temperature increase, the models tend to choose other low-carbon options for the transport sector (hydrogen and/or electricity) and biomass resources are instead allocated to heat and power production in the stationary energy system. There is also a time aspect to this as, in order to meet CO_2 stabilization targets at the end of the century, emission reductions get more stringent over time. This suggests that biofuels could be seen as a bridging technology to other low-carbon options such as hydrogen and/or electricity (Gül *et al.*, 2009).

As already indicated in the above sections, time-related aspects can influence the biofuel utilization. Studies applying a shorter time horizon often obtain higher biofuel utilization than studies applying a longer time horizon (see Figs 1 and 2). This is mainly due to assumptions of development (cost reductions and improvements in technical performance) of new alternative technologies over time.

Finally, we are presenting outcomes of the comparison and analysis of the reviewed modelling studies stressing the quantitative conclusions and with a particular emphasis on the importance of the above-discussed critical factors and assumptions. These outcomes may be summarized in the following six points:

- Only studies assuming high biomass potentials (an annual maximum potential of at least 300 EJ) result in biofuel market shares of 35% or more).

- Five of the six studies assuming low biomass potentials (250 EJ or less) result in low (below 10%) biofuel market shares.

- Only studies resulting in a considerable bio-based H_2 deployment also result in high (at least 40%) biofuel market shares.

- All long-term (end-year around 2100) studies assuming large GHG reduction (atmospheric CO_2 concentration stabilization of 450 ppm CO_2 or less) result in low (below 10%) biofuel market shares.

- Three of the eight long-term (end-year around 2100) studies show that the amount of biofuel utilization passes through a maximum and then decreases towards the end of the modelling period.

- Of the eight long-term (end-year around 2100) modelling studies, the two studies with the lowest climate policy ambition (atmospheric CO_2 concentration stabilization of 550 ppm CO_2) show the highest and most strongly increasing biofuel utilization.

Discussion

The presented review provides insights into levels and characteristics of biofuel futures and on factors influencing biofuel deployment. It demonstrates that energy–economy modelling studies portray a diverse picture in regard to future biofuel utilization with shares in most cases ranging from low levels to medium levels (up to about 40%) at the end of the modelled time horizon.

Not all studies are explicit about the type of chosen biofuel but some trends emerge. Generally, liquid wood-based second-generation biofuels and, more specifically, FT liquids are options highlighted in several of the studies. The possibility of using existing infrastructure and vehicles is, in these cases, probably of high significance, but also the combined production based on the FT process of jet fuels (for the aviation sector) and synthetic diesel/gasoline (for the road transport sector) is pointed out as valuable.

A number of factors influencing the resulting biofuel utilization in the modelling results have been identified. These are mainly climate ambition/policies, the technology representation in the transport sector as well as in the stationary energy sector and the assumed biomass potential. As the models cover long time horizons and the conditions often change over time, there is also a time aspect to many of the mentioned factors (e.g., technology costs, CO_2 reduction requirements and energy prices).

The *climate ambition/policy* (the level of GHG reduction constraints or emission cost penalties) is relevant for how much of the available biomass is used. With higher climate ambition, the proportion of the total biomass potential that is used increases.

The *technology representation*, that is what technologies that are available in the model, to what relative costs and to what potential, determines the allocation of biomass. The relative cost of alternative technologies is complex and varies with scarcity rents and CO_2 penalties, which, in turn, are functions of the climate ambition. This relates to biofuels in relation to other technologies in the transport sector as well as in the stationary energy system, but also between different biofuel options. For example, favourable assumptions regarding non-biomass-based low-carbon electricity generation, such as CCS or nuclear power, imply a low demand for biomass in the stationary system and, in many cases, this means more available biomass for biofuel production. On the contrary, a high (allowed) potential and low costs for hydrogen or electricity-based transportation will decrease the competitiveness of biofuels. A high total *biomass supply potential* can imply that the potential of the most cost-effective biomass usage can be filled and still leave biomass resources to other, less cost-effective, alternatives.

The resulting biofuel utilization depends on several factors and there are considerable differences between the studies. Differences are in many cases due to quantitative assumptions regarding more or less uncertain input data. While this highlights challenges with quantitative long-term future modelling of energy-economic systems, it also demonstrates a strong relevance of the same: without making quantitative statements regarding parameters such as biomass potentials, system-wide CO_2 reduction objectives and cost of alternative technologies, not much can be said about the effective future contribution of biofuels from an overall systems perspective.

In this review of future studies based on global energy systems modelling, we find that the future market penetration of biofuels range from low (0–10%) to high levels (above 40%) in the reviewed model results. Most of the studies show low to intermediate biofuels market shares (below 40%) at the end of the studied time horizons for climate policy scenarios not including sector-specific polices. The total biofuel market share exceeds 40% only in studies resulting in large-scale deployment of bio-based hydrogen.

Factors influencing biofuel utilization in the model results include biomass potential, climate ambition/policies, technology representation in the transport sector and in the stationary energy sector, oil price and energy policies in addition to GHG-related constraints or penalties.

Although biofuels tend to not dominate the transport sector at the end of the modelled time horizons, compared with today's level, many model studies show a significant increase in biofuel use. Besides biofuels, the development and deployment of energy-efficient vehicle technologies, such as hybrids and fuel cell vehicles (in the longer term), are essential in many of the future transport scenarios.

Acknowledgements

This study resulted out of a project within the Swedish Knowledge Centre for Renewable Transportation Fuels (f3), financed by the Swedish Energy Agency, the Region Västra Götaland and the f3 Partners, including universities, research institutes and industry (see www.f3centre.se).

References

Akashi O, Hanaoka T (2012) Technological feasibility and costs of achieving a 50% reduction of global GHG emissions by 2050: mid- and long-term perspectives. *Sustainability Science*, 7, 139–156.

Anandarajah G, McDowall W, Ekins P (2013) Decarbonising road transport with hydrogen and electricity: long term global technology learning scenarios. *International Journal of Hydrogen Energy*, 38, 3419–3432.

Azar C, Lindgren K, Andersson BA (2003) Global energy scenarios meeting stringent CO_2 constraints— cost-effective fuel choices in the transportation sector. *Energy Policy*, 31, 961–976.

EC (European Commission) (2009) Directive 2009/28/EC of 23 April 2009 on the promotion of the use of energy from renewable sources. *Journal of the European Union*, 16–62.

EC (European Commission) (2015) Directive (EU) 2015/1513 of the European Parliament and of the Council of 9 September 2015 amending Directive 98/70/EC relating to the quality of petrol and diesel fuels and amending Directive 2009/28/EC on the promotion of the use of energy from renewable sources.

Erb K-H, Haberl H, Krausmann F et al. (2009) Eating the planet: feeding and fuelling the world sustainably, fairly and humanely - a scoping study. Institute of Social Ecology and PIK Potsdam, Vienna.

Fulton L, Cazzola P, Cuenot F (2009) IEA Mobility Model MoMo and its use in the ETP 2008. *Energy Policy*, 37, 3758–3768.

Gielen DJ, Fujino J, Hashimoto S, Moriguchi Y (2002) Biomass strategies for climate policies? *Climate Policy*, 2, 319–333.

Gielen D, Fujino J, Hashimoto S, Moriguchi Y (2003) Modeling of global biomass policies. *Biomass and Bioenergy*, **25**, 177–195.

Girod B, van Vuuren DP, Grahn M, Kitous A, Kim SH, Kyle P (2013) Climate impact of transportation A model comparison. *Climatic Change*, **118**, 595–608.

Grahn M, Azar C, Lindgren K, Berndes G, Gielen D (2007) Biomass for heat or as transportation fuel? A comparison between two model-based studies. *Biomass and Bioenergy*, **31**, 747–758.

Grahn M, Azar C, Williander MI, Anderson JE, Mueller SA, Wallington TJ (2009a) Fuel and vehicle technology choices for passenger vehicles in achieving stringent CO$_2$ targets: connections between transportation and other energy sectors. *Environmental Science and Technology*, **43**, 3365–3371.

Grahn M, Azar C, Lindgren K (2009b) The role of biofuels for transportation in CO$_2$ emission reduction scenarios with global versus regional carbon caps. *Biomass and Bioenergy*, **33**, 360–371.

Gül T, Kypreos S, Turton H, Barreto L (2009) An energy-economic scenario analysis of alternative fuels for personal transport using the Global Multi-regional MARKAL model GMM. *Energy*, **34**, 1423–1437.

Hedenus F, Karlsson S, Azar C, Sprei F (2010) Cost-effective energy carriers for transport – The role of the energy supply system in a carbon-constrained world. *International Journal of Hydrogen Energy*, **35**, 4638–4651.

IEA (International Energy Agency) (2008) Energy Technology Perspectives 2008 IEA, Paris. Available at: http://www.iea.org (accessed 15 May 2016).

Kitous A, Criqui P, Bellevrat E, Chateau B (2010) Transformation patterns of the worldwide energy system - scenarios for the century with the POLES model. *The Energy Journal*, **31**, 49–82.

Kyle P, Kim SH (2011) Long-term implications of alternative light-duty vehicle technologies for global greenhouse gas emissions and primary energy demands. *Energy Policy*, **39**, 3012–3024.

OECD/IEA (2012) World Energy Outlook 2012, OECD/IEA, Paris.

van Ruijven B, van Vuuren DP (2009) Oil and natural gas prices and greenhouse gas emission mitigation. *Energy Policy*, **37**, 4797–4808.

Takeshita T (2009) A strategy for introducing modern bioenergy into developing Asia to avoid dangerous climate change. *Applied Energy*, **86**, S222–S232.

Takeshita T (2012) Assessing the co-benefits of CO$_2$ mitigation on air pollutants emissions from road vehicles. *Applied Energy*, **97**, 225–237.

Takeshita T, Yamaji K (2008) Important roles of Fischer-Tropsch synfuels in the global energy future. *Energy Policy*, **36**, 2773–2784.

Turton H (2006) Sustainable global automobile transport in the 21st century: an integrated scenario analysis. *Technological Forecasting and Social Change*, **73**, 607–629.

2

Solving the multifunctionality dilemma in biorefineries with a novel hybrid mass–energy allocation method

SYLVESTRE NJAKOU DJOMO[1] (iD), MARIE TRYDEMAN KNUDSEN[1], RANJAN PARAJULI[1], MIKAEL SKOU ANDERSEN[1], MORTEN AMBYE-JENSEN[2], GERFRIED JUNGMEIER[3], BENOÎT GABRIELLE[4] and JOHN ERIK HERMANSEN[1]

[1]Department of Agroecology, Aarhus University, Blichers Allé 20, P.O. Box 50, DK-8830 Tjele, Denmark, [2]Department of Engineering, Aarhus University, Hangovej 2, P.O. Box 239, DK-8200 Aarhus Denmark, [3]Joanneum Research Forschungsgesellschaft mbH, Elisabethstraße 18/II, 8010 Graz, Austria, [4]EcoSys Research Unit, AgroParisTech, INRA, F-78850 Thiverval-Grignon, France

Abstract

Processing biomass into multifunctional products can contribute to food, feed, and energy security while also mitigating climate change. However, biorefinery products nevertheless impact the environment, and this influence needs to be properly assessed to minimize the burden. Life cycle assessment (LCA) is often used to calculate environmental footprints of products, but distributing the burdens among the different biorefinery products is a challenge. A particular complexity arises when the outputs are a combination of energy carrying no mass, and mass carrying no energy, where neither an allocation based on mass nor on energy would be appropriate. A novel hybrid mass–energy (HMEN) allocation scheme for dealing with multifunctionality problems in biorefineries was developed and applied to five biorefinery concepts. The results were compared to results of other allocation methods in LCA. The reductions in energy use and GHG emissions from using the biorefinery's biofuels were also quantified. HMEN fairly distributed impacts among biorefinery products and did not change the order of the products in terms of the level of the pollution caused. The allocation factors for HMEN fell between mass and economic allocation factors and were comparable to energy allocation factors. Where the mass or the energy allocation failed to attribute burdens, HMEN addressed this shortcoming by assigning impacts to non-mass or to nonenergy products. Under the partitioning methods and regardless of the feedstock used, bioethanol reduced GHG by 72–98% relative to gasoline. The GHG savings were 196% under the substitution method, but no GHG savings occurred for sugar beet bioethanol under the surplus method. Bioethanol from cellulosic crops had lower energy use and GHG emissions than from sugar beet, regardless of the allocation method used. HMEN solves multifunctional problems in biorefineries and can be applied to other complex refinery systems. LCA practitioners are encouraged to further test this method in other case studies.

Keywords: allocation, biochemical, biofuels, biomaterial, biorefinery, GHG emissions, HMEN method, life cycle assessment

Introduction

To cope with population growth and the rapid depletion of fossil resources, the EU and the Unites States have proclaimed their interest in strengthening green growth in the bioeconomy (EC, 2012, The White House, 2012). More than 30 countries have expressed their intentions to increase their reliance on biological resources (Bosch et al., 2015), so the share of biomaterials and bioenergy is expected to increase in the coming decades. The development of biorefineries is crucial for achieving the transition to a bioeconomy. Biorefineries convert biomass into food, energy, chemicals, and materials (Sacramento-Rivero et al., 2016). They can contribute to sustainable resources use and so conserve finite resources, while mitigating climate change and other impacts. However, environmental impacts occur during the production and conversion phases of biomass into energy and bio-based materials (Creutzig et al., 2015; Gerssen-Gondelach et al., 2016), and these impacts should be reflected in the environmental assessment of individual biorefinery products.

To support environmental claims about biorefineries, a life cycle assessment (LCA) is used (Cherubini & Jungmeier, 2010; Nuss & Gardner, 2013; Pereira et al., 2015; Silalertruksa et al., 2017), but as only one product is often of interest for the LCA, environmental loads associated with a biorefinery system are often split among all biorefinery products using appropriate

Correspondence: Sylvestre Njakou Djomo
e-mail: sylvestre.njakoudjomo@agro.au.dk

allocation methods. Allocation in multifunctional processes has been extensively discussed in the literature (Azapagic & Clift, 1999), and several methods for solving these problems have been proposed (Jungmeier et al., 2002). The LCAs for petroleum refineries and for bioenergy and biorefinery systems are, in some cases, sensitive to allocation methods (Wang et al., 2004; Börjesson, 2009; Gnansounou et al., 2009; Luo et al., 2009; Cherubini et al., 2011). The choice of an adequate method is still a contentious issue, and an arbitrary choice can lead to incorrect LCA results (Reap et al., 2008) and thus poor decision-making (Weidema, 2000).

Studies on petroleum refineries and bioenergy systems have so far used mass (Gabrielle & Gagnaire, 2008), energy (Huo et al., 2009), and exergy (Cherubini et al., 2008) as parameters to allocate resource use and GHG emissions to the products and coproducts of these systems. Mass, energy, and exergy allocations are often considered to be based on physical parameters because they use measurement units such as weight or energy content as their basis. But the direct application of these methods to biorefineries is challenging because of the complexity of biorefining processes, the large number of end products, and the diversity of their functions. Mass allocation is unsuitable for nonmass products like electricity, while, without further assumptions, the energy allocation would not work for nonenergy products (e.g., fertilizers) that do not have a heating value (Singh et al., 2010). The exergy allocation is complex as a result of difficulties in establishing exergy values for some substances (Cherubini et al., 2011).

The economic allocation (Spirinckx & Ceuterick, 1996; Guinée et al., 2004) and the linear programming method (Azapagic & Clift, 1999; Babusiaux, 2003; Pierru, 2007; Hirshfeld & Kolb, 2012; Elgowainy et al., 2014; Balakrishnan et al., 2015) have also been used to split burdens among petroleum refinery, bioenergy, or biorefinery products. The economic allocation considers the financial incentives, which are the main drivers of production and associated impacts. But the method cannot be applied to systems where coproducts do not yet have a market or where market prices fluctuate (Wang et al., 2011). The linear programming method models the physical and technical relationships between the inputs and outputs and environmental burdens of the system. It follows similar logics of economic allocation (Bredeson et al., 2010) and provides detailed information on the operations within the biorefinery (Balakrishnan et al., 2015). But the approach does not work if there is a fixed ratio between products and coproducts because the functional outputs cannot be varied independently (Azapagic & Clift, 1999). Linear programming is also data intensive, and the data needed may not be easily accessible.

Other studies avoid allocation either via system subdivision (Furuholt, 1995; Wang et al., 2004; Bredeson et al., 2010), substitution (Kim & Dale, 2002; Eriksson et al., 2007), or using the system expansion approach (Njakou Djomo et al., 2015). The subdivision method disaggregates processes into subprocesses within a given system and splits off those that are relevant to the functional output. It captures the differences in environmental loads of producing individual products at the next sublevel (Wang et al., 2004). But biorefinery processes are very integrated and cannot be meaningfully split into subprocesses. The use of the substitution method requires the identification of the main product and the coproducts. The main product is then allocated the entire burden, but also credited with the impacts that the coproducts can avoid by replacing other products on the market. However, in the case of the biorefinery, the identification of the main product is not obvious because the overall idea of biorefining is to utilize synergies in the production with the purpose of obtaining multiple products (Parajuli et al., 2015).

The system expansion method broadens the boundary to the point where allocation is not needed and where the compared systems cover the same functional unit (Ahlgren et al., 2015). However, system expansion is difficult to apply when aggregating different functions. Neither does it show impacts of individual products, as its results refer to a group of functions rather than a single product or function. Finally, a small number of studies have used the surplus method (Fu et al., 2003; Pimentel & Patzek, 2005) to overcome allocation problems with bioenergy systems. This latest method identifies and assigns all burdens to the main product. In this method, coproducts are burden-free and thus considered as waste products. This, however, seems to be a simplification of the reality for the reason that some biorefinery coproducts have well-established markets (e.g., protein and lignin) and cannot thus be regarded as wastes (Maes et al., 2015).

As none of the methods above are without drawbacks when applied to biorefineries, new methods for solving multifunctionality problems in biorefineries could help to settle the debate. In response to this situation, a number of researchers proposed a framework to deal with allocation in biorefining (Cherubini et al., 2011; Sandin et al., 2015). Although appealing, their approaches may be criticized in that they fail to reduce the number of options that LCA practitioners are faced with and thus increase the risk of controversies. For many authors, mass- and energy-based allocation methods are very attractive as they are simple and based on clear and defensible rationales (Pelletier et al., 2015). They provide an engineering perspective to allocation within a given refinery (Wang et al., 2004). However, they are not directly applicable to biorefinery systems for the reasons mentioned above. Therefore, a new method based on physical parameters and which is an improvement over the current mass and energy allocation is needed. Here, we present a new hybrid mass–energy 'HMEN' allocation method and the results from its application to different biorefinery concepts. To test the robustness of this new approach, its results were compared to results of other allocation methods in LCA. Finally, gasoline and natural gas were used to illustrate the effects of replacing conventional fossil fuels with bioethanol and biogas.

Materials and methods

Biorefinery models

Different biorefinery concepts exist today, with different pathways for biomass conversion and different final products. Simple biorefinery concepts use current available technologies to convert, via a platform, a single biomass feedstock into two or three marketable products. Complex biorefinery concepts use novel technologies to convert, via a number of platforms, a variety of biomass feedstocks into several marketable products (IEA, 2009). Both energy (e.g., biofuels) and nonenergy products (e.g., feed, biomaterial, and/or biochemicals) from simple

and complex biorefineries were evaluated from cradle-to-gate in this study (Fig. 1).

Allocation framework

To overcome the limitations associated with allocation based solely on energy content or mass, a new hybrid mass–energy (HMEN) allocation method was developed (Fig. 1). This new allocation method captures well the differences in energy efficiency between biorefinery systems. A dispatch factor for energy $\alpha = \frac{\sum \eta_i}{\eta_0}$ was derived from the following relationship: $\eta_0 = \sum \eta_i + \sum \eta_j$, where η_0 is the overall energy efficiency of the biorefinery, $\sum \eta_i$ is the efficiency of the energy stream, and $\sum \eta_j$ the efficiency of the material stream. As this relation equals to $\frac{\sum \eta_i}{\eta_0} + \frac{\sum \eta_j}{\eta_0} = 1$ when its two sides are divided by η_0, it thus follows that $1 - \alpha$ represents the dispatch factor for the material stream (Table 1). Based on the abovementioned dispatch factors, the classical equations for distributing burdens between energy and nonenergy products can be modified as shown in Eqns (1) and (2).

$$\Phi_i = \alpha \frac{\theta_i}{\sum_i^n \theta_i} \qquad (1)$$

$$\Phi_j = (1 - \alpha) \frac{\lambda_j}{\sum_j^m \lambda_j} \qquad (2)$$

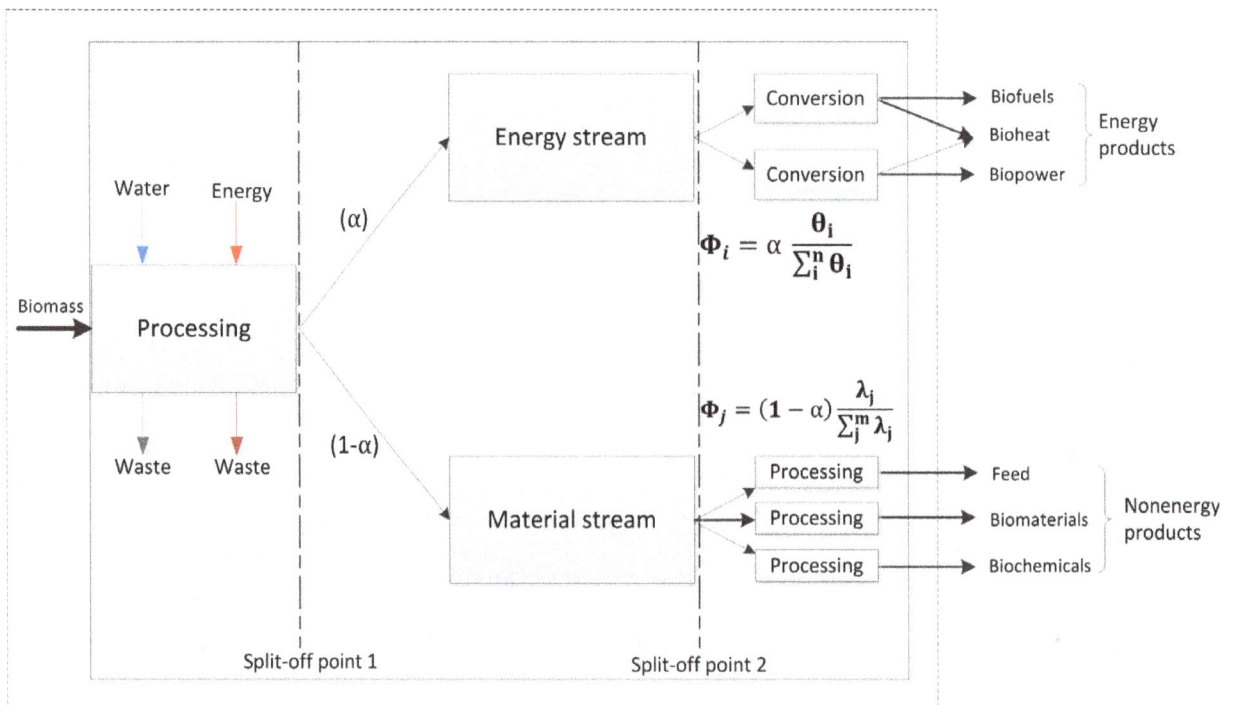

Fig. 1 System boundary and schematic representation of the hybrid mass–energy (HMEN) method. At split-off point 1, a dispatch factor (α) which divides a biorefinery into energy and material streams is computed. At the split-off point 2, the dispatch factor is combined with mass and energy allocation method to derive the allocation coefficients. The dotted line represents the system boundary.

Table 1 Key figures for the processing and conversion of the different feedstocks in the selected biorefinery plants

Biorefinery 1 (BioR₁)

Component	Mass (t h⁻¹)	Energy (GJ h⁻¹)
Inputs		
Straw	3.44	57.6
Enzymes	0.034	0
Outputs		
2G Bioethanol	0.573	16.0
Lignin pellets	0.47	10.1
C5 molasses*	0.965	15.0

Performance	%
Overall efficiency	71.3
Efficiency, energy stream	45.3
Dispatch factor, energy (α)	64.0

Biorefinery 2 (BioR₂)

Component	Mass (t h⁻¹)	Energy (GJ h⁻¹)
Inputs		
Straw	44.1	635.0
Manure	71.3	42.81
Biowaste	43.5	27.13
MSW	21.4	223.6
Enzyme	2.14	0
Outputs		
2G bioethanol	8.58	231.7
Biogas	1.69	30.39
Biomethane	1.68	72.2
Lignin pellets	4.18	92.04
Digestate*	3.65	80.2
Heat	111.5	251.9
Electricity	0	35.81

Performance	%
Overall efficiency	85.5
Efficiency, energy stream	77.0
Dispatch factor, energy (α)	90.0

Biorefinery 3 (BioR₃)

Component	Mass (t h⁻¹)	Energy (GJ h⁻¹)
Inputs		
Rapeseed	2.5	92.4
Water	1.1	0.0
Steam	0.2	0.36
Electricity	0	0.45
Outputs		
1G Bio-oil	1.0	37
Protein products*	0.61	9.6
Soapstock*	0.02	0.4
Fiber*	0.76	12.0
Ash†	0.03	0.0

Performance	%
Overall efficiency	63.4
Efficiency, energy stream	40.0
Dispatch factor, energy (α)	63.0

Biorefinery 4 (BioR₄)

Component	Mass (t h⁻¹)	Energy (GJ h⁻¹)
Inputs		
Sugar beet	103.5	1760
Catch crop & leaves	64.3	1093
Lime (CaO)	7.5	0
Carbon dioxide	11.5	0
Steam	15.6	35.2
Electricity	0	7.3
Outputs		
1G bioethanol	14.5	388
Biogas	45.0	799
Sugar*	43.1	646
Protein products*	12.9	193
Foam earth* (CaCO₃)	8.5	0
Carbon dioxide†	15.0	0

Performance	%
Overall efficiency	70.0
Efficiency, energy stream	41.0
Dispatch factor, energy (α)	59.0

Biorefinery (BioR₅)

Component	Mass (t h⁻¹)	Energy (GJ h⁻¹)
Inputs		
Green grass	0.29	5.2
Silage grass	0.17	3.0
Manure	0.24	4.6
Electricity	0	0.09
Heat	0.14	0.32
Outputs		
Biogas	0.17	3.11
Fiber*	0.09	1.46
Coagulated Protein*	0.07	1.32
Digestate*	0.30	6.6

Performance	%
Overall efficiency	95.0
Efficiency, energy stream	24.0
Dispatch factor, energy (α)	25.0

The overall energy efficiency is calculated as the total energy output of biorefinery products (i.e., energy and nonenergy products) divided by the total energy input.

The efficiency of the energy stream is computed as the sum of energy output of the energy products divided by the total energy input.

The dispatch factor for energy (α) is calculated by dividing the efficiency of the energy stream by the overall energy efficiency.

Note that for the BioR₁ and BioR₂, part of the biomass input is used to produce the energy needed for the process, so no external energy is required.

*Represents the nonenergy products which can be materials or chemicals.

†Represents waste (i.e., products with no economic value).

In Eqns (1) and (2) above, θ_i denotes the energy content of a given energy product i from a multifunctional process, and λ_j is the mass of a given nonenergy product j from a multifunctional process, while $\sum_i^n \theta_i$ and $\sum_j^m \lambda_j$ represent the total energy content and the total mass of the generated energy and nonenergy products, respectively. The indices n and m represent the number of energy and nonenergy products, respectively. If $\alpha = 0$, the biorefinery produces only material or chemical products; if $\alpha = 1$, the biorefinery generates only an energy product, and if $0 < \alpha < 1$, the biorefinery system yields both energy and material/chemical products. Note that Eqns (1) and (2) are reduced to a classical energy or mass allocation method, respectively, when $\alpha = 1$ or $\alpha = 0$.

Finally, the share of environmental burdens of a bioenergy/biofuel product from the biorefinery can be calculated as indicated in Eqn (3). Similarly, the environmental burdens assigned to a biomaterial/biochemical product from the biorefinery can be computed using Eqn (4):

$$E_i = \Phi_i E_T = \left(\alpha \frac{\theta_i}{\sum_i^n \theta_i} \right) \cdot E_T \qquad (3)$$

$$E_j = \Phi_j E_T = \left((1 - \alpha) \frac{\lambda_j}{\sum_i^m \lambda_j} \right) \cdot E_T \qquad (4)$$

where E_i is the share of environmental burdens for a given energy product, E_j represents the share of environmental impacts for a given biomaterial/biochemical product from the biorefinery, and E_T is the total environmental impact of the biorefinery system. Note that the sum of all allocated burdens should equal the total environmental impact generated by the biorefinery (i.e., 100% rule).

Case study selection

The HMEN method was used to estimate and compare the share of energy use and GHG emissions of biorefinery products. Five biorefinery plants were selected, based on the criteria that (i) they generate both energy and nonenergy products, and (ii) there are data available on the overall efficiency of the system. The selected biorefinery plants differed in terms of feedstock used and product outputs (Table 1). The first biorefinery plant (BioR$_1$) converts straw into bioethanol, lignin pellets, and molasses (Larsen et al., 2012). The second plant (BioR$_2$) transforms a mixture of straw, manure, industrial waste and biowaste to bioethanol, biogas, biomethane, heat, power, lignin, and digestate (MEC I/S, 2015). The third plant (BioR$_3$) processes oilseed rape grains into bio-oil, meal cake, and soapstock (Schneider & Finkbeiner, 2013), while the fourth plant (BioR$_4$) converts sugar beet and catch crop leaves into bioethanol, biogas, sugar, protein products, and foam earth (CaCO$_3$) (ECN, 2010). The fifth plant (BioR$_5$) uses grass, grass silage, and manure to produce biogas, fiber, protein products, and fertilizers (ECN, 2010).

Because the impacts from biomass production can be larger than those from conversion (Jungmeier et al., 2002), the study also covers all relevant agricultural operations as well as upstream production of inputs to these operations. Biorefinery inputs such as enzymes, electricity, and heat were included

where relevant in the calculation (Table 1). However, changes in soil carbon stock due to land use changes were excluded from the analysis, as were the storage and end use of biorefinery products. As manure, industrial waste, and biowaste are wastes, only the energy use and GHG emissions related to their transport to the biorefinery plants were considered. A transport distance of 50 km was assumed for all biomass feedstock. Data on energy use and GHG emissions during the production and transport of biomass were derived from the Ecoinvent database (Ecoinvent, 2014), while the data on electricity and heat used for enzyme production were derived from Dunn et al. (2012).

Comparison with other allocation methods

The robustness of the HMEN method was assessed by comparing its results to those of the mass, energy, economic allocation, surplus method, system expansion, and the substitution method. For the substitution approach, it was assumed that biofuels represent the main product of the biorefinery in each case study. The economic allocation was based on market prices, while the energy allocation was based on the lower heating values of the different biorefinery products (Table S1).

Comparison of biorefinery biofuels and conventional fuels

The energy use and GHG emissions of the different biofuels (i.e., bioethanol, biogas, biomethane) were compared to those of conventional fuels. To this end, it was assumed that bioethanol replaces gasoline, while biogas or biomethane replaces natural gas. It was further assumed that biofuels are carbon neutral because the CO_2 emitted during the combustion of biofuels corresponds to the CO_2 uptake by the feedstock during their growth. Gasoline production uses 52 GJ t^{-1} and emits about 624 kg CO_2 t^{-1} (Ecoinvent, 2014), whereas its combustion releases an additional ~3341 kg CO_2 t^{-1} (Cherubini & Jungmeier, 2010). Natural gas production consumes 50 GJ t^{-1} and emits 340 kg CO_2 t^{-1} (Ecoinvent, 2014), while about 2805 kg CO_2 t^{-1} is emitted during natural gas combustion. Assuming an energy content of 43 GJ t^{-1} for gasoline and 27 GJ t^{-1} for bioethanol, 1 t bioethanol can thus replace 0.62 t gasoline. Similarly, 1 t biogas can displace 0.35 t natural gas if assuming an energy content of 18 and 52 GJ t^{-1} for biogas (@ 55% CH$_4$) and natural gas (@ 92% CH$_4$), respectively. The energy content of biomethane (@ 96% CH$_4$) is 43 GJ t^{-1}, so 1 t biomethane can displace 0.83 t natural gas (Table S1).

Results

The allocation coefficients, the energy use, and GHG emissions computed using the HMEN and other allocation procedures are shown in Table 2. Under the mass allocation approach, liquid biofuels (e.g., bioethanol) received a lower allocation factor because of their lower weights relative to the other biorefinery products (Table 2). However, under the economic allocation, liquid biofuels were assigned high allocation coefficients because of the high ratio of price relative to the ratio of differences in mass between liquid biofuels and other cogenerated products. For example, the ratio of the price differences between bioethanol and lignin pellets was 10 times higher than the mass ratio (i.e., 0.33) between these two biorefinery products. The energy allocation and the HMEN method gave similar weights to all energy products, but their allocation factors for nonenergy products differed when these products represented a significant share of the coproducts (Table 2). Importantly, where mass and energy allocation failed to distribute burdens to certain products because of their zero mass value or energy content, the HMEN approach addressed this shortcoming by assigning actual burdens to these products (Table 2). This shows how problematic LCA results can be when mass or energy allocation is applied to biorefineries. Indeed, there is a disproportionate advantage for products without energy content (e.g., foam earth) over products with energy content when the energy allocation was used. The same was true for products with mass over products without mass (e.g., electricity) when the mass allocation was adopted (Table 2). Although the allocation factors differed between the HMEN approach and the other partitioning methods (i.e., mass or economic allocation), the estimates of energy use and GHG emissions per product were in most cases within the same order of magnitude (Table 2). HMEN also had the same prioritization of biorefinery products as the energy allocation in all cases. Its ranking of products was also similar to that of the mass allocation method in nearly all cases, but differed from the economic allocation in some cases. The similarity observed between the HMEN method and the energy allocation in this study reflects the underlying conceptual linkage between the two approaches. In fact, dispatch factors were computed using the energy efficiency, which is, in turn, based on the total energy outputs and energy inputs of the biorefinery systems.

The substitution method credited the main products (i.e., biofuels) with the impacts generated by the alternative goods displaced by their coproducts. No allocation factors were computed here, but because the total avoided burdens exceeded in some cases the overall impacts of the biorefineries, negative estimates for energy use and GHG emissions were obtained (Table 2). This contrasted with the partitioning methods, which all computed a positive total physical energy use and GHG emissions. Such results suggest that when the substitution approach is used, some biorefinery products may become a net sink for energy and/or GHG emissions even before a comparison of these bio-based products with their conventional counterparts.

Figure 2 shows the comparison of the different allocation methods used in this study. Estimates of other allocation schemes were compared to those of the surplus method set as a reference. For simplicity, only biorefineries producing bioethanol were selected (i.e., $BioR_1$, $BioR_2$, and $BioR_4$). The comparison between the mass, energy, economic, and HMEN method showed only small differences among them. Indeed, all estimates of energy use (Fig. 2b) and GHG emissions (Fig. 2a) were within the same order of magnitude. Under the substitution approach, it was clear that avoided impacts by some of the coproducts more than compensated for the overall environmental impacts of some biorefinery products. Figure 2 also shows that the substitution method is not only sensitive to the choice of the conventional good displaced by the coproducts, but also to the number and type of coproducts generated. In fact, large energy and GHG credits were given to the main product when the biorefinery system generated many energy products as in the $BioR_2$ case (Fig. 2).

The variability in specific energy use and GHG emissions of biorefinery products is presented in Fig. 3. These estimates were obtained by dividing the energy use and GHG emissions in Table 2 by the amount of final products of each biorefinery plant in Table 1. Given that substitution provided only results for the main product, the choice of the main product was varied in each case study to obtain estimates of specific energy use and GHG emissions for all products. Figure 3 shows that the energy use and GHG emissions vary widely depending on the allocation method adopted. Under the partitioning approach (i.e., mass, energy, economic, HMEN), the energy use of bioethanol from straw ranged from 1.7 to 2.9 GJ t^{-1} (Fig. 3c), while its GHG emissions varied from 149 to 247 kg CO_2 t^{-1} (Fig. 3a). For straw–bioethanol, the substitution approach computed values of −15.9 GJ t^{-1} for energy use (Fig. 3d) and 155.6 kg CO_2 t^{-1} for GHG emissions (Fig. 3b), while the energy use and GHG emission values calculated using the surplus method were 4.8 GJ t^{-1} and 412.6 kg CO_2 t^{-1}, respectively (Fig. 3b,d).

The energy use of bioethanol from mixed biomass sources ranged from 0.4 to 2.3 GJ t^{-1} (Fig. 3c), while the GHG emissions varied from 32.9 to 181.8 kg CO_2 t^{-1} (Fig. 3a) under partitioning approaches. Under the surplus method, the energy use was 6.3 GJ t^{-1} (Fig. 3d) and GHG emissions were 503.2 kg CO_2 t^{-1} (Fig. 3b). Negative values were obtained for energy use (−66.1 GJ t^{-1}) and GHG emissions (−2358.6 kg CO_2 t^{-1}) when the substitution method was chosen (Fig. 3d,b). Under the partitioning approach, sugar beet bioethanol consumed 1.4–1.9 times more nonrenewable energy and emitted two to 2.8 times more GHGs than straw-based bioethanol (Fig. 3a,c). When the surplus

Table 2 Allocation methods and implications for energy and GHG emissions for biorefinery products

Biorefinery	Products	Partitioning — Mass allocation			Energy allocation			Economic allocation			HMEN approach			Surplus method			Avoiding allocation — System expansion		Substitution	
		Coef (%)	Energy (GJ h^{-1})	GHG (kg CO_2 h^{-1})	Coef (%)	Energy (GJ h^{-1})	GHG (kg CO_2 h^{-1})	Coef (%)	Energy (GJ h^{-1})	GHG (kg CO_2 h^{-1})	Coef (%)	Energy (GJ h^{-1})	GHG (kg CO_2 h^{-1})	Coef (%)	Energy (GJ h^{-1})	GHG (kg CO_2 h^{-1})	Energy (GJ h^{-1})	GHG (kg CO_2 h^{-1})	Energy (GJ h^{-1})	GHG (kg CO_2 h^{-1})
BioR₁	2G bioethanol*	0.36	1.00	85.49	0.39	1.07	92.03	0.60	1.65	141.51	0.39	1.07	92.03	1.00	2.76	236.41	–	–	-9.09	89.18
	Lignin pellets	0.03	0.08	7.01	0.25	0.68	58.10	0.04	0.11	9.59	0.25	0.68	58.10	0	0	0	–	–	-10.83	-56.8
	C₅ molasses	0.61	1.68	143.91	0.36	1.01	86.28	0.36	1.00	85.31	0.36	1.01	86.28	0	0	0	–	–	-1.02	-90.43
	Total	1.00	2.76	236.41	1.00	2.76	236.41	1.00	2.76	236.41	1.00	2.76	236.41	1.00	2.76	236.41	2.76	236.41	2.76	236.41
BioR₂	2G bioethanol*	0.07	3.54	282.19	0.29	15.8	1259.64	0.36	19.57	1559.5	0.29	15.8	1259.64	1.00	54.2	4317.8	–	–	-567.31	-20237.3
	Biogas	0.013	0.70	55.58	0.04	2.07	165.19	0.02	1.33	105.94	0.04	2.07	165.19	0	0	0	–	–	-29.51	-199.4
	Biomethane	0.013	0.69	55.39	0.09	4.92	392.4	0.04	2.10	164.1	0.09	4.92	392.45	0	0	0	–	–	-70.3	-475.1
	Lignin pellets	0.03	1.72	137.48	0.12	6.28	500.29	0.02	0.83	66.48	0.12	6.28	500.29	0	0	0	–	–	-99.36	-520.9
	Heat	0.85	46.01	3667.17	0.32	17.19	1369.71	0.39	21.3	1695.8	0.32	17.19	1369.71	0	0	0	–	–	-302.4	-16732.1
	Electricity	0	0	0	0.05	2.44	194.65	0.13	7.06	562.31	0.05	2.44	194.65	0	0	0	–	–	-105.25	-5083.6
	Digestate	0.03	1.51	120	0.10	5.47	435.93	0.04	2.05	163.75	0.10	5.47	435.94	0	0	0	–	–	-14.66	-1544.0
	Total	1.00	54.2	4317.8	1.00	54.2	4317.8	1.00	54.2	4317.8	1.00	54.2	4317.8	1.00	54.2	4317.8	54.2	4317.8	54.2	4317.8
BioR₃	2G Bio-oil*	0.42	5.59	826.97	0.63	8.37	1238.42	0.81	10.85	1605.4	0.63	8.36	1236.95	1.00	13.4	1976.5	–	–	11.11	1737.7
	Meal cake	0.26	3.51	519.34	0.16	2.18	321.99	0.18	2.42	358.85	0.17	2.26	334.11	0	0	0	–	–	-2.066	-223.32
	Fiber	0.31	4.15	613.61	0.20	2.72	402.66	0.002	0.031	4.55	0.20	2.67	394.76	0	0	0	–	–	-0.100	-8.90
	Soapstock	0.01	0.11	16.54	0.01	0.09	13.39	0.004	0.051	7.62	0.01	0.072	10.64	0	0	0	–	–	-0.084	-6.542
	Total	1.00	13.4	1976.5	1.00	13.4	1976.5	1.00	13.4	1976.5	1.00	13.4	1976.5	1.00	13.4	1976.5	13.4	1976.5	13.4	1976.5
BioR₄	1G bioethanol*	0.12	35.58	4351.7	0.19	56.51	7112.28	0.27	79.34	9986.1	0.19	56.5	7112.3	1.00	295.1	37137.8	–	–	-636.81	20723.4
	Biogas	0.36	107.08	13477.4	0.39	116.37	14646.15	0.29	84.64	10652.2	0.39	116.4	14646.2	0	0	0	–	–	-785.67	-5310.7
	Sugar	0.35	102.49	12899.4	0.32	94.09	11841.57	0.34	101.35	12755.2	0.28	81.6	10274.4	0	0	0	–	–	-120.5	-8407.9
	Protein products	0.10	30.602	3851.55	0.10	28.109	3537.81	0.09	26.3	3309.8	0.08	24.4	3067.7	0	0	0	–	–	-24.2	-2613.2
	Foam earth	0.07	20.322	2557.72	0	0	0	0.01	3.45	434.39	0.05	16.2	2037.2	0	0	0	–	–	-1.54	-82.7
	Total	1.00	295.1	37137.8	1.00	295.1	37137.8	1.00	295.1	37137.8	1.00	295.1	37137.8	1.00	295.1	37137.8	295.1	37137.8	295.1	37137.8
BioR₅	Biogas*	0.27	0.26	24.47	0.25	0.241	20.5	0.60	0.58	49.14	0.25	0.24	20.5	1.00	0.97	82.2	–	–	-0.40	-63.6
	Coagulated protein	0.11	0.11	9.09	0.11	0.10	8.69	0.016	0.02	1.33	0.11	0.11	9.4	0	0	0	–	–	-0.16	-17.78
	Fiber	0.14	0.14	11.69	0.12	0.113	9.61	0.002	0.002	0.153	0.15	0.14	12.1	0	0	0	–	–	-0.01	-1.10
	Digestate	0.47	0.46	39	0.53	0.511	43.4	0.384	0.372	31.60	0.49	0.47	40.3	0	0	0	–	–	-1.2	-126.9
	Total	1.00	0.97	82.2	1.00	0.97	82.2	1.00	0.97	82.2	1.00	0.97	82.2	1.00	0.97	82.2	0.97	82.2	0.97	82.2

*Considered as the main product of the biorefinery.

†Note that in the case of substitution, the total energy use (or GHG emissions) is calculated as the difference between the energy use (or GHG emissions) of the main product and that of the coproducts.

method was selected, the energy used was 20.3 GJ t^{-1} (Fig. 3a), while the GHG emissions were 2555.9 kg CO_2 t^{-1} (Fig. 3b). Estimates of energy use and GHG emissions for sugar beet bioethanol were −43.8 GJ t^{-1} and 1426.2 kg CO_2 t^{-1}, respectively, when the substitution method was adopted (Fig. 3b,d).

Overall, the variability in energy use and GHG emissions across the allocation methods suggested the latter had a strong influence on the environmental performance of biorefinery products, but the degree of the influence depended on the biomass feedstock utilized. Some trends were, however, uncovered: In all cases, bioethanol from mixed biomass sources (i.e., manure + straw + food waste) consumed less nonrenewable energy and emitted less GHGs than both bioethanol from straw and sugar beet. Bioethanol from straw had better environmental performances than sugar beet bioethanol. In general, biofuels from cellulosic crops (i.e., bioethanol, biogas, biomethane) used less energy and emitted less GHGs than bioethanol from sugar beet or bio-oil from rapeseed (Fig. 3). Likewise, protein products from cellulosic crops performed better than those produced from sugar beet and rapeseed. The same observation was true for fiber from cellulosic and sugar beet crops. Finally, digestates from cellulosic biomass had a lower impact both in terms of energy use and GHG emissions than digestate from sugar beet (Fig. 3).

Relative to gasoline and under the partitioning methods, all bioethanol reduced the energy use (82–99%) and GHG emissions (72–98%). The largest reduction in energy used (93–99%) occurred when mixed biomass was the feedstock, while the lowest reduction in energy use (82–93%) was linked to sugar beet. Bioethanol from mixed biomass also achieved the largest reduction in GHG emissions (93–98%), while bioethanol from sugar beet had the lowest reduction in GHG emissions (72–88%). Small savings in energy use (37%) and a small increase in GHG emissions (4%) relative to gasoline were observed when the surplus method was used, but under the substitution method, savings in energy use and reduction in GHG emissions reached 304% and 196%, respectively. Compared to natural gas and under the partitioning methods, all biogas reduced the energy use by 80–98% and GHG emissions by 70–97%. As with bioethanol, little to no reduction in either energy use or GHG emissions was achieved when the surplus method was adopted as allocation method. However, when the substitution approach was chosen, the maximum saving in energy used was 2855%, while the maximum reduction in GHG emissions was 1366%. These results showed that allocation methods influenced the savings of biofuel relative to conventional fuels.

Discussion

The HMEN method presented in this study overcomes the limitations of allocation methods based solely on the mass or energy content of biorefinery products, which both suffer from drawbacks. In particular, this new, hybrid method was able to assign environmental impacts to the products with no mass (such as electricity) or no energy content (e.g., earth-foam), which is a major improvement over classical allocations (Table 2). This means that LCA practitioners using these schemes run the risk of overlooking important environmental burdens for certain biorefinery products, a risk which the new method (HMEN) can handle and mitigate (Table 2). Compared to the economic allocation, another widely used method for biorefineries, HMEN is still based on the physical relationship between products and is therefore independent of price fluctuations. Because it uses only the physical flows within biorefineries to split burdens between products, it can limit the freedom of choice and thus the risks of controversy around the outcomes of the evaluation. The use of dispatch factors (i.e., energy content) as a weighting factor is not only consistent with the hypothesis that energy consumption is tied to the amount of mass transported, but also consistent with the conservation of energy during biorefining processes. However, such use of energy content as a weighting factor does not provide information on the degradation of energy or resources during a process, nor does it quantify the usefulness or quality of the various material streams flowing through a system and exiting as products and/or wastes (Wang et al., 2004). Nevertheless, the HMEN method is more robust than the mass and energy allocation currently used in LCA of biorefineries (Table 2). Like the HMEN approach, the economic allocation method distributed impacts among biorefinery products. Differences in estimates between the economic allocation and the HMEN method were in most cases insignificant; for bioethanol, for example, the differences in allocated energy use and GHG emissions were less than a factor of 1.5 (Table 2). Although the economic allocation captures some underlying motivation to produce different biorefining products, the potential market fluctuations (Malça & Freire, 2006), the lack of a physical basis for using market values as weighting factors (Wang et al., 2004), the difficulty of applying this method when no market experience exits, and the uncertainties inherent in this approach favor the use of the HMEN method for attributional LCAs of biorefining products.

The HMEN method performs at a high level of resolution (i.e., at biorefinery level) with regard to process streams in biorefineries. This means it does not allocate

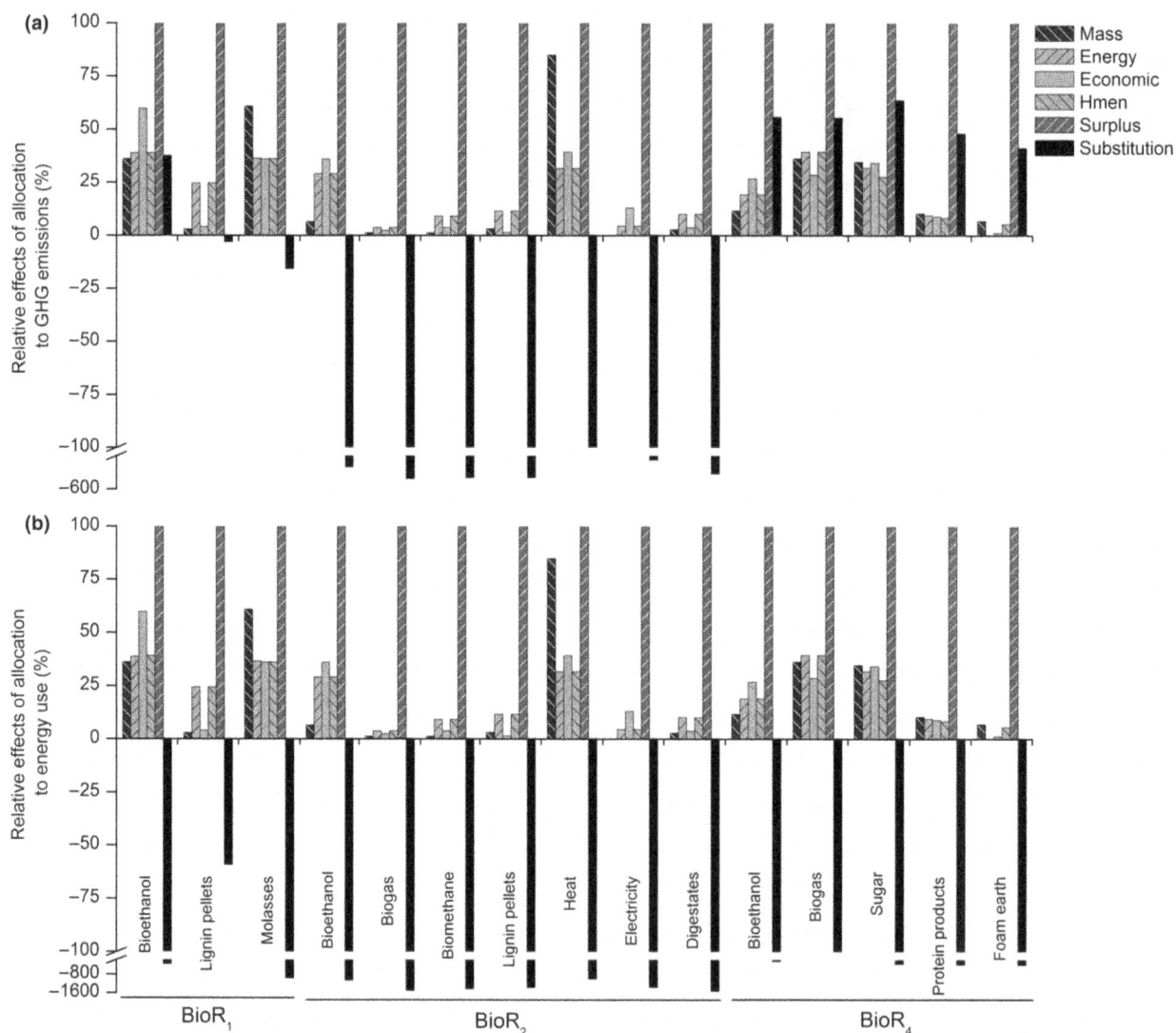

Fig. 2 Comparison of the influence of allocation methods on specific energy use (b) and GHG emissions (a).

emissions to each intermediate product as the subdivision or the LP method does. Although it has been demonstrated that differences in energy use and GHG emissions between allocation at refinery level and the refining at process level can be up to 40% (Furuholt, 1995), our opinion is that intermediate products in biorefineries are often used for internal processes within biorefineries and not sold as final products. Moreover, except for the subdivision and the linear programming, most of the allocation models in LCA perform at a high level of resolution because of a data availability issue. Finally, it has to be acknowledged that the HMEN method does not accommodate the indirect market-mediated effects, a situation which can only be dealt with through substitution. However, this approach is not exempt from shortcomings: Substitution fails to quantify the environmental impacts of a specific

product with sufficient accuracy, as the uncertainties of added system can be overriding (Pawelzik et al., 2013).

We showed that estimates of the energy use and GHG emissions of biorefinery systems varied strongly depending on the allocation method adopted (Fig. 3), which is in line with several previous findings that report effects of allocation method on LCA results and conclusions (Azapagic & Clift, 1999; Heijungs & Guinée, 2007). We found that very few allocation methods arrived at the same prioritization of biorefinery products. This contrasts with the conclusions of Curran (Curran, 2007) that different allocation methods lead to same prioritization of products (Table 2). HMEN results in the same prioritization as the energy allocation, and similar ranking as the economic or mass allocation. Overall, these results reiterate the need for additional and more stringent guidance on allocation issues for

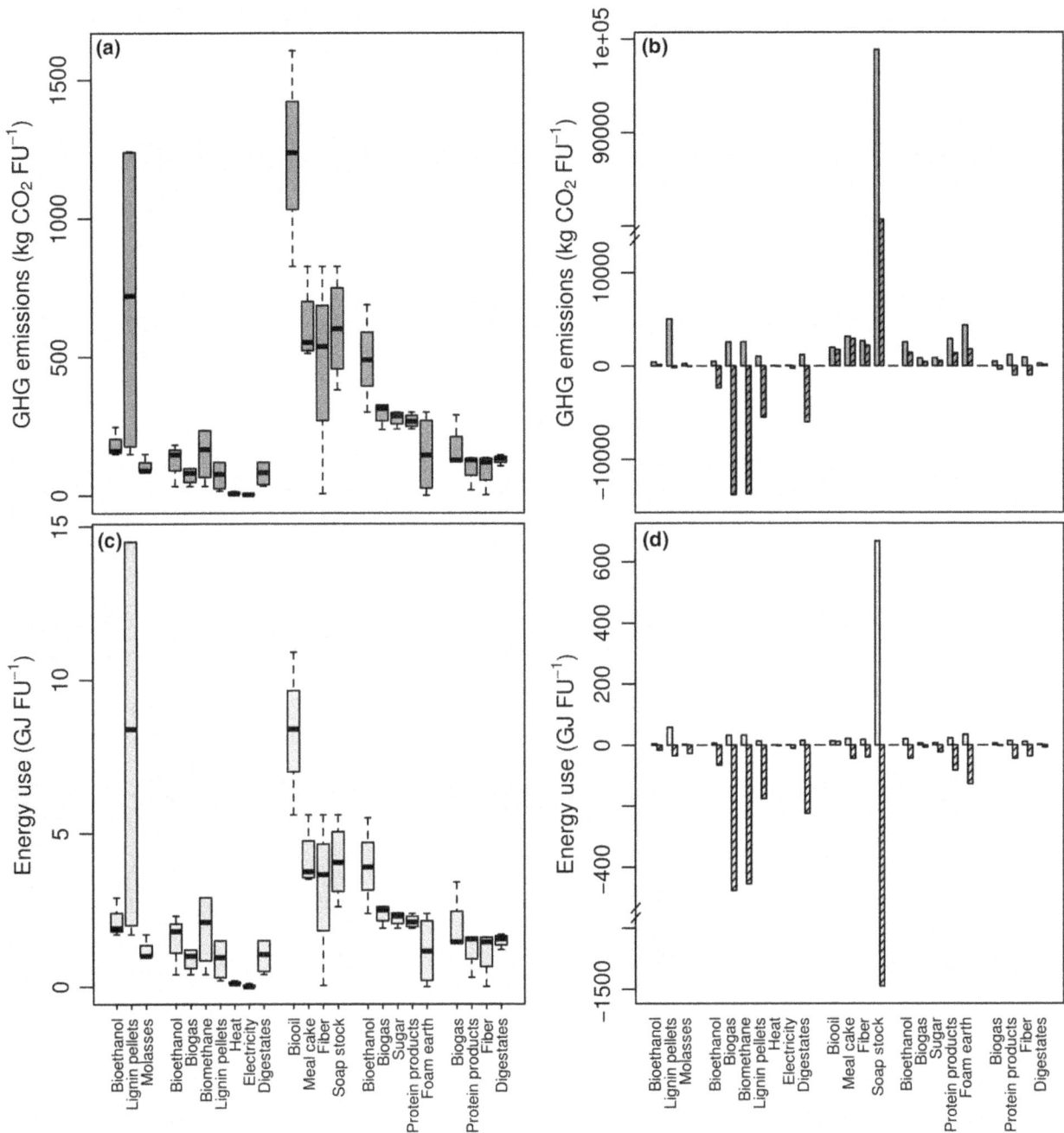

Fig. 3 Variability in specific energy use and GHG emissions of various biorefinery products, depending on type of allocation method applied: (a,c) mass, energy, and economic allocation; and (b,d) surplus (straight) and substitution (striped) methods. Functional unit is 1 t (tonne) for fuels and materials and 1 GJ for heat and electricity.

LCA practices. Without additional guidance, inconsistencies will not be reduced, and this will limit the usability of information from LCA studies in decision-making (Pelletier *et al.*, 2015).

Existing directives are not very helpful for practitioners as they provide different recommendations for assessments of multifunctional systems (Wardenaar *et al.*, 2012). The EU-RED (RED, 2009) suggests the use of energy allocation, the US-EPA (EPA, 2010) and IEA (Jungmeier *et al.*, 2014) adopt the system expansion method, while the PAS2050 (BSI, 2011) suggests the use of economic allocation where system expansion is not possible. The ISO (ISO 14044, 2006) and the ILCD book (EC-JRC, 2010) recommend firstly avoiding allocation, and secondly applying physical causality-based allocation methods when allocation is unavoidable. Allocation problems in biorefining can be solved in many ways depending on the research questions (Reap *et al.*, 2008; Cherubini *et al.*, 2011), but any method used for solving this problem must reduce risks of controversy, have the fewest unintended consequences, and generate solutions that are consistent with the principles of ISO standards for LCA. Although substitution is the preferred method in ISO, we showed that it may result in negative estimates of energy use and GHG emissions for the main products because of the credits from the exported energy products of the biorefinery systems. This observation suggests that substitution should not be applied universally without examining the individual situation (Wang *et al.*, 2011). Moreover, the application of the substitution method requires the identification of the main product, which in the case of the biorefinery is difficult (Parajuli *et al.*, 2015), especially if the products are given equal importance. Guidance on identification of the main products exists (Weidema, 2000), but its importance in biorefineries is limited because they generate several marketable products (Bozell & Petersen, 2010). These, coupled to uncertainties about the displaced products, further limit the suitability of this method for biorefineries.

Direct comparison of results from this work with those of other studies on allocation in biorefineries is difficult due to differences in methods used, system boundaries, feedstock and technologies investigated, generated products, in addition to other assumptions. In their analysis of bioethanol from sugar beet refinery, Malça & Freire (2006) assigned between 28% and 37% of energy use to bioethanol. Allocation factors for energy use attributable to sugar beet bioethanol in this study ranged from 12% to 27%. Pulps and sugar were the coproducts in Malça & Freire (2006), while in our study, biogas, protein products, foam earth, and sugar were produced, and this may explain the difference observed. The allocation factors (63–99%) for GHG emissions assigned to straw–bioethanol in Cherubini *et al.* (2011) are higher than those for GHG emissions (36–60%) attributed to straw-based bioethanol in this study (Table 2). Also, the hybrid method of Cherubini *et al.* (2011) assigned 84% of GHG emissions to bioethanol. This latest allocation factor is more than double that for straw-based bioethanol estimated by the HMEN method (39%). Differences in the conceptual framework as well as the type and amount of coproducts explain the difference in results of the two approaches. The hybrid method of Cherubini *et al.* (2011) is a combination of the partitioning and substitution approaches, while the HMEN in this study is a mass- and energy-based allocation. Electricity, heat, and phenols were the coproducts in Cherubini *et al.* (2011), while molasses and lignin pellets were the coproducts in our study

(Table 2). For switchgrass bioethanol in the United States, allocation factors for energy use (62-98%) and GHG emissions (69–98%) (Wang *et al.*, 2011) were also higher than the calculated factors for energy use (36–329%) and GHG emissions (36–60%) in this study. In the US study, switchgrass was converted into bioethanol and electricity, but in our study, straw was converted into bioethanol, molasses, and lignin pellets. Finally, the comparison of bioethanol to gasoline, like that of biogas and biomethane to natural gas, showed that, in general, these biofuels save GHG emissions relative to their fossil fuel counterparts. Reduction in GHG emissions ranged from 42% to 196% for bioethanol, 58% to 1365% for biogas, and from 98% to 616% for biomethane. This reinforces the generally accepted conclusion that biofuels reduce GHG emissions (Jury *et al.*, 2010; Bühle *et al.*, 2011; Monti *et al.*, 2012; Muñoz *et al.*, 2014). However, for bioethanol, the saving in GHG emissions would be much lower because only 5% bioethanol is blended with gasoline in many countries.

By integrating mass and energy allocation methods, the HMEN approach solves the dilemma of multifunctional allocation in biorefineries. HMEN is based on physical relationships between the products and factors in the amounts of biomass used to produce each biorefinery product. It allows environmental burdens to be distributed to all biorefinery products without significant computational difficulties. The application of HMEN to a set of diverse case studies showed that HMEN surpassed existing mass and energy allocation methods. HMEN is thus preferable for complex systems such as biorefineries. Some weaknesses of the HMEN method were highlighted, and future work will explore the possibility to overcome them. Although we have limited the test to biorefineries, the approach can be used in other systems where multifunctionality problems occur. LCA practitioners are thus encouraged to test the method in other case studies.

References

Ahlgren S, Björklund A, Ekman A et al. (2015) Review of methodological choices in LCA of biorefinery systems – key issues and recommendations. *Biofuels, Bioproducts and Biorefining*, **9**, 606–619.

Azapagic A, Clift R (1999) Allocation of environmental burdens in co-product systems: product-related burdens (Part 1). *The International Journal of Life Cycle Assessment*, **4**, 357–369.

Babusiaux D (2003) Affectation des émissions de CO_2 et de polluants d'une raffinerie aux produits finis pétroliers. *Oil & Gas Science and Technology – Revue IFP*, **58**, 685–692.

Balakrishnan M, Sacia ER, Sreekumar S et al. (2015) Novel pathways for fuels and lubricants from biomass optimized using life-cycle greenhouse gas assessment. *Proceedings of the National Academy of Sciences*, **112**, 7645–7649.

Börjesson P (2009) Good or bad bioethanol from a greenhouse gas perspective – what determines this? *Applied Energy*, **86**, 589–594.

Bosch R, Van De Pol M, Philp J (2015) Policy: define biomass sustainability. *Nature*, **523**, 526–527.

Bozell JJ, Petersen GR (2010) Technology development for the production of bio-based products from biorefinery carbohydrates-the US Department of Energy's "Top 10" revisited. *Green Chemistry*, **12**, 539–554.

Bredeson L, Quiceno-Gonzalez R, Riera-Palou X, Harrison A (2010) Factors driving refinery CO_2 intensity, with allocation into products. *The International Journal of Life Cycle Assessment*, **15**, 817–826.

BSI P (2011) Specification for the assessment of the life cycle greenhouse gas emissions of goods and services. BSI-British Standards Institution, London, UK. Available at: http://www.bsigroup.com/upload/Standards%2&%20Publications/Energy/PAS2050.pdf (accessed 7 February 2016).

Bühle L, Stülpnagel R, Wachendorf M (2011) Comparative life cycle assessment of the integrated generation of solid fuel and biogas from biomass (IFBB) and whole crop digestion (WCD) in Germany. *Biomass and Bioenergy*, **35**, 363–373.

Cherubini F, Jungmeier G (2010) LCA of a biorefinery concept producing bioethanol, bioenergy, and chemicals from switchgrass. *The International Journal of Life Cycle Assessment*, **15**, 53–66.

Cherubini F, Bargigli S, Ulgiati S (2008) Life cycle assessment of urban waste management: energy performances and environmental impacts. The case of Rome, Italy. *Waste Management*, **28**, 2552–2564.

Cherubini F, Strømman AH, Ulgiati S (2011) Influence of allocation methods on the environmental performance of biorefinery products—a case study. *Resources, Conservation and Recycling*, **55**, 1070–1077.

Creutzig F, Ravindranath NH, Berndes G et al. (2015) Bioenergy and climate change mitigation: an assessment. *GCB Bioenergy*, **7**, 916–944.

Curran MA (2007) Studying the effect on system preference by varying coproduct allocation in creating life-cycle inventory. *Environmental Science & Technology*, **41**, 7145–7151.

Dunn JB, Mueller S, Wang M, Han J (2012) Energy consumption and greenhouse gas emissions from enzyme and yeast manufacture for corn and cellulosic ethanol production. *Biotechnology Letters*, **34**, 2259–2263.

EC (2012) Innovating for Sustainable Growth: a bioeconomy for Europe. European Commission, Brussels. Available at: http://ec.europa.eu/research/bioeconomy/pdf/bioeconomycommunicationstrategy_b5_brochure_web.pdf (accessed 15 July 2016).

EC-JRC (2010) International Reference Life Cycle Data System (ILCD) Handbook-General Guide for Life Cycle Assessment-Detailed guidance. First edition March 2010. EUR 24708 EN. Luxembourg. Publication Office of the European Union.

Ecoinvent (2014) Swiss centre for life cycle inventory, Database, Ecoinvent Centre, Dubendorf, 2013.

Elgowainy A, Han J, Cai H, Wang M, Forman GS, Divita VB (2014) Energy efficiency and greenhouse gas emission intensity of petroleum products at U.S. refineries. *Environmental Science & Technology*, **48**, 7612–7624.

Energy Center of the Netherlands (ECN) (2010) *Development of Advanced Biorefinery schemes to be Integrated into existing industrial (fuel producing) complexes*. Report of the BIOREF-INTEG, The LE Petten, Netherlands.

EPA (2010) Renewable Fuel Standard Program (RFS2) Regulatory Impact Analysis, US Environmental Protection Agency, Office of Transportation and Air Quality, Washington, DC.

Eriksson O, Finnveden G, Ekvall T, Björklund A (2007) Life cycle assessment of fuels for district heating: a comparison of waste incineration, biomass- and natural gas combustion. *Energy Policy*, **35**, 1346–1362.

Fu GZ, Chan AW, Minns DE (2003) Life cycle assessment of bio-ethanol derived from cellulose. *The International Journal of Life Cycle Assessment*, **8**, 137–141.

Furuholt E (1995) Life cycle assessment of gasoline and diesel. *Resources, Conservation and Recycling*, **14**, 251–263.

Gabrielle B, Gagnaire N (2008) Life-cycle assessment of straw use in bio-ethanol production: a case study based on biophysical modelling. *Biomass and Bioenergy*, **32**, 431–441.

Gerssen-Gondelach SJ, Wicke B, Faaij APC (2016) GHG emissions and other environmental impacts of indirect land use change mitigation. *GCB Bioenergy*, **9**, 725–742.

Gnansounou E, Dauriat A, Villegas J, Panichelli L (2009) Life cycle assessment of biofuels: energy and greenhouse gas balances. *Bioresource Technology*, **100**, 4919–4930.

Guinée JB, Heijungs R, Huppes G (2004) Economic allocation: examples and derived decision tree. *The International Journal of Life Cycle Assessment*, **9**, 23–33.

Heijungs R, Guinée JB (2007) Allocation and 'what-if' scenarios in life cycle assessment of waste management systems. *Waste Management*, **27**, 997–1005.

Hirshfeld DS, Kolb JA (2012) Analysis of energy use and CO_2 emissions in the U.S. refining sector, with projections for 2025. *Environmental Science & Technology*, **46**, 3697–3704.

Huo H, Wang M, Bloyd C, Putsche V (2009) Life-cycle assessment of energy use and greenhouse gas emissions of soybean-derived biodiesel and renewable fuels. *Environmental Science & Technology*, **43**, 750–756.

IEA (2009) Biorefineries: adding value to the sustainable utilisation of biomass (Task 42).

ISO 14044 (2006) ISO-14044 Environmental management-Life cycle assessment. Requirements and guidelines, Geneva, Switzerland.

Jungmeier G, Werner F, Jarnehammar A, Hohenthal C, Richter K (2002) Allocation in LCA of wood-based products experiences of cost action E9. *The International Journal of Life Cycle Assessment*, **7**, 369–375.

Jungmeier G, Van Ree R, De Jong E et al. (2014) Assessing biorefinery using wood for bioeconomy-Current statut and future perspective of IEA bioenergy Task 42. "Biorefining" 24th European Biomass Conference and Exhibition, 6-9 June 2016. Amsterdam, The Netherlands.

Jury C, Benetto E, Koster D, Schmitt B, Welfring J (2010) Life cycle assessment of biogas production by monofermentation of energy crops and injection into the natural gas grid. *Biomass and Bioenergy*, **34**, 54–66.

Kim S, Dale BE (2002) Allocation procedure in ethanol production system from corn grain – I. System expansion. *International Journal of Life Cycle Assessment*, **7**, 237–243.

Larsen J, Haven MØ, Thirup L (2012) Inbicon makes lignocellulosic ethanol a commercial reality. *Biomass and Bioenergy*, **46**, 36–45.

Luo L, Van Der Voet E, Huppes G, Udo De Haes HA (2009) Allocation issues in LCA methodology: a case study of corn stover-based fuel ethanol. *The International Journal of Life Cycle Assessment*, **14**, 529–539.

Maes D, Van Dael M, Vanheusden B, Goovaerts L, Reumerman P, Márquez Luzardo N, Van Passel S (2015) Assessment of the sustainability guidelines of EU Renewable Energy Directive: the case of biorefineries. *Journal of Cleaner Production*, **88**, 61–70.

Malça J, Freire F (2006) Renewability and life-cycle energy efficiency of bioethanol and bio-ethyl tertiary butyl ether (bioETBE): assessing the implications of allocation. *Energy*, **31**, 3362–3380.

MEC I/S MEC (2015) *The Dawning of Danish Bio-Economy*. Statut Report, Holstebro, Denmark.

Monti A, Barbanti L, Zatta A, Zegada-Lizarazu W (2012) The contribution of switchgrass in reducing GHG emissions. *GCB Bioenergy*, **4**, 420–434.

Muñoz I, Flury K, Jungbluth N, Rigarlsford G, I Canals LM, King H (2014) Life cycle assessment of bio-based ethanol produced from different agricultural feedstocks. *The International Journal of Life Cycle Assessment*, **19**, 109–119.

Njakou Djomo S, Witters N, Van Dael M, Gabrielle B, Ceulemans R (2015) Impact of feedstock, land use change, and soil organic carbon on energy and greenhouse gas performance of biomass cogeneration technologies. *Applied Energy*, **154**, 122–130.

Nuss P, Gardner KH (2013) Attributional life cycle assessment (ALCA) of polyitaconic acid production from northeast US softwood biomass. *The International Journal of Life Cycle Assessment*, **18**, 603–612.

Parajuli R, Dalgaard T, Jørgensen U et al. (2015) Biorefining in the prevailing energy and materials crisis: a review of sustainable pathways for biorefinery value chains and sustainability assessment methodologies. *Renewable and Sustainable Energy Reviews*, **43**, 244–263.

Pawelzik P, Carus M, Hotchkiss J et al. (2013) Critical aspects in the life cycle assessment (LCA) of bio-based materials – reviewing methodologies and deriving recommendations. *Resources, Conservation and Recycling*, **73**, 211–228.

Pelletier N, Ardente F, Brandão M, De Camillis C, Pennington D (2015) Rationales for and limitations of preferred solutions for multi-functionality problems in LCA: is increased consistency possible? *The International Journal of Life Cycle Assessment*, **20**, 74–86.

Pereira LG, Chagas MF, Dias MOS, Cavalett O, Bonomi A (2015) Life cycle assessment of butanol production in sugarcane biorefineries in Brazil. *Journal of Cleaner Production*, **96**, 557–568.

Pierru A (2007) Allocating the CO_2 emissions of an oil refinery with Aumann–Shapley prices. *Energy Economics*, **29**, 563–577.

Pimentel D, Patzek TW (2005) Ethanol production using corn, switchgrass, and wood; biodiesel production using soybean and sunflower. *Natural Resources Research*, **14**, 65–76.

Reap J, Roman F, Duncan S, Bras B (2008) A survey of unresolved problems in life cycle assessment. *The International Journal of Life Cycle Assessment*, **13**, 290–300.

RED (2009) Directives 2009/28/EC of the European Parliament and of the Council of 23 April 2009 on the promotion of the use of energy from renewable sources and amending and subsequently repealing Directives 2001/77/EC and 2003/30/EC vol 2013.

Sacramento-Rivero JC, Navarro-Pineda F, Vilchiz-Bravo LE (2016) Evaluating the sustainability of biorefineries at the conceptual design stage. *Chemical Engineering Research and Design*, **107**, 167–180.

Sandin G, Røyne F, Berlin J, Peters GM, Svanström M (2015) Allocation in LCAs of biorefinery products: implications for results and decision-making. *Journal of Cleaner Production*, **93**, 213–221.

Schneider L, Finkbeiner M (2013) *Life Cycle Assessment of EU Oilseed Crushing and Vegetable Oil Refining*. Report, Technische Universitat Berlin, Berlin, Germany.

Silalertruksa T, Pongpat P, Gheewala SH (2017) Life cycle assessment for enhancing environmental sustainability of sugarcane biorefinery in Thailand. *Journal of Cleaner Production*, **140**(Part 2), 906–913.

Singh A, Pant D, Korres NE, Nizami A-S, Prasad S, Murphy JD (2010) Key issues in life cycle assessment of ethanol production from lignocellulosic biomass: challenges and perspectives. *Bioresource Technology*, **101**, 5003–5012.

Spirinckx C, Ceuterick D (1996) Biodiesel and fossil diesel fuel: comparative life cycle assessment. *The International Journal of Life Cycle Assessment*, **1**, 127–132.

The White House (2012) National bioeconomy blueprint. The White House Office of Science and Technology, Washington, DC. Available at: https://www.whitehouse.gov/sites/default/files/microsites/ostp/national_bioeconomy_blueprint_april_2012.pdf (accessed 5 October 2016).

Wang M, Lee H, Molburg J (2004) Allocation of energy use in petroleum refineries to petroleum products. *The International Journal of Life Cycle Assessment*, **9**, 34–44.

Wang M, Huo H, Arora S (2011) Methods of dealing with co-products of biofuels in life-cycle analysis and consequent results within the U.S. context. *Energy Policy*, **39**, 5726–5736.

Wardenaar T, Van Ruijven T, Beltran AM, Vad K, Guinée J, Heijungs R (2012) Differences between LCA for analysis and LCA for policy: a case study on the consequences of allocation choices in bio-energy policies. *The International Journal of Life Cycle Assessment*, **17**, 1059–1067.

Weidema B (2000) Avoiding co-product allocation in life-cycle assessment. *Journal of Industrial Ecology*, **4**, 11–33.

3

Evaluation of Agricultural Production Systems Simulator as yield predictor of *Panicum virgatum* and *Miscanthus* x *giganteus* in several US environments

JONATHAN J. OJEDA[1,2], JEFFREY J. VOLENEC[3], SYLVIE M. BROUDER[3], OCTAVIO P. CAVIGLIA[1,2,4] and MÓNICA G. AGNUSDEI[5]

[1]National Research Council (CONICET), Oro Verde, Argentina, [2]Facultad de Ciencias Agropecuarias, Universidad Nacional de Entre Ríos, Ruta 11, km 10.5 (3101), Oro Verde, Entre Ríos, Argentina, [3]Department of Agronomy, Purdue University, West Lafayette, IN, USA, [4]Instituto Nacional de Tecnología Agropecuaria (INTA), Estación Experimental Agropecuaria Paraná, Ruta 11, km 12.5 (3101), Oro Verde, Entre Ríos, Argentina, [5]Instituto Nacional de Tecnología Agropecuaria (INTA), Estación Experimental Agropecuaria Balcarce, Ruta 226, km 73.5 (7620), Balcarce, Buenos Aires, Argentina

Abstract

Simulation models for perennial energy crops such as switchgrass (*Panicum virgatum* L.) and *Miscanthus* (*Miscanthus* x *giganteus*) can be useful tools to design management strategies for biomass productivity improvement in US environments. The Agricultural Production Systems Simulator (APSIM) is a biophysical model with the potential to simulate the growth of perennial crops. APSIM crop modules do not exist for switchgrass and *Miscanthus*, however, re-parameterization of existing APSIM modules could be used to simulate the growth of these perennials. Our aim was to evaluate the ability of APSIM to predict the dry matter (DM) yield of switchgrass and *Miscanthus* at several US locations. The *Lucerne* (for switchgrass) and *Sugarcane* (for *Miscanthus*) APSIM modules were calibrated using data from four locations in Indiana. A sensitivity analysis informed the relative impact of changes in plant and soil parameters of APSIM *Lucerne* and APSIM *Sugarcane* modules. An independent dataset of switchgrass and *Miscanthus* DM yields from several US environments was used to validate these re-parameterized APSIM modules. The re-parameterized modules simulated DM yields of switchgrass [0.95 for CCC (concordance correlation coefficient) and 0 for SB (bias of the simulation from the measurement]] and *Miscanthus* (0.65 and 0% for CCC and SB, respectively) accurately at most locations with the exception of switchgrass at southern US sites (0.01 for CCC and 2% for SB). Therefore, the APSIM model is a promising tool for simulating DM yields for switchgrass and *Miscanthus* while accounting for environmental variability. Given our study was strictly based on APSIM calibrations at Indiana locations, additional research using more extensive calibration data may enhance APSIM robustness.

Keywords: Agricultural Production Systems Simulator, bioenergy, biomass, *Miscanthus*, model re-parameterization, switchgrass, United States

Introduction

Many studies throughout US have reported the extraordinary potential for high biomass production of switchgrass (*Panicum virgatum* L.) (Vogel *et al.*, 2002; Kiniry *et al.*, 2012; Burks, 2013; Arundale *et al.*, 2014; Trybula *et al.*, 2014) and *Miscanthus* (*Miscanthus* x *giganteus*) (Heaton *et al.*, 2004, 2008; Khanna *et al.*, 2008; Jain *et al.*, 2010; Kiniry *et al.*, 2012; Mishra *et al.*, 2013; Trybula *et al.*, 2014), both perennial rhizomatous grasses with C4 photosynthesis. Stakeholders involved in developing biomass crops for bioenergy are therefore increasingly interested in estimating potential yields of both species over large geographical domains (Clifton-Brown *et al.*, 2004). Direct measurements of dry matter (DM) yields of these species are scarce relative to corn (*Zea mays* L.), soybean (*Glycine max* [L.] Merr.) and other grain crop species, and this lack of data over large geographies and at a fine spatial resolution remains a limitation to informed decision making (Clifton-Brown *et al.*, 2004).

Satisfactory predictions of switchgrass biomass production were achieved with models like ALMANAC in Texas, Arkansas and Louisiana (Kiniry *et al.*, 2005) and SWAT in Indiana (Trybula *et al.*, 2014). Stampfl *et al.* (2007) achieved satisfactory simulations of *Miscanthus* biomass production across diverse climate and soil

Correspondence: Jonathan J. Ojeda
E-mails: ojeda.jonathan@conicet.gov.ar and ojeda.jonathan@inta.gob.ar

conditions in Europe using the MISCANMOD model developed by Clifton-Brown *et al.* (2000, 2004). Likewise, European studies for renewable energy used a FORTRAN version of MISCANMOD (Hastings *et al.*, 2008) and showed satisfactory simulation of *Miscanthus* biomass production derived by model improvements in the drought stress function, temperature effect in radiation use efficiency (RUE) and the inclusion of photoperiodism effects (Hastings *et al.*, 2009). Parameterization of WINOWAC was also performed for *Miscanthus* (Miguez, 2007). Other examples of modelling growth/adaptation of these species include the use of the STELLA software (Pallipparambil *et al.*, 2015) to identify Ohio, Missouri, Arkansas and Illinois as suitable locations for *Miscanthus*, as well as to determine sensitive parameters for biomass production. In a recent study (Strullu *et al.*, 2015) the STICS crop-soil model accurately predicted *Miscanthus* biomass production and environmental impacts in various environments in France and the UK. Despite this important progress in the calibration, development, and modification of several simulation models, the ability to predict DM yield both of switchgrass and *Miscanthus* by a single model has not yet been achieved.

In this context, a model scaled for a large geographic region and demonstrating adequate performance to predict DM yield is needed. The Agricultural Production Systems Simulator (APSIM) (Keating *et al.*, 2003) is a biophysical model with potential to simulate growth of annual and perennial crops. The APSIM model has been developed in Australia to simulate, on a daily time step, the main biophysical processes of a generic plant in response to management and weather (Keating *et al.*, 2003; Holzworth *et al.*, 2014). However, without preexisting APSIM crop modules to simulate switchgrass and *Miscanthus*, the re-parameterization of other APSIM crop modules such as the APSIM *Lucerne* (Robertson *et al.*, 2002) and APSIM *Sugarcane* modules (Keating *et al.*, 1999) could act as alternatives to simulate growth of both crops. In order to allow the use of APSIM for this purpose, a supervised calibration with a detailed data base and an evaluation of its predictive ability over a broad range of soils and environments is required. Our objectives were to (i) calibrate APSIM *Lucerne* module for switchgrass and APSIM *Sugarcane* module for *Miscanthus* using experimental field data collected in several locations across Indiana and (ii) validate these re-parameterized APSIM modules with independent data from numerous US locations where the accuracy and biases were evaluated.

Materials and methods

The calibration of the APSIM *Lucerne* and APSIM *Sugarcane* modules was made using the following steps: (i) data on

climate, soil, and management were collected for model inputs; (ii) soil parameterization by location; (iii) adaptation of original plant modules to model switchgrass and *Miscanthus* growth using actual data from literature or field experiments and (iv) sensitivity analysis to evaluate parameter influence on LAI and the DM yield. Outcomes of the sensitivity analysis by successive iterations directed the compilation of existing data and additional field measurements used to develop model parameters. The model was calibrated through graphical comparison and statistical analyses of observed and modelled leaf area index (LAI) and DM yield data from IN locations with the objective to increase the concordance correlation coefficient (CCC, Tedeschi, 2006) and decrease the bias of the simulation from the measurement (SB, Kobayashi & Us Salam, 2000). These data included not only detailed measurements of LAI and DM yields at the final harvest, but also during crop growth and development. Model validation was made by using graphical comparisons and statistical analyses of observed and modelled DM yield data from 35 locations across the US. Data for switchgrass were grouped by region (southern vs. northern locations) and ecotype (upland vs. lowland). A complete description of datasets used for calibration and validation are provided in the supplementary information (Tables S3 and S4).

Data for model simulations

The data used for model calibration were obtained from field trials across IN (Table 1). For switchgrass model calibration, data from the Water Quality Field Station at Purdue University Agronomy Center for Research and Education (ACRE) near West Lafayette (40°28′11.99″N; 87°0′36.00″W) and Throckmorton Purdue Agricultural Center (TPAC) five miles south of Lafayette in Tippecanoe County (40°17′59.99″N; 86°54′0.00″W) (Table 1). The *Miscanthus* calibration included two additional IN locations: Northeast Purdue Agricultural Center (NEPAC) in Whitley County between Fort Wayne and Columbia City (41°8′24.00″N; 85°29′23.99″W), and the Southeast Purdue Agricultural Center (SEPAC) six miles east of North Vernon in Jennings County (39° 1′48.00″N; 85°31′11.99″W) (Table 1). A complete description of datasets used for calibration and validation of the model is shown in the supplementary information (Tables S3 and S4 for switchgrass and *Miscanthus*, respectively). Subsequent model validation used data of DM yields gathered across the US, which were collected from published and unpublished studies from 34 dryland locations and one irrigated location (Davis, CA) in 16 states (Fig. 1; Table 1).

Climate data sources

Daily meteorological data for each location were derived from two data sources. Maximum and minimum air temperatures and rainfall were obtained from National Climatic Data Center (NOAA, http://www.ncdc.noaa.gov), while daily solar radiation was obtained from the NASA Prediction of Worldwide Energy Resource (POWER) - Climatology Resource for Agroclimatology (http://power.larc.nasa.gov). This long-term database also was used as a secondary source of maximum and minimum air temperatures to replace missing values from the

Table 1 Soil and climate characterization of locations used for the calibration/validation of the Agricultural Production Systems Simulator (APSIM)

State	Location	Latitude, longitude	Rainfall (mm)*	Jan	Feb	Mar	Apr	May	Jun	Jul	Aug	Sep	Oct	Nov	Dec	Series	Taxonomic classification
								Tmax/Tmin (°C)*								Soil description	
Calibration																	
IN	Columbia City‡	41.14, −85.49	938	−0.8/−8.5	0.9/−6.5	7.2/−1.0	14.9/5.1	21.5/11.0	26.5/15.7	28.4/17.6	27.6/16.8	23.6/12.6	16.4/6.9	8.7/1.2	1.5/−5.1	Boyer	Typic Hapludalfs
IN	West Lafayette†,‡	40.47, −87.01	978	0.3/−7.4	2.6/−5.1	9.0/0.3	16.7/6.5	22.9/12.7	27.8/17.2	29.6/19.0	28.9/18.0	24.7/13.8	17.9/8.2	9.9/2.0	2.5/−4.5	Drummer	Typic Haplaquolls
IN	Lafayette†,‡	40.30, −86.90	969	0.4/−7.2	2.4/−5.1	8.7/0.1	16.3/6.3	22.5/12.4	27.2/16.9	29.0/18.6	28.4/17.8	24.4/13.5	17.5/7.8	9.8/1.9	2.6/−4.4	Lauramie Octagon	Mollic Hapludalfs Mollic Hapludalfs
IN	Butlerville‡	39.03, −85.52	1117	2.0/−5.3	3.9/−3.8	9.6/1.0	17.0/7.0	22.9/12.8	27.3/17.2	29.0/19.1	28.4/18.4	24.8/14.3	18.2/8.4	10.8/2.6	4.0/−3.0	Cobbsfork	Typic Glossaqualf
Validation																	
ND	Munich†	48.76, −98.34	452	−10.7/−19.7	−8.2/−17.5	−1.6/−10.9	10.1/−0.6	18.8/7.3	23.9/12.4	27.2/15.1	25.9/14.3	19.6/9.3	9.8/1.6	−1.2/−8.1	−8.7/−16.6	Tonka	Argiaquic Argialbolls
ND	Streeter†	46.73, −99.48	471	−7.9/−16.9	−5.7/−14.4	1.2/−7.8	11.2/−0.1	19.2/7.3	24.2/12.3	27.7/15.0	26.2/13.7	20.3/9.0	11.0/1.8	0.7/−6.7	−6.2/−14.1	Barnes	Udic Haploborolls
SD	Bristol†	45.27, −97.83	560	−7.6/−16.9	−5.1/−14.0	2.0/−6.9	12.0/0.9	20.0/8.3	25.2/13.4	28.4/15.8	26.9/14.5	21.2/9.8	12.0/2.8	1.8/−5.5	−5.5/−13.5	Buse	Udorthentic Haploboroll
SD	Huron†	44.39, −98.22	586	−5.2/−14.2	−2.6/−11.5	4.3/−5.1	13.1/1.6	20.6/8.7	25.6/13.9	29.2/16.4	27.7/14.9	22.3/10.2	13.4/3.2	3.5/−4.6	−3.5/−11.5	Dudley	Typic Natrustolls
SD	Highmore†	44.38, −100.28	472	−3.2/−12.3	−0.9/−9.9	5.5/−4.6	13.3/1.4	20.5/8.4	25.4/13.5	28.9/16.2	27.7/14.6	22.2/9.7	13.6/2.9	4.1/−4.6	−2.2/−10.4	Glenham	Typic Argiustoll
SD	Ethan†	43.65, −97.79	618	−4.3/−13.2	−1.7/−10.5	5.4/−4.1	13.8/2.4	21.0/9.3	26.1/14.5	29.2/16.9	27.8/15.4	22.8/10.6	14.2/3.8	4.5/−3.9	−2.5/−10.5	Houdek	Typic Haplustolls
NE	Crofton†	42.73, −97.50	729	−2.7/−11.7	−0.1/−9.0	6.9/−3.0	14.6/3.1	21.3/9.8	26.4/14.8	29.3/17.2	28.2/15.8	23.2/11.0	15.1/4.4	5.8/−3.0	−1.1/−9.1	Crofton	Typic Ustorthent
NE	Atkinson†	42.46, −98.65	626	−1.9/−10.9	0.5/−8.5	7.3/−3.0	14.5/2.8	21.1/9.4	26.0/14.6	29.2/17.3	27.9/15.8	23.1/10.9	15.1/4.1	6.0/−3.2	−0.6/−8.9	Pivot	Entic Haplustoll
NY	Ithaca†	42.46, −76.46	930	−2.4/−10.0	−1.0/−9.3	4.5/−4.3	12.3/2.2	19.4/8.0	24.2/12.9	26.2/15.3	25.1/14.6	21.0/10.8	13.9/4.9	6.8/−0.4	0.0/−6.4	Erie	Aeric Fragiaquepts
IA	Ames†	41.98, −93.69	925	−2.3/−10.6	0.3/−8.0	7.8/−1.2	15.9/5.1	22.2/11.6	27.4/16.8	29.9/19.1	29.0/17.8	24.0/12.7	16.5/6.4	7.4/−0.6	−0.1/−7.4	Clarion	Typic Hapludoll
IL	Dekalb†,‡	41.85, −88.85	897	−1.4/−9.6	0.8/−6.9	7.8/−0.6	15.8/5.6	22.4/12.0	27.7/16.8	29.7/18.7	28.7/17.6	24.3/13.1	17.1/7.4	8.6/1.0	0.9/−6.0	Flanagan	Typic Endoaquoll
NE	Mead‡	41.17, −96.46	743	−1.6/−10.1	0.9/−7.6	8.2/−1.4	16.0/4.6	22.3/11.2	27.3/16.2	30.1/18.5	29.2/17.3	24.1/12.4	16.3/6.0	7.2/−1.4	−0.1/−7.7	Tomek	Pachic Argiudolls
IA	Lucas†	40.92, −93.38	939	−0.5/−8.7	2.0/−6.3	9.4/0.0	17.0/6.0	22.8/12.2	27.8/17.4	30.4/19.8	29.7/18.6	24.8/13.4	17.6/7.2	8.8/0.4	1.5/−5.9	Grundy	Aquertic Argiudoll
IA	Wayne†	40.83, −93.25	934	−0.5/−8.7	2.0/−6.3	9.4/0.0	17.0/6.0	22.8/12.2	27.8/17.4	30.4/19.8	29.7/18.6	24.8/13.4	17.6/7.2	8.8/0.4	1.5/−5.9	Clarinda	Vertic Argiaquoll
NE	Douglas†	40.68, −96.19	789	−0.1/−8.6	2.4/−6.2	9.7/−0.3	17.1/5.6	22.9/11.9	27.8/16.9	30.7/19.4	30.0/18.3	24.7/13.2	17.4/7.0	8.6/−0.3	1.2/−6.4	Wymore	Aquertic Argiudolls
IN	Lafayette†	40.30, −86.90	969	0.4/−7.2	2.4/−5.1	8.7/0.1	16.3/6.3	22.5/12.4	27.2/16.9	29.0/18.6	28.4/17.8	24.4/13.5	17.5/7.8	9.8/1.9	2.6/−4.4	Lauramie	Mollic Hapludalfs

Table 1 (continued)

State	Location	Latitude, longitude	Rainfall (mm)*	Jan	Feb	Mar	Apr	May	Jun	Jul	Aug	Sep	Oct	Nov	Dec	Series	Taxonomic classification
IL	Havana‡	40.30, −89.94	999	0.0/−7.8	2.4/−5.4	9.4/0.4	17.1/6.7	23.3/12.9	28.4/17.7	30.6/19.7	29.6/18.7	25.1/14.1	18.2/8.3	9.7/1.6	2.2/−4.9	Watseka	Typic Endoaquolls
NE	Lawrence†	40.29, −98.26	679	3.4/−5.8	5.8/−4.2	12.3/0.7	18.6/6.0	23.7/12.0	28.4/17.5	31.1/20.1	30.3/19.3	25.7/14.4	19.0/8.1	10.8/0.7	4.1/−4.4	Hastings	Udic Argiustolls
NJ	Adelphia‡	40.22, −74.25	1200	2.5/−4.6	3.8/−3.8	8.6/0.1	15.0/5.6	20.9/11.1	25.9/16.4	28.4/19.3	27.4/18.7	24.0/15.1	17.7/8.8	11.2/3.4	5.2/−1.5	Holmdel	Aquic Hapludults
IL	Urbana†,‡	40.08, −88.23	1024	0.0/−7.7	2.4/−5.3	9.2/0.4	16.9/6.6	23.2/12.9	28.1/17.6	30.1/19.4	29.3/18.3	24.9/14.0	18.0/8.3	9.8/1.8	2.3/−4.8	Flanagan	Typic Endoaquoll
IL	Brownstown‡	38.95, −88.96	978	3.2/−4.3	5.6/−2.3	11.6/2.4	18.7/8.3	24.3/14.2	28.6/18.7	30.3/20.5	29.9/19.7	25.9/15.4	19.8/9.7	12.1/3.5	5.2/−2.3	Cisne	Mollic Albaqualfs
CA	Davis‡	38.50, −121.70	453	13.2/3.8	14.1/4.1	16.5/5.2	19.7/7.1	24.3/10.7	29.0/14.2	32.3/16.4	31.4/15.7	29.2/14.6	24.0/11.6	17.3/7.0	13.0/4.0	Brentwood	Typic Xerochrepts
KY	Lexington‡	38.13, −84.50	1197	3.5/−3.8	5.5/−2.5	11.0/1.7	18.0/7.3	23.3/12.8	27.6/17.4	29.1/19.4	28.7/18.8	25.3/14.9	19.0/8.8	11.8/3.3	5.4/−1.7	Maury	Typic Paleudalfs
IL	Dixon Spring‡	37.45, −88.67	1246	4.7/−3.0	7.0/−1.2	12.7/3.3	19.5/9.0	24.8/14.7	29.0/19.3	30.6/21.0	30.3/20.3	26.5/16.1	20.6/10.2	13.1/4.2	6.5/−1.2	Grantsburg	Oxyaquic Fragiudalfs
VA	Gretna†,‡	36.93, −79.39	1141	6.8/−1.8	9.0/−0.7	14.1/2.9	20.2/8.2	25.0/13.4	28.8/18.0	30.2/20.2	29.2/19.5	26.0/15.9	20.6/9.5	14.2/4.2	8.6/0.0	Mayodan	Typic Hapludult
TN	Milan†	35.93, −88.71	1340	7.6/−0.4	10.0/1.2	15.2/5.3	21.1/10.2	25.9/15.6	29.7/19.9	30.9/21.6	30.8/21.1	27.5/17.4	22.2/11.4	15.2/5.7	9.5/1.2	Grenada / Vicksburg	Oxyaquic Fraglossudalfs / Typic Udifluvents
OK	Muskogee†	35.74, −95.64	1074	7.5/−1.1	10.2/0.8	15.6/5.2	21.3/10.0	25.7/15.4	29.8/20.1	32.4/21.9	32.4/21.6	27.9/17.7	22.0/12.0	14.8/5.5	8.6/0.1	Parsons	Mollic Albaqualf
AR	Lewisville†	33.40, −93.58	1233	10.9/1.4	13.3/3.4	18.0/7.1	23.2/11.6	27.4/16.9	31.2/21.2	32.5/22.6	32.5/22.2	29.2/18.9	23.9/13.1	17.4/7.5	12.1/2.9	Bowie	Plinthic Paleudult
TX	Dallas†	32.58, −97.26	865	11.7/1.9	14.3/3.8	19.0/7.8	23.9/12.4	27.2/17.2	30.8/21.1	32.5/22.7	33.1/22.8	29.1/19.6	24.3/14.5	17.6/8.0	12.4/2.9	Houston Black	Udic Calciusterts
TX	Stephenville†	32.13, −98.20	756	11.5/1.3	14.0/3.2	18.9/7.4	23.9/11.9	26.8/16.6	30.2/20.6	32.1/22.1	32.5/22.1	28.5/18.7	23.7/13.8	17.1/7.3	12.0/2.1	Windthorst	Udic Paleustalfs
LA	Clinton†	30.51, −90.05	1224	15.4/5.8	17.4/7.4	20.9/10.4	24.7/14.0	28.1/18.5	30.4/22.0	30.9/23.1	31.0/23.0	29.0/20.8	25.2/15.5	20.5/10.8	16.6/7.2	Dexter	Ultic Hapludalf
TX	College Station†	30.36, −96.35	966	14.4/5.1	16.8/6.8	20.7/10.0	25.1/14.1	28.1/18.7	30.9/22.0	31.8/23.0	32.1/23.3	29.6/20.9	25.6/16.3	19.8/10.7	15.3/6.3	Norwood	Fluventic Eutrochrepts

*Rainfall, mean annual rainfall in long term; Tmax/Tmin, average maximum and minimum air temperatures in long term. The long-term data used for calculate Rainfall and Tmax/Tmin were based on period 1984 to 2014 for all locations.
†Switchgrass data were evaluated in this location.
‡*Miscanthus* data were evaluated in this location.

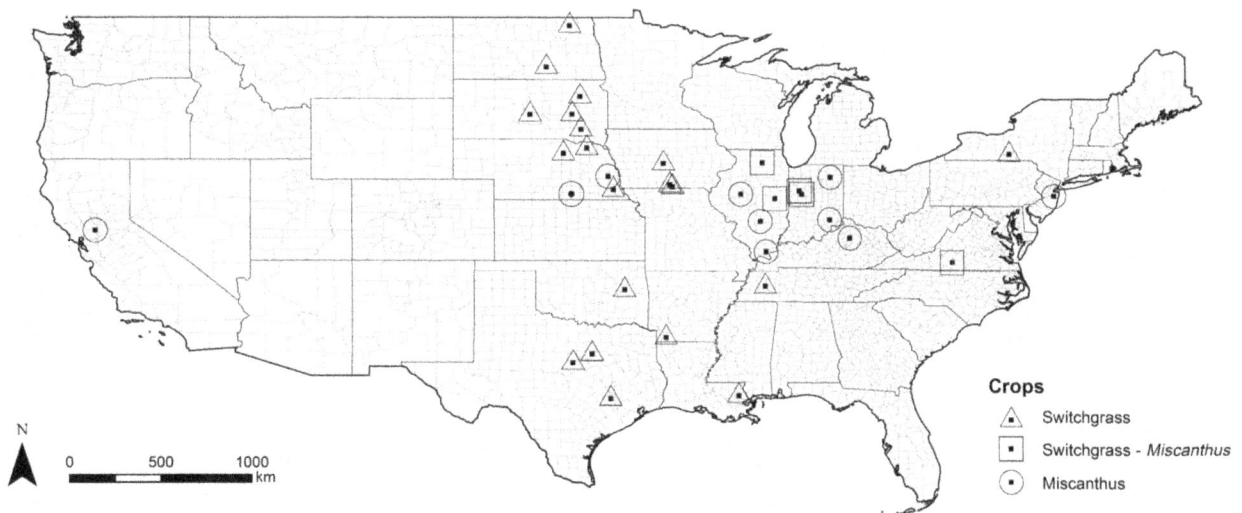

Fig. 1 Experimental sites included in the dry matter yield database for APSIM calibration/validation for switchgrass (△), switchgrass-*Miscanthus* (⊡) and *Miscanthus* (⊙). The data used for switchgrass validation were grouped in northern and southern locations. Northern locations: Indiana (IN), Illinois (IL), Tennessee (TN), Nebraska (NE), Iowa (IA), South Dakota (SD), New York (NY) and North Dakota (ND). Southern locations: Texas (TX), Virginia (VA), Oklahoma (OK), Louisiana (LA) and Arkansas (AR).

NOAA database. Interestingly, recent evaluations of the NASA-POWER solar radiation data indicate very good agreement with measured solar radiation data in areas with flat topography (White *et al.*, 2011; Wart *et al.*, 2013) and with maximum and minimum air temperatures across the US (White *et al.*, 2008). Our evaluations demonstrated a similar fit for daily solar radiation (n = 59031 daily observations) and maximum and minimum air temperatures (n = 69505 daily observations) using measured data from 19 weather stations near the experimental locations used in this study (Fig. 2; Table S1). The number of air temperature data corrections/filled data was always lower than 2% for all variables.

Long-term monthly mean minimum air temperature ranged from −19.7 to 23.3 °C and the monthly mean maximum air temperature between −10.7 to 32.5 °C. The mean annual rainfall varied from 452 to 1340 mm for Munich, ND and Milan, TN respectively. A summary of climate information by location is reported in Table 1.

Soil parameterization

APSIM requires several soil parameters to adequately reflect the variability among locations (Probert *et al.*, 1998; www.apsim.info). As the soil database of both the 7.3 and 7.5 release versions do not include the soils where these biophysical experiments were conducted, new APSIM soil profiles were created using the following process. First, dominant soil series were identified for each location based on data provided in the literature and in consultation with agronomists and local scientists (Table 1). Second, for each soil series actual soil data (texture, organic carbon [OC] and pH) were obtained from the National Cooperative Soil Survey Soil Characterization Database (NCSS, http://ncsslabdatamart.sc.egov.usda.gov) (see actual data in Table 2). Estimates of the drained upper limit (DUL) and the drained lower limit (LL) were estimated using

the HYDRAULIC PROPERTIES CALCULATOR Software developed by Saxton & Rawls (2006) based on soil texture and OC data obtained from NCSS. The estimating equations reported by Saxton & Rawls (2006) were developed by correlation of an extensive data set (1722 samples) provided by the USDA/NRCS National Soil Survey Laboratory. As measured data of soil water parameters were not available for each evaluated location, the accuracy of the HYDRAULIC PROPERTIES CALCULATOR Software (Saxton & Rawls, 2006) to predict soil water parameters was gauged using the observed data of LL (mm mm^{-1}) and DUL (mm mm^{-1}) from soil series near the locations used in this study. An example of the soil parameterization for Drummer soil series at Water Quality Field Station, West Lafayette IN is presented in Table 2. A complete description of actual and estimated soil parameters used for the calibration/validation of APSIM are provided in the supplementary information for all location evaluated in this study (Table S5).

The APSIM modules were configured for soil N and C (APSIM *SoilN*), crop residue dynamics (APSIM *Surface Organic Matter*) and soil water (APSIM *SoilWat*). Actual OC (%) values were used for initialization (Table S5). To initialize the soil nitrogen pool for the simulations a 10-year simulation of previous management at the experimental locations (corn-soybean rotations), the location-specific climate and soil physical data were used. Crop growth data from these simulations were excluded from subsequent analysis.

For each soil, organic matter (OM, %, OM = OC*1.72; Dalgliesh & Foale, 1998); soil pH 1 : 5 (pH measured for a ratio of 1 part soil and 5 parts water solution according to GlobalSoilMap, 2012; estimated by Libohova *et al.*, 2014); texture class; air dry (AD, mm mm^{-1}) corresponding to the moisture limit for dry evaporation of the soil; saturated volumetric water (SAT, mm mm^{-1}); bulk density (BD, Mg m^{-3}); hydraulic conductivity (ks, mm day^{-1}); total porosity (PO, 0–1 calculated as 1-BD/2.6); drainage coefficient (SWCON, day^{-1}) were estimated (Table S5).

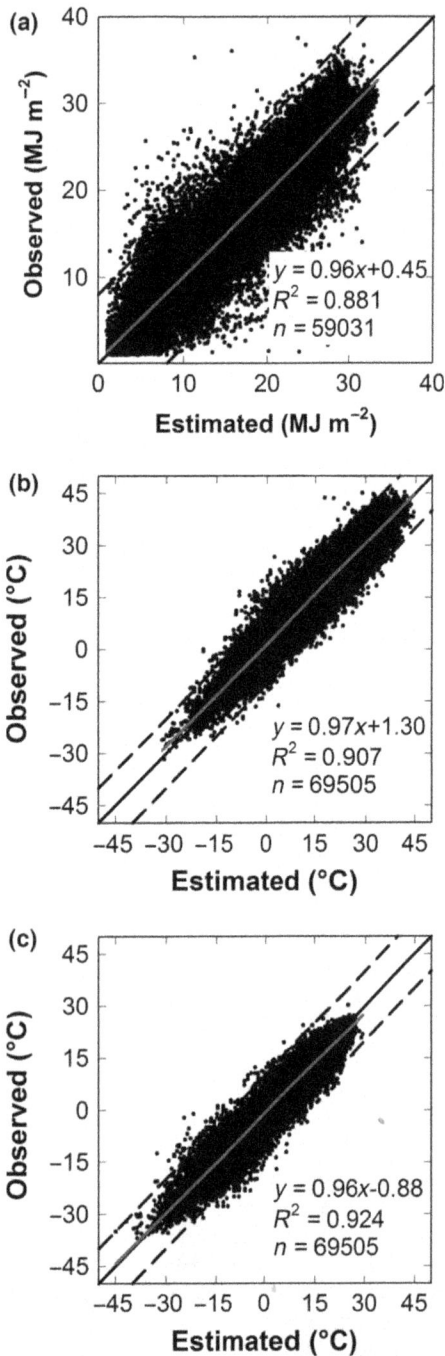

Fig. 2 Observed (a) daily incident solar radiation, (b) daily maximum air temperature and (c) daily minimum air temperature measured at 19 meteorological stations across United States plotted against daily data estimated from NASA-POWER. Solid black line represents the function $y = x$ (i.e. 1 : 1 relationship), dotted line represents ± 20% of 1 : 1 relationship and solid grey line represent the linear fit to the data.

Saturated water content was calculated from BD as described by Dalgliesh & Foale (1998). The parameter AD was estimated as 0.5 × LL in 0–0.15 m soil layer, 0.9 × LL in 0.15–0.3 m soil layer and

equal to LL at depths >0.3 m (Cresswell et al., 2009). The SWCON, the rate at which water drains, was estimated from DUL and BD (Jones & Kiniry, 1986). For each soil layer within each soil series the water extraction coefficient (KL, mm day^{-1}) was set at 0.08 mm day^{-1} (Robertson et al., 1993a,b; Dardanelli et al., 1997, 2004). The root exploration factor (XF, 0–1) was set to 1 for up to 1 m depth and then decreased exponentially to 0.6 at the maximum soil depth (Monti & Zatta, 2009). The maximum rooting depth was set according to maximum soil depth when there were no impediments to crop rooting. A complete description of actual and estimated soil parameters used for the calibration/validation of APSIM are provided in the supplementary information for each location (Table S5).

Initial soil water values were not available at most locations. Hence, an analysis of soil moisture data at sowing in some locations was performed. The data of seven climate monitoring stations (Lincoln, NE; Bedford and Lafayette, IN; Ithaca, NY; Crossville, TN; Bronte and Palestine, TX) were obtained from US Climate Reference Network (USCRN, http://www.ncdc.noaa.gov [Diamond et al., 2013; Bell et al., 2013]). The average of soil moisture at 0.2 m depth from these locations, from January to late April, was compared with the DUL of the 36 soils used in this study. With the exception of the TX sites, the initial soil water moisture in spring was always close to DUL. Based on this analysis, 100% of the initial soil water content was used at the onset of all simulations (not shown).

APSIM configuration

Without pre-existing APSIM modules for simulating switchgrass and Miscanthus we re-parameterized the APSIM Lucerne (Robertson et al., 2002) and Sugarcane (Keating et al., 1999) modules to simulate the growth of switchgrass (Table 4) and Miscanthus (Table 5) respectively. Switchgrass and Miscanthus simulations were undertaken using a daily time-step of the APSIM Version 7.5 and 7.3 respectively (Keating et al., 2003; Holzworth et al., 2014). After exhaustive and comparative analysis of plant modules, the re-parameterized APSIM Lucerne module best simulated switchgrass growth in terms of phenological and physiological functions (Table 4). Similarly, the phylogenetic proximity between Saccharum officinarum and Miscanthus, and similarity in physiology, phenology and growth, was the main justification for using the Sugarcane module as a starting point for re-parameterizing APSIM for predicting DM yield of Miscanthus. All the changes in the APSIM Lucerne and Sugarcane modules were implemented through changing parameterization in the initialisation file in extensible mark-up language (XML) format. Different management rules (i.e. sowing, harvesting, fertilization, irrigation, plant density, row spacing, etc.) were created according to practices used in the field and are reported in detail in the supplementary information for switchgrass (Table S3) and Miscanthus (Table S4). The harvesting rules were set to remove the biomass up to 0.03 m (Ojeda et al., 2016). When the dates of management interventions were not available, local average dates for the application of these practices were used. A complete description of management practices used in the simulations is reported in the supplementary information (Tables S3 and S4 for switchgrass and

Table 2 Soil parameters for Drummer soil series at Water Quality Field Station, West Lafayette IN. Estimated data were obtained from actual data using pedotransfer functions

| | Actual data | | | | | | | Estimated data | | | | | | | |
| | Texture | | | | | | | | | | | | | | |
Depth cm	Sand %	Silt %	Clay %	OC %	pH 1 : 1	OM %	pH 1 : 5	AD mm mm^{-1}	LL	DUL	SAT	BD Mg m^{-3}	ks mm day^{-1}	PO (0–1)	SWCON day^{-1}
0–23	15	56	30	2.49	6.0	4.3	5.8	0.095	0.189	0.361	0.483	1.24	165	0.52	0.277
23–43	12	56	32	0.96	6.3	1.7	6.2	0.180	0.200	0.371	0.490	1.22	152	0.53	0.270
43–71	9	60	31	0.71	6.6	1.2	6.5	0.194	0.194	0.373	0.493	1.21	165	0.53	0.268
71–94	14	62	23	0.34	7.2	0.6	7.1	0.152	0.152	0.342	0.478	1.25	232	0.52	0.292
94–109	65	26	9	–	8.0	–	8.0	0.075	0.075	0.174	0.441	1.35	1317	0.48	0.575
109–190	64	29	6	–	8.0	–	8.0	0.075	0.075	0.177	0.441	1.35	1292	0.48	0.565
190–244	41	41	18	–	8.1	–	8.1	0.126	0.126	0.269	0.447	1.33	439	0.49	0.372
244–330	42	42	16	–	8.1	–	8.1	0.126	0.126	0.267	0.446	1.33	445	0.49	0.375

OC, organic carbon; pH (1 : 1), pH in a 1 : 1 suspension of soil in water; pH (1 : 5), pH in a 1 : 5 suspension of soil in water; OM, organic matter; AD, air dry; LL, lower limit; DUL, drained upper limit or field capacity; SAT, saturated volumetric water content; ks, hydraulic conductivity; BD, bulk density; PO, total porosity; SWCON, drainage coefficient.

Miscanthus, respectively). The used model output from each simulation was crop DM yield (kg ha^{-1}). The simulations in West Lafayette IN included the additional analysis of LAI as another model output. The cultivars used in the field experiments were created as two generic switchgrass genotypes (generic lowland and generic upland) and one Miscanthus genotype (generic) differing in thermal time requirements needed to attain specific phenological stages (Table S6).

Sensitivity analysis

A sensitivity analysis enables users to determine the responses of key model outputs (e.g., harvestable biomass, hereafter DM yield) to variations in selected input parameters. Hence, as part of model calibration, a sensitivity analysis of the APSIM Lucerne and Sugarcane module's to plant and soil parameters (Table 3) was performed using the one-at-a-time method to evaluate parameter influence on LAI and DM yield. Three soil datasets were chosen to represent a range in relevant soil textures (silty, loamy and sandy). We used these soils to analyse the sensitivity of the model parameters through a large range of plant available water capacity (PAWC).

Based on an exhaustive review of the literature, and field-measured data, and in order to adequately predict the growth of switchgrass and Miscanthus with the APSIM Lucerne and Sugarcane modules, the most sensitive plant parameters for switchgrass (Fig. 3; Fig. 6) and Miscanthus (Fig. 3; Fig. 9) were identified (Tables 4 and 5 for switchgrass and Miscanthus, respectively). Thereafter, in extensible mark-up language (XML) format, these parameters were modified (Tables 4 and 5 for switchgrass and Miscanthus, respectively). In all cases, the modified parameters were calculated as an average of reported values in the literature or field measurement based on the range of each parameter. For all locations, we used the same values of parameters to simulate the DM yield in the re-parameterized modules. We followed the same parameterization process for

both crops, although there were more sensitive parameters for switchgrass than for Miscanthus, which explain the differences in the number of parameters listed in Tables 4 and 5. It should be noted that we only showed the modified parameters, since default (original) values might be easily obtained from the XML file available in the free APSIM version.

In the APSIM Sugarcane module crop growth is divided into two sections, plant and ratoon crop. The parameters in the plant and ratoon crop sections determine the crop growth for the first and second harvests onwards. Hence, the model modifications were made in both sections of XML file (plant and ratoon crop). To assess potential errors in soil datasets, after plant model modifications, a sensitivity analysis was undertaken for PAWC (Fig. 4a for switchgrass and Fig. 4b for Miscanthus). The maximum variation (%) in the parameters that determine the maximum PAWC in the soil - AD, LL, DUL and SAT - was determined based on the 36 soils used in this study (Table S2). Therefore, for the sensitivity analysis AD, LL, DUL and SAT were modified in ±29%, ±23%, ±10% and ±5%, respectively, in order to provide realistic boundaries. Second, sensitivity of KL, XF and initial OC was evaluated by modifying the range ±50% of initial values (Fig. 4c,e,g for switchgrass and Fig. 4d, f, h for Miscanthus). Using the same approach explained previously, the maximum pH variation (%) was determined (Table S2). Model sensitivity to pH change was evaluated by increasing and decreasing soil pH by 14% of the actual soil values of switchgrass (Fig. 4i) and Miscanthus (Fig. 4j). The total number of simulations necessary to complete the sensitivity analysis of soil parameters was 958.

Re-parameterization of switchgrass plant module

Crop phenology in APSIM is controlled by the sum of heat units from sowing to maturity. Accordingly, the parameter y_tt (thermal time requirements needed to attain specific phenological stages) was set to the growth habit of switchgrass (Kiniry et al.,

Table 3 Plant and soil parameters evaluated through the sensitivity analysis for the APSIM re-parameterization to simulate the DM yield of switchgrass and *Miscanthus* with their description, acronym/abbreviation. Note that some parameters vary according with the crop

| | | Acronym/Abbreviation | |
	Definition	Switchgrass	*Miscanthus*
Plant	Thermal time calculation	y_tt	y_tt
	Stem reduction effect on phenology	stage_stem_reduction_harvest	–
	Radiation use efficiency	y_rue	y_rue
	Transpiration efficiency coefficient	transp_eff_cf	transp_eff_cf
	Temperature response of photosynthesis – RUE*	y_stress_photo	y_stress_photo
	Water stress on phenology	y_swdef_leaf	swdf_pheno_limit
		y_swdef_pheno	y_swdef_pheno
		y_swdef_pheno_flowering	y_swdef_pheno_flowering
		y_swdef_pheno_start_grain_fill	y_swdef_pheno_start_grain_fill
	Water stress on photosynthesis	–	swdf_photo_limit
	Water stress during photosynthesis to leaf senescence rate	–	sen_rate_water
	Frosting stress	frost_fraction	–
	Extinction coefficient	y_extinct_coef	extinction_coef
	Biomass partitioning	ratio_root_shoot	ratio_root_shoot
Soil	Plant available water capacity	PAWC	PAWC
	Water extraction coefficient	KL	KL
	Root exploration factor	XF	XF
	Initial organic carbon	OC	OC
	pH	pH	pH

*RUE, radiation use efficiency.

2005). Similarly, the *stage_stem_reduction_harvest* parameter was modified so that, after harvest, switchgrass starts a new regrowth. In order to achieve this initial point of growth, *stage_stem_reduction_harvest* was reduced from 4 to 3 (Table 4).

The LAI and, hence, DM yield in APSIM are defined directly by the RUE and the transpiration efficiency coefficient (Kc) parameters both fixed in each phenological stage. Modifications of the physiological parameters reported elsewhere for the APSIM *Lucerne* module (Dolling *et al.*, 2005; Brown *et al.*, 2006; Chen *et al.*, 2008) were also used here to simulate switchgrass DM yields. After sensitivity analysis, RUE (coded by *y_rue*; Madakadze *et al.*, 1998; Kiniry *et al.*, 1999, 2012; Heaton *et al.*, 2008; Jain *et al.*, 2010; Trybula *et al.*, 2014) and Kc (coded by *transp_eff_cf*; Byrd & May, 2000) were set based on switchgrass values obtained in the literature (Table 4). For all locations, we used the same RUE and Kc value. In addition, two other parameters (the temperature response of photosynthesis, *y_stress_photo*, and the extinction coefficient, *y_extinct_coef*) directly associated with DM yield were modified as follows. The temperature response of photosynthesis was modified based on previous corn studies (Andrade *et al.*, 1993; Louarn *et al.*, 2008) validated for switchgrass (Grassini *et al.*, 2009) (Table 4). Similarly, *y_extinct_coef* was modified based on the differences in leaf structure between lucerne vs. switchgrass (Kiniry *et al.*, 1999; Trybula *et al.*, 2014).

Re-parameterization of Miscanthus plant module

Several researchers have reported wide differences in RUE values for *Miscanthus*. Kiniry *et al.* (2012) reported a low value of

RUE (1.3 g MJ^{-1}) in central TX. In contrast, the same authors reported a value of 3.7 g MJ^{-1} in the north-eastern MO whereas Heaton *et al.* (2008) reported a high value of RUE (4.1 g MJ^{-1}) in IL. Other studies reported RUE values of 2.2, 2.4 and 2.3–3.0 g MJ^{-1} in Italy (Cosentino *et al.*, 2007), UK (Clifton-Brown *et al.*, 2001) and IL (Dohleman & Long, 2009), respectively. Hence, given these discrepancies in RUE values among studies, the parameter *y_rue* was modified from 1.8 g MJ^{-1} to 3.0 g MJ^{-1} for all phenological stages (Table 5). This used value of RUE was calculated from these studies as an average of the ratio between accumulated yield (from emergence to peak biomass) and total annual incident radiation. For all locations, we used the same RUE value (3.0 g MJ^{-1}).

The light extinction coefficient (coded by *y_extinct_coef*) through the leaf cover of the crop provides a measurement of the absorption of light by leaves (Zub & Brancourt-Hulmel, 2010). *Miscanthus* achieves *y_extinct_coef* values between 0.45 (Trybula *et al.*, 2014) to 0.68 (Clifton-Brown *et al.*, 2000). Based on the insensitivity to the changes of this parameter in the range reported in the literature (Fig. 3d), the *y_extinct_coef* default value for *Miscanthus* (0.38) was unchanged. *Miscanthus* partition biomass has been parameterized for the WIMOVAC model using data from Beale & Long (1997) and has been validated using data from European studies (Miguez, 2007). Based on the data collected by Burks (2013) and Trybula *et al.* (2014) in West Lafayette IN, the *ratio_root_shoot* parameter was modified in APSIM for all phenological stages (Table 5). These authors measured the *Miscanthus* aboveground and root biomass in different crop growth stages. Therefore, we used these data to re-parameterize the *ratio_root_shoot* into the *Sugarcane*

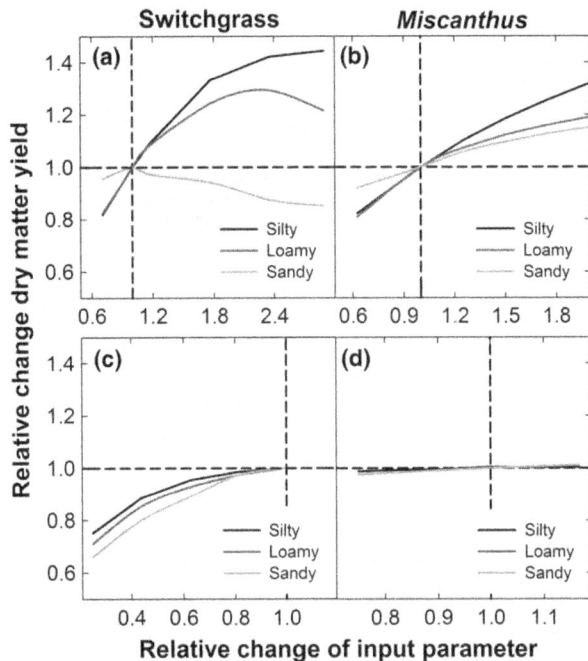

Fig. 3 Relative change in predicted dry matter yield of switch-grass and *Miscanthus* vs. relative change of plant parameters for three contrasting soil textures. Switchgrass and *Miscanthus* were modelled using APSIM *Lucerne* and APSIM *Sugarcane* modules, respectively. The plant parameters analysed were: *y_rue*, radiation use efficiency (a, b) and *y_extinct_coef*, extinction coefficient (c, d). The value 1 on the *x*-axis corresponds to the model default values used in the sensitivity analysis. Broken lines indicate the baseline parameter and no changes in dry matter yield, respectively.

APSIM module for emergence, juvenile and flowering stages. A complete description of *ratio_root_shoot* values obtained by these authors are provided in the supplementary information (Table S7).

Evaluation of model performance

Initially, model performance was visually assessed by comparing scatter plots of observed values in the *y*-axis vs. modelled values in the *x*-axis (Piñeiro *et al.*, 2008). When multiple data points were available for a particular treatment in an experiment, standard deviations are included as an estimate of error. The evaluation of model performance described in Tedeschi (2006) and Kobayashi & Us Salam (2000) were used to statistically evaluate model performance. The parameters used were: observed and modelled mean and standard deviation of the DM yield, the concordance correlation coefficient (CCC), and mean square error (MSE). The CCC integrates precision through Pearson's correlation coefficient, which represents the proportion of the total variance in the observed data that can be explained by the model, and accuracy by bias which indicates how far the regression line deviates from the concordance ($y = x$) line. Similarly, the MSE was partitioned into bias (SB, %, the bias of the simulation from the measurement) and mean

square variation (MSV, %, the difference between the simulation and the measurement with respect to the deviation from the means), using IRENE software (Fila *et al.*, 2003). Bias and MSV are orthogonal and, consequently, can be analysed independently (Kobayashi & Us Salam, 2000). Model calibration was deemed complete when the CCC and SB were higher than 0.7 and <30%, respectively, for the LAI and DM yield of both crops.

In both crops, the growth period from sowing was simulated including the establishment phase, during which time rhizome biomass, root depth, and DM yield are increasing, and the post-establishment phase, in which perennial organs and root system are fully developed and the DM yield is fairly constant. This is influenced more by variation in weather than changes in plant establishment/underground organ development. However, only the observed DM yield from the post-establishment phase was included in this analysis to evaluate the accuracy of the model to predict DM yield with the established crop. The duration of the establishment phase varied from two to four years, depending on the experimental site (Tables S3 and S4). For switchgrass validation, the data sets were grouped by northern locations (IN, Indiana; IL, Illinois; TN, Tennessee; NE, Nebraska; IA, Iowa; SD, South Dakota; NY, New York; ND, North Dakota) and southern locations (TX, Texas; VA, Virginia; OK, Oklahoma; LA, Louisiana; AR, Arkansas). The same grouping was not applied to *Miscanthus*, because DM yields in southern US locations are extremely low and difficult to find in the literature. Therefore, the capability of APSIM to simulate the *Miscanthus* DM yield was not evaluated in southern US locations.

Results

Switchgrass

The most sensitive parameters of plant module were RUE and the light extinction coefficient coded by *y_rue* and *y_ extinct_coef* parameters, respectively. The sensitivity of the model to the modification of these parameters (Fig. 3a,c) was high in the selected soils. The largest change in DM yield (29% and 44%) occurred when *y_rue* was increased from 1.7 to 4 g MJ^{-1} and 4.9 g MJ^{-1} in the loamy and silty soils, respectively (Fig. 3a). In contrast, increasing *y_rue* to 4.9 g MJ^{-1} reduced switchgrass DM yield by ~15% in the sandy soil (Fig. 3a). The trend in DM yield to decreased *y_extinct_-coef* was similar for these soils with declines 5–11% and 25–34% when *y_extinct_coef* was decreased from 0.8 to 0.5 and 0.8 to 0.2, respectively (Fig. 3c).

The sensitivity analysis carried out to identify possible effects of changing soil parameters in the re-parameterized model (Fig. 4) on switchgrass DM yields also showed soil-specific responses with the greatest responses in DM yield occurring in the sandy soil. For example, when PAWC was decreased, predicted DM yield declined 1%, 7% and 11% for silty, loamy and

Table 4 Parameterization of APSIM *Lucerne* plant module for switchgrass simulation. List of the modified parameters with their section into the XML file, definitions, acronym, units, default (original), used values (modified) and range of values found in the literature and references

Definition	Acronym	Units	Default value/s	Used values	Range/References
Crop phenology					
Thermal time calculation	x_temp	°C	1 5 10 15 30 40	12 25 45	Kiniry et al., 2005
	y_tt	°Cd	0 3 6.5 10 25 0	0 13 0	
Stem reduction	stage_code	0–11	1 2 3 4 5 6 7 8 9 10 11	1 2 3 4 5 6 7 8 9 10 11	
Stem reduction effect on phenology	stage_stem_reduction_harvest	0–11	1 2 3 4 4 4 4 4 4 11	1 2 3 3 3 3 3 3 3 11	
Photosynthesis, biomass growth and partition					
Stage dependent RUE*,†	stage_code	0–11	1 2 3 4 5 6 7 8 9 10 11	1 2 3 4 5 6 7 8 9 10 11	**1.7–4.9**/Madakadze et al., 1998; Kiniry et al., 1999, 2012 Heaton et al., 2008; Jain et al., 2010
	y_rue	g MJ^{-1}	0 0.1 0.4 0.7 1.7 1.7 1.7 1.7 0 0	0 0 0 1.7 1.7 1.7 1.7 0 0	
Stage dependent transpiration efficiency coefficients,†	stage_code	0–11	2 3 4 5 6 7 8 9 10 11	2 3 4 5 6 7 8 9 10 11	Trybula et al., 2014
	transp_eff_cf	Pa	0.001 0.003 0.005 0.006 0 0 0 0 0 0	0.0055 0.0055 0.0055 0.0055 0.0055 0.0055 0 0 0 0	Byrd & May, 2000
Temperature response of photosynthesis - RUE	x_axe_temp	°C	0 8 20 25 40	0 6 9 12 17 21 42	Andrade et al., 1993; Louarn et al., 2008
	y_stress_photo	0–1	0 1 1 1 0	0 0.2 0.4 0.7 1 1 0	
Extinction coefficient	x_row_spacing	mm	100 1000	100 1000	Grassini et al., 2009; **0.20–0.55**/Kiniry et al., 1999
	y_extinct_coef	0–1	0.8 0.8	0.5 0.5	Trybula et al., 2014

*RUE, radiation use efficiency.

†The modification shown in this table made in the plant section for transpiration efficiency coefficient also was made in the lucerne regrowth section of the XML file. The meaning of bold values is the range of parameters values found in the literature.

Table 5 Parameterization of APSIM *Sugarcane* module plant module for *Miscanthus* simulation. List of the modified parameters with their section into the XML file, definitions, acronym, units, default (original), used values (modified) and range of values found in the literature and references

Definition	Acronym	Units	Default value/s							Used value/s							Range/References
Plant and ratoon crop*																	
Stage dependent RUE†	stage_code	0–6	1	2	3	4	5	6		1	2	3	4	5	6		**1.3–4.1**/Clifton-Brown
	rue	g MJ⁻¹	0	0	1.8	1.8	1.8	0		0	0	0	3	3	3	0	*et al.*, 2001 Cosentino *et al.*, 2007 Heaton *et al.*, 2008 Dohleman & Long, 2009 Jain *et al.*, 2010 Kiniry *et al.*, 2012 Trybula *et al.*, 2014
Biomass partitioning	stage_code	0–6	1	2	3	4	5	6		1	2	3	4	5	6		**0.2–0.8**/Burks, 2013 Trybula *et al.*, 2014
	ratio_root_shoot	0–1	0	0	0.2	0.2	0.1	0		0	0	0	0.85	0.37	0.22	0	

*The modifications shown in this table made in the plant section into the XML file also were made in the ratoon crop section.
†RUE, radiation use efficiency.

sandy soils, respectively (Fig. 4a). When PAWC was increased, the DM yield was enhanced 3%, 4% and 5% for silty, loamy and sandy soils, respectively (Fig. 4a). The highest DM yield response to changes in XF, KL and pH also occurred in the sandy soil (13%, 14% and 21%, respectively; Fig. 4e,c,i). By comparison, the model was less sensitive (<4%) to the changes in the initial OC (Fig. 4g).

The model with the default settings demonstrated a poor ability to simulate LAI and DM accumulation of switchgrass. Summary statistics comparing observed and modelled LAI from original and modified model parameters at the Water Quality Field Station in West Lafayette IN demonstrated the improvement in LAI predictions, as indicated by the increased CCC values (0 to 0.81) and a reduction in the SB (93 to 30%) (Fig. 5a; Table 6). Similarly, and as expected, when modified plant parameters (Fig. 6; Table 4) were introduced into the APSIM *Lucerne* module, prediction of switchgrass DM yield at the same location was improved, as indicated by the increase CCC (0.11 to 0.96) and the reduction in SB (57 to 4%) (Table 6).

The APSIM *Lucerne* module showed excellent accuracy for predicting the accumulated DM yield at IN locations used for switchgrass model calibration (Fig. 10a) as evidenced by the values of 0.93 for the CCC and 0% for the SB (Table 7). The APSIM *Lucerne* module also predicted DM yields when validated using yield data from trials conducted at northern locations (Fig. 11a), but model accuracy at southern locations was unsatisfactory. In fact, the CCC = 0.95 from comparisons using data from northern locations contrasted with the CCC = 0.01 for comparisons using data from southern locations (Table 7). Remarkably, SBs obtained for northern and southern locations were similar, 0 vs. 2%. The observed switchgrass DM yield during

validation ranged from 5329 kg DM ha⁻¹ in SD to 10668 kg DM ha⁻¹ in IN, with the average discrepancy in DM yield being 1%. The modelled DM yield ranged from 1391 kg ha⁻¹ in VA to 10786 kg ha⁻¹ in TN. The better DM yield predictions in the northern locations were in IL, TN, NE, IA and SD. In contrast, in NY, IN, and ND the switchgrass DM yield was simulated with less precision (Table 7). By comparison, the modelled DM yields at southern locations were, on average, 10% less than observed values. When data were clustered by ecotype at northern locations, the DM yield was better predicted for upland ecotypes (0.96 for the CCC and 0% for the SB) than for lowland ecotypes (0.64 for the CCC and 9% for the SB). The variation of DM yield was well predicted by the re-parameterized model irrespective of stage of establishment of the crop.

Better estimates of soil water parameters (LL and DUL) were obtained from the HYDRAULIC PROPERTIES CALCULATOR Software (Saxton & Rawls, 2006) for northern locations (CCC = 0.92–0.98 and SB = 0–14%) than for southern locations (CCC = 0.75–0.92 and SB = 0–37%) (Table 8; Fig. 7).

Results of regression analysis of observed DM yields on accumulated annual rainfall revealed a poor fit at southern US locations ($R^2 = 0.18$ and slope regression −4.43; Fig. 8a). In contrast, the rainfall regression at northern locations showed a greater R^2 value (0.43) than the southern locations, and a positive slope (6.18) (Fig. 8b).

Miscanthus

As with the *Lucerne* plant module, the most sensitive parameter of the *Sugarcane* plant module was y_rue. However, unlike y_rue, the model was not sensitive to changes in y_extinct_coef (Fig. 3d). Model sensitivity to modification of y_rue (Fig. 3b,d) varied depending on

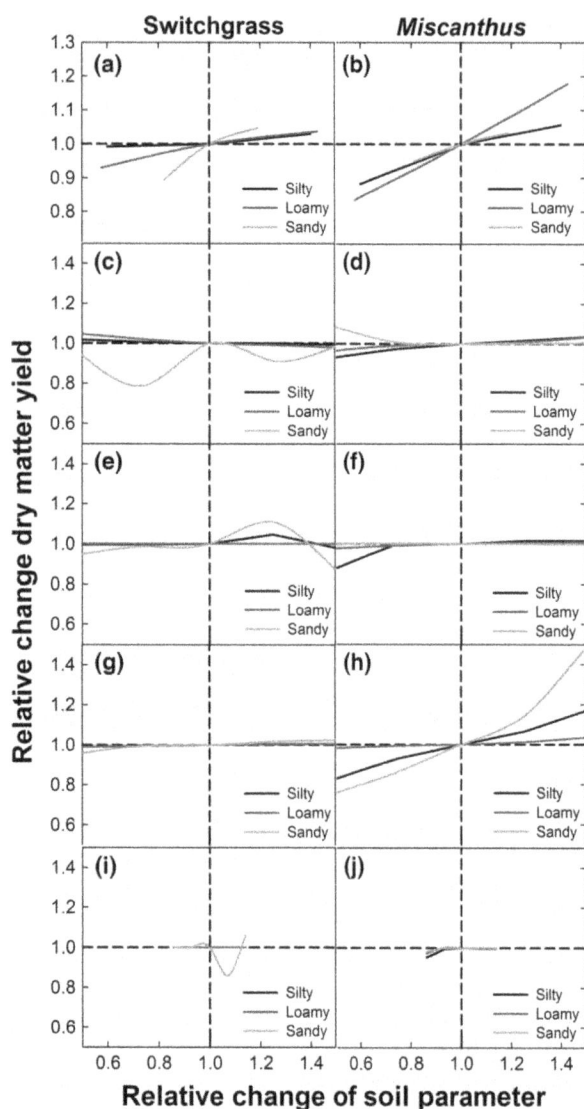

Fig. 4 Relative change in dry matter yield of switchgrass and *Miscanthus* vs. relative change of soil parameters using APSIM *Lucerne* and APSIM *Sugarcane* modules, respectively, for three contrasting soil textures. The soil parameters analysed were PAWC, plant available water capacity (a, b); KL, water extraction coefficient (c, d); XF, root exploration factor (e, f); OC, initial organic carbon (g, h) and pH (i, j). The value 1 on the x-axis corresponds to the default values used in the sensitivity analysis. Broken lines indicate the baseline parameter and no changes in dry matter yield, respectively.

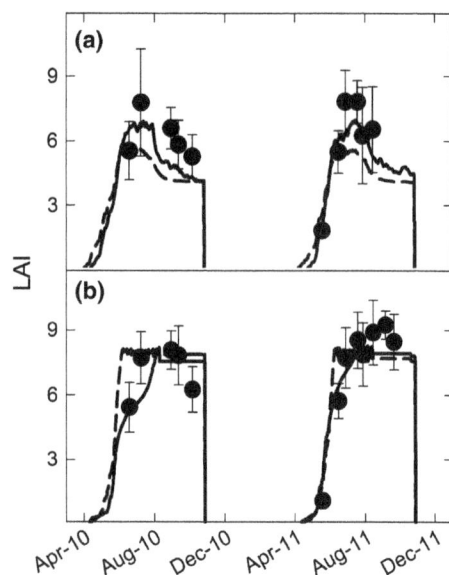

Fig. 5 Modelled pre- (---), post-APSIM modification (—) and observed (●) LAI of (a) switchgrass and (b) *Miscanthus* at Water Quality Field Station (West Lafayette, IN) during two seasons. Vertical bars represent the standard deviation of observed values.

soil type. The largest change in DM yield (increment of 15–32%) occurred when y_rue was increased from 1.8 to 4 g MJ^{-1} (Fig. 3b). Reducing y_rue from 1.8 to 1.25 g MJ^{-1} resulted in 8–19% lower DM yields when compared to initial model conditions (Fig. 3b).

The sensitivity analysis carried out to identify possible effects of soil parameters (Fig. 4) on *Miscanthus* DM yield showed differential responses depending on soil

type and parameter. When PAWC was increased, changes in DM yield were higher for the loamy soil (18%) than for the silty and sandy soils (6% and 3% respectively; Fig. 4b). Similarly, when PAWC was decreased, the reductions in DM yield were greater for the loamy soil (17%) than for the silty and sandy soils (12% and 5%, respectively; Fig. 4b). When initial OC was increased by 50% from default values, predicted increases in DM yield on sandy soil (48%) were higher than the silty soil (17%) and the loamy soil (4%; Fig. 4h). In contrast, the model exhibited low sensitivity of DM yield to soil pH, KL and XF with changes in DM yield predicted to be no >5%, 9% and 12% for the respective parameters (Fig. 4j,d,f).

The original APSIM *Sugarcane* model with the default plant parameters could not accurately predict *Miscanthus* LAI and accumulated DM yield. Summary statistics comparing observed to predicted LAI with the re-parameterized model using data from the Water Quality Field Station in West Lafayette, IN demonstrated improvement in LAI predictions as indicated by the high CCC (0.69) and low SB (<30%) (Fig. 5b; Table 9). Similarly, and as expected, when modifications of plant parameters (Fig. 9; Table 5) were introduced into the model, the prediction of *Miscanthus* DM yield, at the same location was improved as indicated by the excellent CCC (0.94) and low SB (<30%) (Table 9).

The re-parameterized APSIM *Sugarcane* module showed excellent accuracy for predicting *Miscanthus*

Table 6 Summary statistics indicating the cumulative improvement that resulted from re-parameterization of the APSIM *Lucerne* model for predicting LAI (n = 11) and dry matter yield (n = 20) of switchgrass at Water Quality Field Station, West Lafayette, IN. The parameters modified were *y_tt*, thermal time requirements needed to attain specific phenological stages; *y_rue*, radiation use efficiency; *transp_eff_cf*, transpiration efficiency coefficient; *y_stress_photo*, temperature response of photosynthesis and *y_extinct_coef*, extinction coefficient. CCC, SB and MSV are the concordance correlation coefficient, bias of the simulation from the measurement and mean square variation, respectively

	Original model	y_tt	y_rue	transp_eff_cf	y_stress_photo	y_extinc_coef
LAI						
Mean (Observed)	6.1	6.1	6.1	6.1	6.1	6.1
Mean (Modelled)	0.0	4.7	5.2	5.4	5.5	5.5
SD (Observed)	1.7	1.7	1.7	1.7	1.7	1.7
SD (Modelled)	0.0	0.9	1.6	1.4	1.3	1.5
Testing parameters						
CCC	0.00	0.66	0.60	0.67	0.76	0.81
SB (%)	93	60	26	23	23	30
MSV (%)	7	40	74	77	77	70
Dry matter yield (kg ha^{-1})						
Mean (Observed)	5908	5908	5908	5908	5908	5908
Mean (Modelled)	411	2324	5503	7435	6263	6154
SD (Observed)	4939	4939	4939	4939	4939	4939
SD(Modelled)	1774	1920	4094	4929	4543	4530
Testing parameters						
CCC	0.11	0.65	0.85	0.94	0.96	0.96
SB (%)	57	58	3	45	7	4
MSV (%)	43	42	97	55	93	96

DM yield at IN locations used for model calibration (Fig. 10b) as evidenced by the values of 0.92 and 13% for the CCC and SB respectively (Table 10). The model validation was acceptable for most locations (0.65 and 0% for CCC and SB, respectively; Fig. 11b). However, the model accuracy at KY and NJ was unacceptable as indicated by the low CCC values of 0.38 and 0.46, respectively (Table 10). However, the SB obtained during validation was similar and <30%. The observed DM yield of *Miscanthus* for validation ranged from 8398 kg DM ha^{-1} in VA to 33980 kg DM ha^{-1} under irrigated conditions in CA (Fig. 11b). The modelled DM yield for calibration was 9% higher than observed DM yield. This difference, however, was negligible (0.5%) when compared to the observed and modelled DM yield association done for validation (Table 10). The modelled DM yield ranged from 13883 kg DM ha^{-1} in VA to 20518 kg DM ha^{-1} in NE.

Discussion

The main objective of this study was to evaluate the ability of APSIM to simulate the growth and DM yields of switchgrass (using the re-parameterized *Lucerne* module) and *Miscanthus* (using the re-parameterized *Sugarcane* module) at several locations across the US. The modelling approach was based on an exhaustive sensitivity analysis of plant and soil parameters using the

one-at-a-time method followed by a detailed model calibration using field data from experiments in IN, and ending with a model validation using data from numerous US locations. Results indicate that these re-parameterized APSIM *Lucerne* and *Sugarcane* modules can accurately simulate growth and yield of switchgrass and *Miscanthus* respectively. Further considerations, specific to each crop, are discussed below.

Switchgrass

The original APSIM *Lucerne* module was developed and extensively tested in many environments for its ability to predict the phenology and DM yield of lucerne (Robertson *et al.*, 2002; Dolling *et al.*, 2005; Chen *et al.*, 2008; Pembleton *et al.*, 2011; Moot *et al.*, 2015; Ojeda *et al.*, 2016). However, in its original format with thermal parameters for a C3 species, the module is not able to adequately simulate switchgrass DM yield. Therefore, several modifications in plant module parameters were needed to improve the prediction of switchgrass DM yield. The range of modelled DM yield in this study for northern locations (5392 to 10 668 kg ha^{-1}) was coincident with modelled DM yields of the upland ecotype (Wang *et al.*, 2015) in the marginally saline soil of northeast Fort Collins, CO (5200 to 9600 kg ha^{-1}), as well as with the observed DM yield described by Schmer *et al.* (2008) on marginal cropland on ten farms in ND, SD

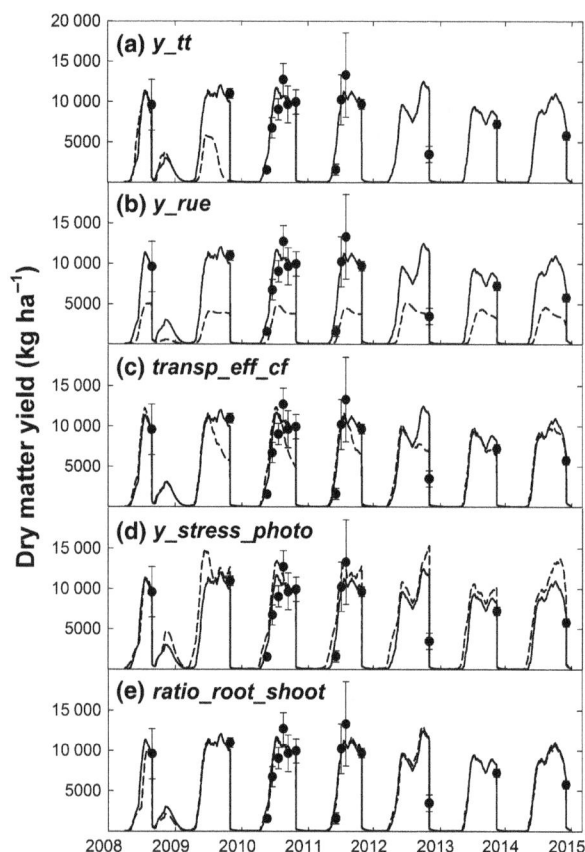

Fig. 6 Modelled pre- (---), post-APSIM *Lucerne* modification (——) and observed dry matter yield (●) of switchgrass cultivar Shawnee at Water Quality Field Station (West Lafayette, IN). The modified parameters shown in each panel are *y_tt*, thermal time requirements needed to attain specific phenological stages (a); *y_rue*, radiation use efficiency (b); *transp_eff_cf*, transpiration efficiency coefficient (c); *y_stress_photo*, temperature response of photosynthesis (d) and *ratio_root_shoot*, ratio root/shoot (e). The effect on dry matter yield was only due to the modification of each individual parameter. Vertical bars represent the standard deviation of observed values.

and NE (5200 to 11 100 kg ha^{-1}) and by Wullschleger *et al.* (2010) for 25 upland cultivars in the northern US.

The re-parameterization was based on sensitivity of DM yield when parameters were modified to values obtained in published studies (Table 4). In addition, differential effects of soil parameters on switchgrass DM yield were observed (Fig. 4). Soil water availability is one of the key soil parameters that explained most of the differences in switchgrass growth and yield (Fig. 4a). Similarly, the low PAWC due to high sand contents in the soil (Saxton & Rawls, 2006) reduced the canopy expansion decreasing the light interception and photosynthesis, thus, reduced plant growth (Durand *et al.*, 1995) in tall fescue. In addition, our results showed highest sensitivity of DM yield to parameter

changes in the sandy soil that had the lowest PAWC. Although the re-parameterized model substantially improved the prediction of DM yield at northern locations, a poor DM yield prediction at southern US locations was found (Figs 10 and 11a). This poor validation was associated with difficulty in accurately estimating PAWC at southern locations (Fig. 7), specifically estimates of LL and DUL by the HYDRAULIC PROPERTIES CALCULATOR Software (Saxton & Rawls, 2006) (Table 8). This was evidenced by the statistical analysis performed on observed and estimated LL and DUL at ten soil series near selected southern and northern locations evaluated in this study (Table 8). For example, the over and under estimation of PAWC at OK and VA (Fig. 7) respectively, would explain the over and under prediction in DM yield at both locations. In contrast, good agreement was found between observed and estimated LL and DUL in two soil series at northern sites in IN and IL (Table 8; Fig. 7). This observation suggests a new line of research that should be addressed to clarify to what extent the under or over estimation of these soil water parameters affects the outcome of predicted DM yield in APSIM.

Previous modelling efforts for predicting DM yield of switchgrass were reported. While Grassini *et al.* (2009) demonstrated similar trends in the DM yield predictions (CCC = 0.77), these results were obtained based on a limited number of observations (8) from two northern US environments (Ames, IA and Mead, NE). Additionally, the accuracy of the ALMANAC model (Kiniry *et al.*, 2005) and APSIM to predict DM yield were similar differing by <7%. However, yield values reported by these authors was nearly double what we observed in our study (ca. 17000 vs. 8000 kg DM ha^{-1}) despite comparable dryland conditions. While ALMANAC accounted for 47% of the variability in observed DM yields (Kiniry *et al.*, 2005), when the CCC was calculated from the published results, both models (APSIM and ALMANAC) were poor predictors of DM yield (CCC<0.50) at southern locations (with the exception of Stephenville TX). These authors also observed high year-to-year variability in measured yields at southern locations in the US (TX, LA and AR) and reported that this was not closely associated with variation in rainfall. The lack of fit for the southern locations was evaluated here using our complete dataset. The results showed that the southern locations showed poor fits for observed DM yield as a function of accumulated annual rainfall (Fig. 8a), in contrast with northern locations (Fig. 8b). An additional explanation for the low fit between observed and modelled DM yield at these locations is that the observed DM yields used to validate the model in TX, AR, and LA were derived from the mean of nine cultivars (Cassida *et al.*, 2005; Table S3) in each location. The absence of genotypic parameters for

Table 7 Summary statistics indicating the performance of the Agricultural Production Systems Simulator (APSIM) model in predicting the dry matter yield (kg ha^{-1}) of switchgrass. The data were divided in model calibration and validation subsets. The used data for validation were further split in groups by location (northern and southern) and ecotype (lowland and upland). Northern locations: IN, Indiana; IL, Illinois; TN, Tennessee; NE, Nebraska; IA, Iowa; SD, South Dakota; NY, New York; ND, North Dakota. Southern locations: TX, Texas; VA, Virginia; OK, Oklahoma; LA, Louisiana; AR, Arkansas. The CCC, SB and MSV are the concordance correlation coefficient, bias of the simulation from the measurement and mean square variation, respectively

	Calibration	Validation																	
		North								South									
	IN	IL	TN	NE	IA	SD	NY	IN	ND	TX	VA	OK	LA	AR	South	North	Upland*	Lowland*
Observations #	41	16	14	12	19	23	12	12	7	15	12	12	5	3	47	115	101	14
Mean (Observed)	8029	7913	10518	5392	6317	5329	8938	10668	5586	7958	6310	6874	6724	8988	7195	7439	7013	10518
Mean (Modelled)	8055	7400	10786	4960	6360	5371	9999	10570	5988	6654	1391	9357	9015	9879	6458	7496	7040	10786
SD (Observed)	4408	3177	1525	2331	1708	2059	1059	1565	1078	2991	2122	2655	3736	1145	2722	2845	2722	1525
SD (Modelled)	4076	2684	1755	2535	1503	2392	1614	1745	1126	4206	316	2961	2391	524	4288	2987	2836	1755
Testing parameters																		
CCC	0.93	0.93	0.86	0.83	0.73	0.63	0.63	0.50	0.44	−0.07	0.20	−0.55	0.08	0.68	0.01	0.95	0.96	0.64
SB (%)	0	20	9	9	0	0	47	0	12	6	88	22	27	70	2	0	0	9
MSV (%)	100	80	91	91	100	100	53	100	88	94	12	78	73	30	98	100	100	91

*For this dataset the statistical analysis was calculated using only the Northern locations.

Table 8 Summary statistics indicating the performance of HYDRAULIC PROPERTIES CALCULATOR Software (Saxton & Rawls, 2006) in predicting the soil water parameters of ten soil series from different states. Southern locations: Virginia (VA), Texas (TX), Kentucky (KY), Arkansas (AR), Oklahoma (OK) and Louisiana (LA). Northern locations: New Jersey (NJ), Illinois (IL), Indiana (IN) and New York (NY)

	Southern locations						Northern locations			
State	VA	TX	KY	AR	OK	LA	NJ	IL	IN	NY
Series	Cecil	Windthorst	Maury	Bowie	Parsons	Dexter	Holmdel	Flanagan	Chalmers	Collamer
Observations #	16	16	16	18	20	22	20	16	14	14
Testing parameters										
CCC	0.75	0.77	0.80	0.87	0.91	0.92	0.92	0.96	0.97	0.98
SB (%)	37	0	30	0	3	19	13	7	0	14
MSV (%)	63	100	70	100	97	81	87	93	100	86

each cultivar of switchgrass used by these authors, did not allow us to re-parameterize/calibrate/validate the model at the cultivar level. Although the model predicted DM yield of upland switchgrass cultivars better than that of lowland cultivars, the limited number of observations and locations evaluated for lowland ecotypes in this study did not allow us to demonstrate differences in APSIM accuracy by ecotype.

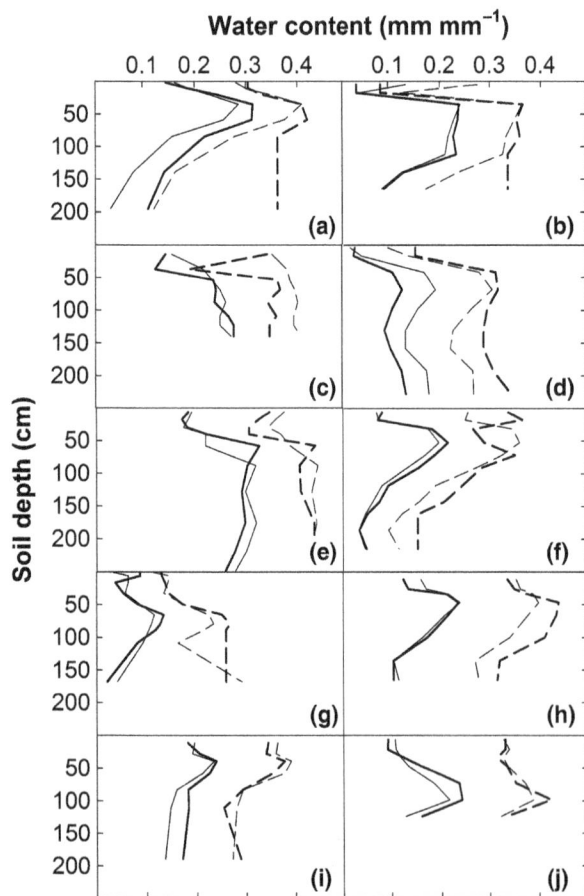

Fig. 7 Drained lower limit (LL, solid lines) and drained upper limit (DUL, dotted lines) observed (thick lines) and estimated (hair lines) by the HYDRAULIC PROPERTIES CALCULATOR Software (Saxton & Rawls, 2006) for the soil series (a) Cecil in VA, (b) Windthorst in TX, (c) Maury in KY, (d) Bowie in AR, (e) Parsons in OK, (f) Dexter in LA, (g) Holmdel in NJ, (h) Flanagan in IL, (i) Chalmers in IN and (j) Collamer in NY.

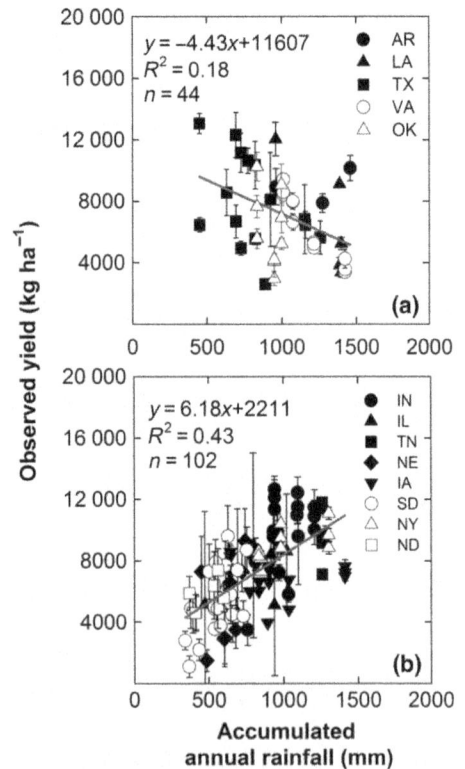

Fig. 8 Relationship between observed dry matter yield of switchgrass vs. accumulated annual rainfall for (a) southern locations and (b) northern locations in US. Northern locations: IN, Indiana; IL, Illinois; TN, Tennessee; NE, Nebraska; IA, Iowa; SD, South Dakota; NY, New York; ND, North Dakota. Southern locations: TX, Texas; VA, Virginia; OK, Oklahoma; LA, Louisiana; AR, Arkansas. Solid grey lines represent linear equation fit to the data. Vertical bars represent the standard deviation in observed values.

Table 9 Summary statistics indicating the cumulative improvement that resulted from re-parameterization of the APSIM *Sugarcane* model for predicting LAI (n = 12) and dry matter yield (n = 20) of *Miscanthus* at Water Quality Field Station, West Lafayette, IN. The parameters modified were *y_rue*, radiation use efficiency and *ratio_root_shoot*, biomass partitioning. The CCC, SB and MSV are the concordance correlation coefficient, bias of the simulation from the measurement and mean square variation, respectively

	Original model	*y_rue*	*ratio_root_shoot*
LAI			
Mean (Observed)	7.6	7.6	7.6
Mean (Modelled)	7.9	7.8	7.3
SD (Observed)	1.2	1.2	1.2
SD (Modelled)	0.2	0.2	0.9
Testing parameters			
CCC	−0.15	−0.03	0.69
SB (%)	5	2	17
MSV (%)	95	98	83
Dry matter yield (kg ha^{-1})			
Mean (Observed)	10 825	10 825	10 825
Mean (Modelled)	9252	11992	9337
SD (Observed)	10 213	10 213	10 213
SD (Modelled)	7970	10 185	7952
Testing parameters			
CCC	0.88	0.90	0.94
SB (%)	12	7	27
MSV (%)	88	93	73

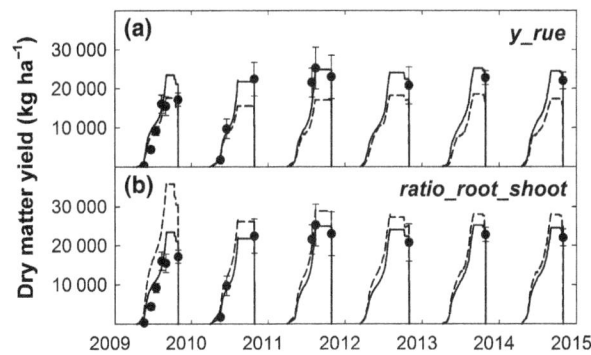

Fig. 10 Scatter plot showing observed vs. modelled dry matter yield for the Indiana sites with calibration data resulting from the re-parameterization of the APSIM model for (a) switchgrass and (b) *Miscanthus*. WQFS, Water Quality Field Station; TPAC, Throckmorton Purdue Agricultural Center; NEPAC, Northeast Purdue Agricultural Center and SEPAC, Southeast Purdue Agricultural Center. Solid black line, dotted line and solid grey line represent 1 : 1 fit (i.e. $y = x$), ± 20% of curve 1 : 1 value and linear equation fit to the data, respectively. Vertical bars represent the standard deviation in observed values where such data were available. The CCC is the concordance correlation coefficient.

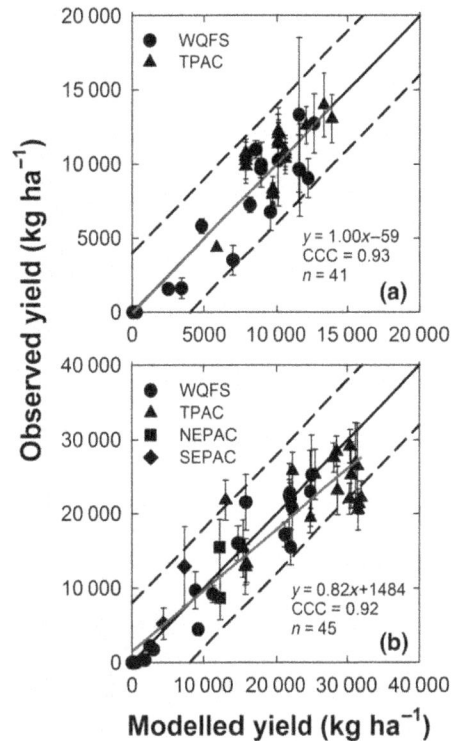

Fig. 9 Modelled pre- (- - -), post-APSIM *Sugarcane* modification (—) and observed dry matter yield (●) of *Miscanthus* at Water Quality Field Station (West Lafayette, IN). The modified parameters shown in each panel are *y_rue*, radiation use efficiency (a) and *ratio_root_shoot*, ratio root/shoot (b). The effect on dry matter yield was due to the independent modification of each individual parameter. Vertical bars represent the standard deviation in observed values.

Miscanthus

Accurate prediction of *Miscanthus* DM yield using the APSIM *Sugarcane* module required fewer model re-parameterizations when compared to changes made in the

APSIM *Lucerne* module parameters to predict DM yield of switchgrass. Sugarcane (*Saccharum officinarum*) shares phenological and physiological attributes with *Miscanthus* due to their close polyphyletic relationship at the subtribe level (Hodkinson *et al.*, 2002), so it is not surprising that this APSIM module predicted the DM yield of *Miscanthus*. As with switchgrass, the main plant parameter modified was the RUE. Model DM yield prediction improved when *y_rue* was increased (Fig. 9a). Unlike switchgrass, no change occurred in *Miscanthus* DM yield prediction from changes in *y_extinct_coef* in the three soils evaluated (Fig. 3d). Similarly, DM yield was not sensitive to change in *y_extinct_coef* using SWAT in IN (Trybula *et al.*, 2014). In addition, Davey *et al.* (2016) demonstrated that the time period in which increases in the *y_extinct_coef* value had a greatest impact on light interception, and consequently on DM yield, is at the beginning of the growing season before

Table 10 Summary statistics indicating the performance of the Agricultural Production Systems Simulator (APSIM) model in predicting the dry matter yield (kg ha^{-1}) of *Miscanthus*. The data was divided in calibration and validation datasets. IN, Indiana; CA, California; NE, Nebraska; IL, Illinois; VA, Virginia; NJ, New Jersey; KY, Kentucky. The CCC, SB and MSV are the concordance correlation coefficient, bias of the simulation from the measurement and mean square variation, respectively

| | Calibration | Validation | | | | | | |
	IN	CA	NE	IL	VA	NJ	KY	Total
Observations #	45	8	15	44	12	17	12	108
Mean (Observed)	15 475	20 421	20 352	18 354	14 926	16 087	17 269	17 927
Mean (Modelled)	17 032	20 101	20 518	18 195	13 883	17 536	16 079	17 841
SD (Observed)	9631	8548	4865	4540	3747	3074	3054	4789
SD (Modelled)	10 924	5213	5321	4025	3469	3488	5442	4653
Testing parameters								
CCC	0.92	0.85	0.69	0.58	0.57	0.46	0.38	0.65
SB (%)	13	1	0	0	10	16	6	0
MSV (%)	87	99	100	100	90	84	94	100

LAI≥4. Therefore, after this stage, further increases in *y_extinct_coef* have little effect on DM yield. Thus, the low sensitivity to this parameter in our study may be based on the constant values of *y_extinct_coef* used for all crop stages.

In a recent study (Zhao *et al.*, 2014) the modelled root biomass of wheat (*Triticum aestivum* L.) was improved trough re-parameterization of the *ratio_root_shoot* parameter using APSIM in China. Likewise, biomass partitioning between roots, rhizomes and shoots for *Miscanthus* has been parameterized for the WIMOVAC model using data from Beale & Long (1997) and this trait has been validated using data from Europe (Miguez, 2007). Based on the mentioned studies, and using data collected by Burks (2013) and Trybula *et al.* (2014) in West Lafayette, IN, the *ratio_root_shoot* parameter was changed in APSIM for all stages, which led to an accurate prediction of DM yield (Fig. 9b; CCC = 0.94).

Similar to switchgrass, sensitivity analysis demonstrated definite trends associated with soil PAWC changes. However, this response differed from switchgrass in that DM yield was greater for the loamy soil than the silty and sandy soils. While the cause is not clear at this time, one plausible explanation is genotypic differences in root exploration between species depending on soil type (Monti & Zatta, 2009). These authors found that *Miscanthus* roots were more concentrated in the top layers of the soil profile as compared with switchgrass, which led to the crop water capture was close and negatively related to root distribution.

Most *Miscanthus* studies for US locations have predicted peak autumn yield (17 500–48 000 kg DM ha^{-1}) and assumed adequate soil moisture and nutrient availability (Heaton *et al.*, 2004; Khanna *et al.*, 2008; Jain *et al.*, 2010; Mishra *et al.*, 2013). However, our predicted DM yields from the validation work for this same region (17 000 kg ha^{-1} at IN to 20 500 kg ha^{-1} at NE) are lower because *Miscanthus* was grown under water and/or nutrient-limiting dryland conditions, (except in CA).

The APSIM *Sugarcane* module was able to satisfactorily predict *Miscanthus* DM yields at IN locations (calibration) and for most locations evaluated (validation) with the exception of NJ and KY. As was discussed for switchgrass, the poor ability of the model to predict DM yield of *Miscanthus* at these two locations was associated with the inaccurate estimation of PAWC. An over-estimation of DUL was found in the Maury soil series at KY (Fig. 7c). In fact, the estimated soil water parameters were not in a satisfactory range as compared with the observed values (0.80 for CCC and 30% for SB) at this site when compared with the soils at IL, IN and NY (0.96–0.98 for CCC and 0–14% for SB). Similarly, a poor fit was found between observed and estimated DUL in the soil layers from 0.5 to 1.5 m in the Holmdel soil series in NJ (Fig. 7g); however, the prediction level of the LL and DUL using the HYDRAULIC PROPERTIES CALCULATOR Software developed by Saxton & Rawls (2006) was acceptable in this soil (0.92 for CCC and 13% for SB).

APSIM model: a promising tool to simulate DM yield for switchgrass and Miscanthus in several US environments

This work was the first attempt to re-parameterize two current APSIM plant modules (*Lucerne* and *Sugarcane*) for predicting the DM yield of switchgrass and *Miscanthus*. Such re-parameterization was conducted based on an extensive literature review and using detailed experimental datasets. We initially focused on the re-parameterization of plant and soil modules and on predicting

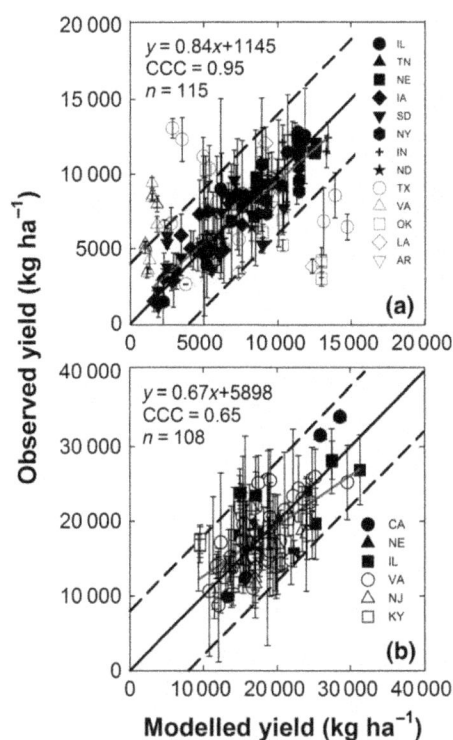

Fig. 11 Scatter plots of observed vs. modelled dry matter yield for all sites with validation data using the re-parameterized APSIM model for (a) switchgrass and (b) *Miscanthus* in different states. ND, North Dakota, NE, Nebraska, IL, Illinois, NY, New York, SD, South Dakota, IA, Iowa, IN, Indiana, TN, Tennessee, AR, Arkansas, TX, Texas, OK, Oklahoma, LA, Louisiana, VA, Virginia, CA, California, KY, Kentucky and NJ, New Jersey. Solid black line, dotted line and solid grey line represent 1 : 1 fit (i.e. $y = x$), ± 20% of curve 1 : 1 value and linear equation fit to the data, respectively. The linear equation, CCC (concordance correlation coefficient) and number of observations (n) correspond to northern locations data and complete dataset for switchgrass and *Miscanthus*, respectively. Vertical bars represent the standard deviation in observed values.

the direction and the magnitude of the DM yield responses.

The study demonstrates:

- The simulation of switchgrass DM yield in northern locations of the US using the re-parameterized APSIM *Lucerne* module had greater accuracy than in southern ones. The improved predictions were associated with a strong, positive association between DM yield and accumulated annual rainfall.

- The original version of the APSIM *Sugarcane* module can be used to accurately simulate the growth and yield of *Miscanthus* in a broad range of geographies and ecosystems within the US that differ in local weather, soil characteristics, and crop management.

- The predictions of the DM yield for *Miscanthus* improved substantially when the physiological

parameters (*rue* and *ratio_root_shoot*) of the model were modified.

- PAWC parameterization in a soil profile was critical for explaining DM yield differences for both crops.

This study represents an advance with respect to previous ones to simulate switchgrass and *Miscanthus* because: (i) the DM yield predictions were carried out with the same model (ii) the re-parameterization was started from two existing APSIM plant modules, (iii) the modelled DM yields have been compared against independent datasets, which include contrasting cultivars of switchgrass and environments, and (iv) the average errors associated with the predictions of DM yield at northern locations of switchgrass and *Miscanthus* were extremely low for both the calibration and the validation (26–57 kg ha^{-1} and 1557–86 kg ha^{-1}, respectively). To improve the APSIM accuracy under these environments additional agronomic studies are needed, since only a limited number of locations were utilized for each species. In addition, as our study was based on APSIM calibrations at IN locations, further calibrations of the model using data obtained from other environments is recommended. Nevertheless, these re-parameterized APSIM modules hold promise as tools for predicting switchgrass and *Miscanthus* yields in several US environments.

Acknowledgements

The authors thank the different contributors to the dataset which allowed model validation: partners of the Sun Grant project (Dr. T. Voigt, University of Illinois; Dr. D. Williams, University of Kentucky; Dr. R. Gaussoin, University of Nebraska; Dr. S. Bonos, Rutgers University; Dr. J. Fike, Virginia Polytechnic Institute and State University) and Dr. E. Heaton, Iowa State University. Thanks also to K. Scheeringa (Indiana State Climate Office) for providing climatic data and Dr. M. Castro Franco (CONICET) for his help in creating the map in Fig. 1. The IN experiments at ACRE, TPAC, PPAC and SEPAC were funded by the Purdue University College of Agriculture and Competitive Grant No. 2011-68005-30411 from USDA-NIFA AFRI (CenUSA). The senior author wishes also to acknowledge financial support from the Fulbright Program. The present work is a part of the thesis submitted by J.J. Ojeda to the Postgraduate program of Unidad Integrada Balcarce (UNMdP-INTA) in partial fulfilment of the requirement of the Doctor's degree. J.J. Ojeda held a scholarship and O.P. Caviglia is a member of CONICET, the National Research Council of Argentina.

References

Andrade FH, Uhart SA, Cirilo A (1993) Temperature affects radiation use efficiency in maize. *Field Crops Research*, **32**, 17–25.

Arundale RA, Dohleman FG, Heaton EA, Mcgrath JM, Voigt TB, Long SP (2014) Yields of *Miscanthus* x *giganteus* and *Panicum virgatum* decline with stand age in the Midwestern USA. *Global Change Biology Bioenergy*, **6**, 1–13.

Beale CV, Long SP (1997) Seasonal dynamics of nutrient accumulation and partitioning in the perennial C4-grasses *Miscanthus* x *giganteus* and *Spartina cynosuroides*. *Biomass and Bioenergy*, **12**, 419–428.

Bell JE, Palecki MA, Baker CB et al. (2013) US Climate Reference Network soil moisture and temperature observations. *Journal of Hydrometeorology*, **14**, 977–988.

Brown HE, Moot DJ, Teixeira EI (2006) Radiation use efficiency and biomass partitioning of lucerne (*Medicago sativa*) in a temperate climate. *European Journal of Agronomy*, **25**, 319–327.

Burks JL (2013) Eco-physiology of three perennial bioenergy systems. PhD Thesis. Purdue University, USA.

Byrd GT, May PA (2000) Physiological comparisons of switchgrass cultivars differing in transpiration efficiency. *Crop Science*, **40**, 1271–1277.

Cassida KA, Muir JP, Hussey MA, Read JC, Venuto BC, Ocumpaugh WR (2005) Biomass yield and stand characteristics of switchgrass in south central US environments. *Crop Science*, **45**, 673–681.

Chen W, Shen YY, Robertson MJ, Probert ME, Bellotti WD (2008) Simulation analysis of lucerne-wheat crop rotation on the Loess Plateau of Northern China. *Field Crops Research*, **108**, 179–187.

Clifton-Brown JC, Neilson B, Lewandowski I, Jones MB (2000) The modelled productivity of *Miscanthus* x *giganteus* (Greef et Deu) in Ireland. *Industrial Crops and Products*, **12**, 97–109.

Clifton-Brown JC, Long SP, Jørgensen U (2001) *Miscanthus* productivity. In: *Miscanthus for Energy and Fibre* (eds Jones MB, Walsh M), pp. 46–67. James and James, London.

Clifton-Brown JC, Stampfl PF, Jones MB (2004) *Miscanthus* biomass production for energy in Europe and its potential contribution to decreasing fossil fuel carbon emissions. *Global Change Biology*, **10**, 509–518.

Cosentino SL, Patane C, Sanzone E, Copani V, Foti S (2007) Effects of soil water content and nitrogen supply on the productivity of *Miscanthus* x *giganteus* Greef et Deu. in a Mediterranean environment. *Industrial Crops and Products*, **25**, 75–88.

Cresswell HP, Hume IH, Wang E, Nordblom TL, Finlayson JD, Glover M (2009) Catchment response to farm scale land use change. CSIRO Land and Water Science Report 9 September 2009, CSIRO Land and Water and NSW Department of Primary Industries, Australia.

Dalgliesh NP, Foale MA (1998) Soil Matters: monitoring soil water and nitrogen in dryland farming. In: *Technical Manual-ISBN 0 64306375 7*, CSIRO/Agricultural Production Systems Research Unit, Toowoomba, Australia.

Dardanelli JL, Bacheier OA, Sereno R, Gil R (1997) Rooting depth and soil water extraction patterns of different crops in a silty loam Haplustoll. *Field Crops Research*, **54**, 29–38.

Dardanelli JL, Ritchie JT, Calmon M, Andriani JM, Collino DJ (2004) An empirical model for root water uptake. *Field Crops Research*, **87**, 59–71.

Davey CL, Jones LE, Squance M et al. (2016) Radiation capture and conversion efficiencies of *Miscanthus sacchariflorus*, *M. sinensis* and their naturally occurring hybrid *M.* x *giganteus*. *Global Change Biology Bioenergy*. doi:10.1111/gcbb.12331.

Diamond HJ, Karl TR, Palecki MA et al. (2013) US Climate Reference Network after one decade of operations: status and assessment. *Bulletin of the American Meteorological Society*, **94**, 485–498.

Dohleman FG, Long SP (2009) More productive than maize in the Midwest: how does *Miscanthus* do it? *Plant Physiology*, **150**, 2104–2115.

Dolling PJ, Robertson MJ, Asseng S, Ward PR, Latta RA (2005) Simulating lucerne growth and water use on diverse soil types in a Mediterranean-type environment. *Crop and Pasture Science*, **56**, 503–515.

Durand JL, Onillon B, Schnyder H, Rademacher I (1995) Drought effects on cellular and spatial parameters of leaf growth in tall fescue. *Journal of Experimental Botany*, **46**, 1147–1155.

Fila G, Bellocchi G, Acutis M, Donatelli M (2003) Irene: a software to evaluate model performance. *European Journal of Agronomy*, **18**, 369–372.

GlobalSoilMap (2012) Specifications, Version 1 GlobalSoilMap. Net Products. Release 2.2.

Grassini P, Hunt E, Mitchell RB, Weiss A (2009) Simulating switchgrass growth and development under potential and water-limiting conditions. *Agronomy Journal*, **101**, 564–571.

Hastings A, Clifton-Brown J, Wattenbach M, Stampfl P, Mitchell CP, Smith P (2008) Potential of *Miscanthus* grasses to provide energy and hence reduce greenhouse gas emissions. *Agronomy for Sustainable Development*, **28**, 465–472.

Hastings A, Clifton-Brown J, Wattenbach M, Mitchell CP, Smith P (2009) The development of MISCANFOR, a new *Miscanthus* crop growth model: towards more robust yield predictions under different climatic and soil conditions. *Global Change Biology Bioenergy*, **1**, 154–170.

Heaton EA, Long SP, Voigt TB, Jones MB, Clifton-Brown JC (2004) *Miscanthus* for renewable energy generation: european Union experience and projections for Illinois. *Mitigation and Adaptation Strategies for Global Change*, **9**, 433–451.

Heaton EA, Dohleman FG, Long SP (2008) Meeting US biofuel goals with less land: the potential of *Miscanthus*. *Global Change Biology*, **14**, 2000–2014.

Hodkinson TR, Chase MW, Lledó DM, Salamin N, Renvoize SA (2002) Phylogenetics of *Miscanthus*, *Saccharum* and related genera (Saccharinae, Andropogoneae, Poaceae) based on DNA sequences from ITS nuclear ribosomal DNA and plastid *trnL* intron and *trnL-F* intergenic spacers. *Journal of Plant Research*, **115**, 381–392.

Holzworth DP, Huth NI, Zurcher EJ et al. (2014) APSIM-evolution towards a new generation of agricultural systems simulation. *Environmental Modelling and Software*, **62**, 327–350.

Jain AK, Khanna M, Erickson M, Huang H (2010) An integrated biogeochemical and economic analysis of bioenergy crops in the Midwestern United States. *Global Change Biology Bioenergy*, **2**, 217–234.

Jones CA, Kiniry KR (1986) *CERES-Maize, A Simulation Model of Maize Growth and Development*, 1st edn. Texas University Press, College Station, TX.

Keating BA, Robertson MJ, Muchow RC, Huth NI (1999) Modelling sugarcane production systems I. Development and performance of the sugarcane module. *Field Crops Research*, **61**, 253–271.

Keating BA, Carberry PS, Hammer GL et al. (2003) An overview of APSIM, a model designed for farming systems simulation. *European Journal of Agronomy*, **18**, 267–288.

Khanna M, Dhungana B, Clifton-Brown JC (2008) Costs of producing *Miscanthus* and switchgrass for bioenergy in Illinois. *Biomass and Bioenergy*, **32**, 482–493.

Kiniry JR, Tischler CR, Van Esbroeck GA (1999) Radiation use efficiency and leaf CO_2 exchange for diverse C4 grasses. *Biomass and Bioenergy*, **17**, 95–112.

Kiniry JR, Cassida KA, Hussey MA et al. (2005) Switchgrass simulation by the ALMANAC model at diverse sites in the southern US. *Biomass and Bioenergy*, **29**, 419–425.

Kiniry JR, Johnson MVV, Bruckerhoff SB, Kaiser JU, Cordsiemon RL, Harmel RD (2012) Clash of the titans: comparing productivity via radiation use efficiency for two grass giants of the biofuel field. *BioEnergy Research*, **5**, 41–48.

Kobayashi K, Us Salam M (2000) Comparing simulated and measured values using mean squared deviation and its components. *Agronomy Journal*, **92**, 345–352.

Libohova Z, Wills S, Odgers NP et al. (2014) Converting pH 1 : 1 H_2O and 1 : 2 $CaCl_2$ to 1 : 5 H_2O to contribute to a harmonized global soil database. *Geoderma*, **213**, 544–550.

Louarn G, Chenu K, Fournier C, Andrieu B, Giauffret C (2008) Relative contributions of light interception and radiation use efficiency to the reduction of maize productivity under cold temperatures. *Functional Plant Biology*, **35**, 885–899.

Madakadze IC, Stewart K, Peterson PR, Coulman BE, Samson R, Smith DL (1998) Light interception, use-efficiency and energy yield of switchgrass (*Panicum virgatum* L.) grown in a short season area. *Biomass and Bioenergy*, **15**, 475–482.

Miguez FE (2007) *Miscanthus* x *giganteus* production: Meta-analysis, field study and mathematical modeling. PhD Thesis. University of Illinois, USA.

Mishra U, Torn MS, Fingerman K (2013) *Miscanthus* biomass productivity within US croplands and its potential impact on soil organic carbon. *Global Change Biology Bioenergy*, **5**, 391–399.

Monti A, Zatta A (2009) Root distribution and soil moisture retrieval in perennial and annual energy crops in Northern Italy. *Agriculture, Ecosystems and Environment*, **132**, 252–259.

Moot DJ, Hargreaves J, Brown E, Teixeira EI (2015) Calibration of the APSIM-Lucerne model for 'Grasslands Kaituna' lucerne crops grown in New Zealand. *New Zealand Journal of Agricultural Research*, **58**, 190–202.

Ojeda JJ, Pembleton KG, Islam MR, Agnusdei MG, Garcia SC (2016) Evaluation of the agricultural production systems simulator simulating Lucerne and annual ryegrass dry matter yield in the Argentine Pampas and south-eastern Australia. *Agricultural Systems*, **143**, 61–75.

Pallipparambil GR, Raghu S, Wiedenmann RN (2015) Modeling the biomass production of the biofuel crop *Miscanthus* x *giganteus*, to understand and communicate benefits and risks in cultivation. *Energy for Sustainable Development*, **27**, 63–72.

Pembleton KG, Rawnsley RP, Donaghy DJ (2011) Yield and water-use efficiency of contrasting lucerne genotypes grown in a cool temperate environment. *Crop and Pasture Science*, **62**, 610–623.

Piñeiro G, Perelman S, Guerschman JP, Paruelo JM (2008) How to evaluate models: observed vs. predicted or predicted vs. observed? *Ecological Modelling*, **216**, 316–322.

Probert ME, Dimes JP, Keating BA, Dalal RC, Strong WM (1998) APSIM's water and nitrogen modules and simulation of the dynamics of water and nitrogen in fallow systems. *Agricultural Systems*, **56**, 1–28.

Robertson MJ, Fukai S, Ludlow MM, Hammer GL (1993a) Water extraction by grain sorghum in a sub-humid environment. I. Analysis of the water extraction pattern. *Field Crops Research*, **33**, 81–97.

Robertson MJ, Fukai S, Ludlow MM, Hammer GL (1993b) Water extraction by grain sorghum in a sub-humid environment. II. Extraction in relation to root growth. *Field Crops Research*, **33**, 99–112.

Robertson MJ, Carberry PS, Huth NI *et al.* (2002) Simulation of growth and development of diverse legume species in APSIM. *Crop and Pasture Science*, **53**, 429–446.

Saxton KE, Rawls WJ (2006) Soil water characteristic estimates by texture and organic matter for hydrologic solutions. *Soil Science Society of America Journal*, **70**, 1569–1578.

Schmer MR, Vogel KP, Mitchell RB, Perrin RK (2008) Net energy of cellulosic ethanol from switchgrass. *Proceedings of the National Academy of Sciences*, **105**, 464–469.

Stampfl PF, Clifton-Brown JC, Jones MB (2007) European-wide GIS-based modelling system for quantifying the feedstock from *Miscanthus* and the potential contribution to renewable energy targets. *Global Change Biology*, **13**, 2283–2295.

Strullu L, Ferchaud F, Yates N *et al.* (2015) Multisite yield gap analysis of *Miscanthus x giganteus* using the STICS model. *BioEnergy Research*, **8**, 1735–1749.

Tedeschi LO (2006) Assessment of the adequacy of mathematical models. *Agricultural Systems*, **89**, 225–247.

Trybula EM, Cibin R, Burks JL, Chaubey I, Brouder SM, Volenec JJ (2014) Perennial rhizomatous grasses as bioenergy feedstock in SWAT: parameter development and model improvement. *Global Change Biology Bioenergy*, **7**, 1185–1202.

Vogel KP, Brejda JJ, Walters DT, Buxton DR (2002) Switchgrass biomass production in the Midwest USA. *Agronomy Journal*, **94**, 413–420.

Wang L, Qian Y, Brummer JE, Zheng J, Wilhelm S, Parton WJ (2015) Simulated biomass, environmental impacts and best management practices for long-term switchgrass systems in a semi-arid region. *Biomass and Bioenergy*, **75**, 254–266.

Wart J, Grassini P, Cassman KG (2013) Impact of derived global weather data on simulated crop yields. *Global Change Biology*, **19**, 3822–3834.

White JW, Hoogenboom G, Stackhouse PW, Hoell JM (2008) Evaluation of NASA satellite-and assimilation model-derived long-term daily temperature data over the continental US. *Agricultural and Forest Meteorology*, **148**, 1574–1584.

White JW, Hoogenboom G, Wilkens PW, Stackhouse PW, Hoel JM (2011) Evaluation of satellite-based, modeled-derived daily solar radiation data for the continental United States. *Agronomy Journal*, **103**, 1242–1251.

Wullschleger SD, Davis EB, Borsuk ME, Gunderson CA, Lynd LR (2010) Biomass production in switchgrass across the United States: database description and determinants of yield. *Agronomy Journal*, **102**, 1158–1168.

Zhao Z, Wang E, Xue L *et al.* (2014) Accuracy of root modelling and its impact on simulated wheat yield and carbon cycling in soil. *Field Crops Research*, **165**, 99–110.

Zub HW, Brancourt-Hulmel M (2010) Agronomic and physiological performances of different species of *Miscanthus*, a major energy crop. A review. *Agronomy for Sustainable Development*, **30**, 201–214.

4

Predicting soil C changes over sugarcane expansion in Brazil using the DayCent model

DENER M. S. OLIVEIRA[1,2] (iD), STEPHEN WILLIAMS[1], CARLOS E. P. CERRI[2] and KEITH PAUSTIAN[1,3]

[1]*Natural Resource Ecology Laboratory, Colorado State University, Fort Collins, CO, 80523-1499, USA,* [2]*Department of Soil Science, Luiz de Queiroz College of Agriculture, University of São Paulo, Piracicaba, 13418-900, Brazil,* [3]*Department of Soil and Crop Sciences, Colorado State University, Fort Collins, CO, 80523-1170, USA*

Abstract

In recent years, the increase in Brazilian ethanol production has been based on expansion of sugarcane-cropped area, mainly by the land use change (LUC) pasture–sugarcane. However, second-generation (2G) cellulosic-derived ethanol supplies are likely to increase dramatically in the next years in Brazil. Both these management changes potentially affect soil C (SOC) changes and may have a significant impact on the greenhouse gases balance of Brazilian ethanol. To evaluate these impacts, we used the DayCent model to predict the influence of the LUC native vegetation (NV)–pasture (PA)–sugarcane (SG), as well as to evaluate the effect of different management practices (straw removal, no-tillage, and application of organic amendments) on long-term SOC changes in sugarcane areas in Brazil. The DayCent model estimated that the conversion of NV-PA caused SOC losses of 0.34 ± 0.03 Mg ha^{-1} yr^{-1}, while the conversion PA-SG resulted in SOC gains of 0.16 ± 0.04 Mg ha^{-1} yr^{-1}. Moreover, simulations showed SOC losses of 0.19 ± 0.04 Mg ha^{-1} yr^{-1} in SG areas in Brazil with straw removal. However, our analysis suggested that adoption of some best management practices can mitigate these losses, highlighting the application of organic amendments ($+0.14 \pm 0.03$ Mg C ha^{-1} yr^{-1}). Based on the commitments made by Brazilian government in the UNFCCC, we estimated the ethanol production needed to meet the domestic demand by 2030. If the increase in ethanol production was based on the expansion of sugarcane area on degraded pasture land, the model predicted a SOC accretion of 144 Tg from 2020 to 2050, while increased ethanol production based on straw removal as a cellulosic feedstock was predicted to decrease SOC by 50 Tg over the same 30-year period.

Keywords: best management practices, biofuels, land use change, second-generation ethanol, soil organic matter, straw removal

Introduction

Bioenergy is critical for environmental security and climate change mitigation. Future projections suggest that 30% of the world's fuel supply might be bio-based by 2050 (Macedo *et al.*, 2015). However, the C balance in the agricultural phase still raises uncertainties about the environmental feasibility of biofuels expansion. Land use change (LUC) due to biofuel crop establishment may be associated with soil C (SOC) losses that negatively impact the biofuel's greenhouse gases (GHG) balance (Fargione *et al.*, 2008; Mello *et al.*, 2014). The relevance of LUC has been emphasized by several authors, especially in relation to political decisions made for increasing biofuel production (Lapola *et al.*, 2010; Hudiburg *et al.*, 2016).

In Brazil, the negative effects of LUC brought out concerns about the efficiency of the sugarcane ethanol as a climate change mitigation option (Fargione *et al.*, 2008; Lapola *et al.*, 2010). However, sugarcane ethanol shows the largest average net GHG mitigation (including LUC effects) compared to other first-generation ethanol feedstocks (Renouf *et al.*, 2008). Nowadays, Brazil is considered to have developed the world's first sustainable biofuel economy and in many respects is the biofuel industry leader (Souza *et al.*, 2014). This reputation is largely based on its sugarcane industry.

Between 2004 and 2012, Brazil's GDP increased by 32% (IPEA, 2016), while GHG emissions decreased by 52% (MCTI, 2014), breaking the link between economic growth and GHG emissions. Despite these advances, the Brazilian government announced ambitious goals in the last UNFCCC: reduce GHG emissions by 43% below 2005 levels by 2030 (iNDC Brazil, 2015). To do so, among other strategies, the government established that

Correspondence: Dener M. S. Oliveira
e-mail: denermsoliveira@gmail.com

the sugarcane contribution to the energy supply in Brazil by 2030 must be around 16%. Meeting this mandate probably will require a substantial increase in sugarcane production area.

Previous studies using the Century model evaluated the effects of green harvest management (GM – harvest without burning) and organic amendments on SOC changes in sugarcane areas in Brazil (Galdos *et al.*, 2009; Brandani *et al.*, 2015). As concluded by these studies, the high crop residue inputs in areas under GM is the main factor associated with increments on SOC in sugarcane areas in Brazil. However, the sugarcane residues have become an attractive source of biomass for bioelectricity and second-generation (2G) ethanol production in Brazil (Walter & Ensinas, 2010). Crop residue removal is associated with decreases on SOC (Wilhelm *et al.*, 2007; Wortmann *et al.*, 2010), but the adoption of some best management practices can mitigate these losses (Paustian *et al.*, 2016).

In recent years, almost all the sugarcane expansion in Brazil has been done under pasture areas (Dias *et al.*, 2016). Using the Century model, Silva-Olaya *et al.* (2016) studied the impact of LUC from native vegetation and pasture to sugarcane cultivation on SOC dynamics in Brazil. The site-level data used in Silva-Olaya *et al.* (2016) were those reported by Mello *et al.* (2014), where most of sugarcane areas were either still harvested with burning or this practice had just been stopped for 3 years or less before the sampling time. In this sense, the longer term effects of the conversion pasture–sugarcane on SOC remain unclear for areas under GM. Moreover, there are not published papers on the effects of straw removal on SOC in sugarcane areas in Brazil. Simulation models provide a feasible and cost-effective option to predict the long-term potential impacts of LUC and management practices on SOC. Furthermore, predictions on SOC changes are a useful tool to encourage decision makers and planners to develop sustainable land use strategies and soil management systems in areas to biofuel production (Campbell & Paustian, 2015). In this study, we used the DayCent model to predict the impact of unburnt sugarcane expansion into pasture areas, as well as to evaluate the effect of different management practices, such as straw removal, no-tillage, and application of organic amendments (vinasse and filter cake), on long-term SOC changes in sugarcane areas in Brazil.

Materials and methods

Description of study sites

For the field data used in this research (Cherubin *et al.*, 2015; Oliveira *et al.*, 2016b), we sampled three land uses – native vegetation (NV), pasture (PA) and sugarcane (SG) – at sites across south-central Brazil, the largest sugarcane region in the world, accounting for 93.4% of Brazilian ethanol production (UNICA, 2015). The climate at all the sites has rainfall concentrated in the spring and summer (October–April), while the dry season is in the autumn and winter (May–September). The soils are typical of the Brazilian tropical region, well-drained and highly weathered, with a predominance of kaolinite, Fe oxides (goethite and hematite), and Al oxide (gibbsite) in the clay-size fraction.

The first site, Lat_17S, is located in Jataí, southwestern region of Goiás state (Lat.: 17°56′16″ S; Long.: 51°38′31″ W) with a mean altitude of 800 m and a predominance of clayey Acrudox soils (USDA, 2014). The climate classification is Awa type (Köppen) mesothermal tropical, with a mean annual temperature of 24.0 °C and an annual precipitation of 1600 mm. The second site, Lat_21S, is located in Valparaíso, west region of São Paulo state (Lat.: 21°14′48″ S; Long.: 50°47′04″ W) with a mean altitude of 425 m and predominance of loamy Hapludalf soils (USDA, 2014). The climate classification is Aw type (Köppen) humid tropical. The area has a mean annual temperature of 23.4 °C and an annual precipitation of 1240 mm. The third site, Lat_23S, is located in Ipaussu, south-central region of the São Paulo state (Lat.: 23°05′08″ S; Long.: 49°37′52″ W), with a mean altitude of 630 m and predominance of clayey Hapludox soils (USDA, 2014). The climate classification is Cwa type (Köppen) tropical. The annual mean temperature is 21.7 °C, and the annual precipitation is 1470 mm. A general description of each land use is shown in Fig. 1. For more information about soil parent material and soil classification, LUC sequence, sampling, and laboratory procedures, see Cherubin *et al.* (2015), Oliveira *et al.* (2016a,b).

The DayCent Model

We used the most recent version of DayCent model (DD14-centEVI) to simulate changes in soil organic matter (SOM) dynamics in areas under LUC to sugarcane expansion in Brazil. DayCent (Parton *et al.*, 1998; Del Grosso *et al.*, 2001) is a modified, daily time step version of the biogeochemical ecosystem Century model (Parton *et al.*, 1987). Both Century and DayCent simulates fluxes of C and N between the atmosphere, vegetation, and soil, including the dynamics of multiple C and N soil organic matter pools, but DayCent also includes other processes such as greenhouse gases emissions.

In DayCent, phenology, net primary productivity, shoot:root ratio, and the C:N ratio of biomass in plant components are species-specific. Moreover, the model calculates potential plant growth as a function of water, light, and soil temperature and limits actual plant growth based on specific plant nutrient requirements. The type and timing of each management event can be specified, including tillage, fertilization, organic matter addition, harvest, burning, and grazing intensity. Litter decomposition and SOM turnover are determined by the amount and quality of residue returned to the soil, the size of the SOM pools, and temperature and water controls (Del Grosso *et al.*, 2001). These aspects allow DayCent to generate accurate simulations for multiple vegetation types under a wide range of management practices at diverse sites, which make the model particularly useful for simulating land use change.

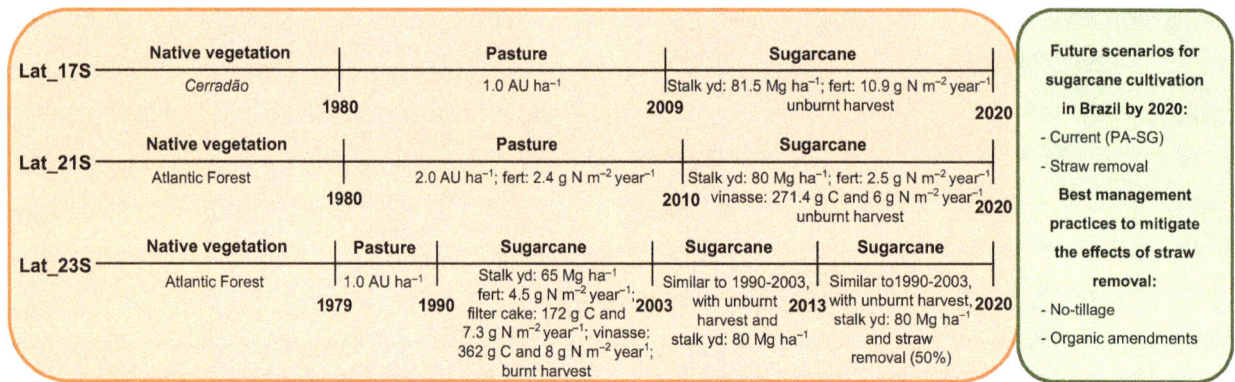

Fig. 1 Land use transitions and brief description of the management practices of studied sites in south-central Brazil and future scenarios to sugarcane cultivation in Brazil. AU, animal units; Fert, fertilizers applied; PA-SG, pasture to sugarcane land use conversion; Stalk yd, stalk yield.

Accordingly, DayCent has been used and validated across a range of land use and management scenarios (Del Grosso *et al.*, 2009; Duval *et al.*, 2013; Hudiburg *et al.*, 2016). The Century model was widely used for simulations in pastures (Cerri *et al.*, 2004, 2007) and sugarcane areas (Galdos *et al.*, 2009, 2010; Brandani *et al.*, 2015; Silva-Olaya *et al.*, 2016) in Brazil. However, there is no published research using the DayCent model for simulations in Brazil so far.

Modeling procedures

The DayCent model requires input of climate and soil data. In this study, we used climate data (daily maximum and minimum average temperature and precipitation) from 1901 to 2015, provided by MsTMIP (Wei *et al.*, 2014). We opted to use this gridded global product because is the only long-term and daily weather data available for these sites. Others weather data available for Brazil (e.g., INMET, Cepagri) were restricted to more recent periods or in a monthly basis. The site-specific soil attributes used to the initialization of the model are available in Cherubin *et al.* (2015).

To initialize the model prior to simulating forest clearing and pasture establishment, we used the forest submodel to estimate equilibrium SOM levels and plant productivity under native forest conditions, over a 7000-year simulation period. Two kinds of native vegetation were simulated using the parameterization developed by Silva-Olaya *et al.* (2016): Cerradão Forest (Lat_17S) and Atlantic Forest (Lat_21S and Lat_23S; Fig. 1). In our simulations, the main difference between these forest types is the N input by biological fixation in Cerradão Forest (Bustamante *et al.*, 2012). The disturbances on these areas were fire events and tree mortality (Cerri *et al.*, 2004). After simulating the equilibrium condition in native vegetation, the model was set to simulate the deforestation process following the slash-and-burn procedure. Those events were parameterized using similar calibration procedures as those developed by Cerri *et al.* (2004) for the Century model.

As for most pastures in Brazil, the pastures evaluated in our assessment are to some degree degraded and do not achieve the level of productivity characteristic of well-managed pastures. To simulate this condition, we adjusted the potential

aboveground production, based on the biomass production for degraded pastures in Brazil reported by Lilienfein & Wilcke (2003). Regarding the grazing management, as the areas presented different stocking rates (Fig. 1), we specified different levels of grazing according to the options currently available in the model. We assumed continuous grazing through all the year, including the dry season period.

The simulations for sugarcane areas were performed using parameterization for the sugarcane crop developed by Galdos *et al.* (2009, 2010), Campbell (2015), and Silva-Olaya *et al.* (2016). The potential biomass production was adjusted to match the field data for south-central Brazil (UNICA, 2015), assuming the biomass partitioning develop for sugarcane by Galdos *et al.* (2010). Sugarcane renovation was performed every 6 years, and the tillage operations (plowing, disking, and subsoiling) were simulated using the intensive default tillage parameters specified at the model. Organic amendments (vinasse and filter cake) are currently applied on two of our sites (Lat_21S and Lat_23S). The composition of the filter cake used in this study was 228 g C kg^{-1}, 12 g N kg^{-1} and 160 g lignin kg^{-1} (Galdos *et al.*, 2009). The composition of vinasse used was based on the analysis reported by Prado *et al.* (2013), with 11.56 g C L^{-1} and 0.42 g N L^{-1}.

Currently, all the sites evaluated are under GM. However, in Lat_23S, the sugarcane was harvested with burning during a 13-year period (Fig. 1). In this specific case, we used the parameters for burning events developed by Galdos *et al.* (2009), in which 85% of the dry matter of the trash (leaves and tops) is removed by the fire, and 80% of the N in the residue material is lost to the atmosphere. For the GM, the model was set to remove 99% of aboveground biomass, with 94% of dry matter in tops and leaves and 1% of stalks returned to the system as litter after the harvest (rates reported by mills in Brazil).

Model outputs and statistical analysis

Usually, DayCent model is set up for simulations of SOM dynamics in the top 0.2 m soil depth (Parton *et al.*, 1998). For this study, DayCent was parameterized to simulate SOM dynamics to a depth of 0.3 m, by decreasing the decay rate of all SOM pools by 15% (W. Parton and M. Hartman, pers.

comm.). Simulation output variables evaluated were total soil C and N stocks, and natural isotopic abundance of ^{13}C. The proportion of soil C derived from native vegetation (native-C) or from pasture and sugarcane (modern-C) was calculated with the equations for soil C-partitioning proposed by Bernoux et al. (1998) using the simulated δ^{13}C values. The rates of soil C change associated with land use/management shifts are the average of the three sites ($n = 3$).

Statistical analyses of model results were performed in accordance with tests proposed by Smith et al. (1997) to assess goodness-of-fit of the DayCent model to the measured C stocks, N stocks, and soil C-partitioning. The statistical metrics were as follows: correlation coefficient (r), root mean square error (RMSE), mean difference (M), relative error (E), and lack of fit (LOFIT).

Future scenarios

Based on feasible management strategies for future sugarcane cultivation in Brazil, we simulated the soil C changes in areas of sugarcane under five management scenarios:

Scenario I: green management, without burning or straw removal (GM)
Scenario II: Straw removal
Scenario III: Straw removal with no-tillage
Scenario IV: Straw removal with organic amendments
Scenario V: Straw removal with no-tillage and organic amendments

We assumed a maximum rate of 75% of straw removal for sugarcane areas in Brazil, as reported by Cardoso et al. (2013). No-tillage operations were simulated using the default files available in the DayCent model. The organic amendments were vinasse and filter cake, applied in rates commonly used in sugarcane areas in Brazil, that is, 200 m^3 ha^{-1} yr^{-1} and 25 Mg ha^{-1} yr^{-1}, respectively (Prado et al., 2013). Moreover, based on the commitments made by Brazilian government in the UNFCCC (iNDC Brazil, 2015), we estimated the sugarcane area expansion according to the projected increased ethanol production needed to meet the domestic demand by 2030 in two scenarios: with and without the contribution of 2G technologies to ethanol production. To reach the estimated production of ethanol in 2030, we assumed linear rates of increment in planted area. Using these two scenarios of expansion and the simulated rates of soil C change under different management practices, we estimated soil C changes in sugarcane areas in Brazil over the next decades, without assuming any biophysical or economic basis for expansion allocation across south-central Brazil.

Results

Model performance

The DayCent model estimates were consistent with the field-observed SOM changes in areas undergoing LUC for sugarcane expansion in Brazil (Table 1, Fig. 2). The measured and simulated SOC were well correlated ($r = 0.98$; $P < 0.05$), with the model underestimating the SOC by 2.1 ± 4.6, 4.8 ± 1.3, and 2.7 ± 8.6% in NV, PA, and SG areas, respectively (Table 1). Despite the correlation between the measured and simulated values ($r = 0.91$; $P < 0.05$), the model showed a tendency to overestimate the N stocks in these areas, with values 23.4 ± 18.2% greater than the measured N stocks in PA and SG areas (Table 1, Fig. 2). The DayCent model also accurately simulated our measured results for C-partitioning, with simulated soil native-C underestimated by 7.7 ± 2.1% and the simulated soil modern-C 9.7 ± 19.7% greater than the measured values.

Goodness-of-fit measures show that the DayCent model represented well the changes of SOM for the NV-PA-SG conversions evaluated. With exception of the N stocks, values for RMSE indicated a small difference between measured and simulated values (Table 2). Values for M and E showed an absence of significant bias in the simulated soil C and N stocks, and C-partitioning. However, LOFIT pointed to lack of fit between the measured and simulated N stocks and soil C-partitioning (Table 2).

Long-term SOC changes undergoing NV-PA-SG conversions in Brazil

The DayCent model estimated that the conversion of NV-PA is associated with SOC losses of 0.34 ± 0.03 Mg C ha^{-1} yr^{-1} in areas of south-central Brazil (Fig. 3). After the conversion of these pastures to sugarcane under GM, we observed the partial recovery of the SOC, at a rate of 0.16 ± 0.04 Mg C ha^{-1} yr^{-1}. We did not include the SOC changes for Lat_23S between 1990 and 2003, when the sugarcane was harvested with burning (Fig. 1). In this case, the SOC losses simulated by the DayCent model were 1.04 Mg C ha^{-1} yr^{-1} (Fig. 3c). Moreover, the simulated SOC losses in the year right after sugarcane crop renovation were 1.14 ± 0.46 Mg C ha^{-1} (Fig. 3). Normalizing the SOC values relative to those under native vegetation (NV=100) at each site, we observed that the simulated SOC changes after LUC showed a very similar pattern across sites, with a consistent SOC loss after the LUC NV-PA and SOC increases within the transition PA-SG (Fig. 3d). By 2050, under the current management practices, the SOC in SG areas was predicted to be 86.1 ± 2.8% of those observed in NV.

The C-partitioning using the simulated δ^{13}C values of SOM also showed a clear pattern in areas undergoing the LUC NV to PA to SG in Brazil (Fig. 3). In PA areas, native-C losses were 0.93 ± 0.41 Mg C ha^{-1} yr^{-1}, coupled with modern-C gains of 0.48 ± 0.20 Mg C ha^{-1} yr^{-1}. For SG, native-C losses and modern-C increases were 0.39 ± 0.17 and 0.56 ± 0.22 Mg C ha^{-1} yr^{-1}, respectively (Fig. 3).

Table 1 Measured and simulated C, N, native-C, and modern-C soil stocks (Mg ha^{-1}) at 0–0.3 m layer of areas under different land uses – native vegetation (NV), pasture (PA), sugarcane (SG) – in three sites (Lat_17S, Lat_21S, Lat_23S) of south-central Brazil

	Soil C stocks (Mg ha^{-1})		Soil N stocks (Mg ha^{-1})		Native-C (Mg ha^{-1})		Modern-C (Mg ha^{-1})	
	Measured	Simulated	Measured	Simulated	Measured	Simulated	Measured	Simulated
Lat_17S								
NV	49.1 ± 3.5*	50.9	3.0 ± 0.3	3.5	49.1 ± 3.5	50.9	–	–
PA	37.2 ± 3.2	38.7	1.7 ± 0.2	2.7	28.6 ± 2.4	27.4	8.6 ± 1.3	11.3
SG	38.2 ± 2.1	39.0	2.3 ± 0.2	3.0	25.8 ± 1.3	25.8	12.4 ± 1.0	13.2
Lat_21S								
NV	48.6 ± 3.2	47.4	4.7 ± 0.6	3.9	48.6 ± 3.2	47.4	–	–
PA	37.2 ± 2.8	36.5	2.6 ± 0.2	3.0	27.5 ± 1.9	26.6	9.7 ± 0.9	9.9
SG	40.1 ± 3.1	37.5	3.1 ± 0.3	3.3	28.2 ± 4.3	26.5	12.5 ± 1.0	11.1
Lat_23S								
NV	89.9 ± 8.5	86.2	7.4 ± 0.5	7.9	89.9 ± 8.5	86.2	–	–
PA	76.9 ± 6.7	74.2	6.1 ± 0.4	7.0	54.1 ± 5.7	49.4	22.8 ± 3.3	24.8
SG	60.5 ± 4.2	68.5	4.5 ± 0.3	7.1	44.8 ± 3.6	44.9	15.8 ± 1.8	23.6

*Standard deviation from the mean values ($n = 9$).

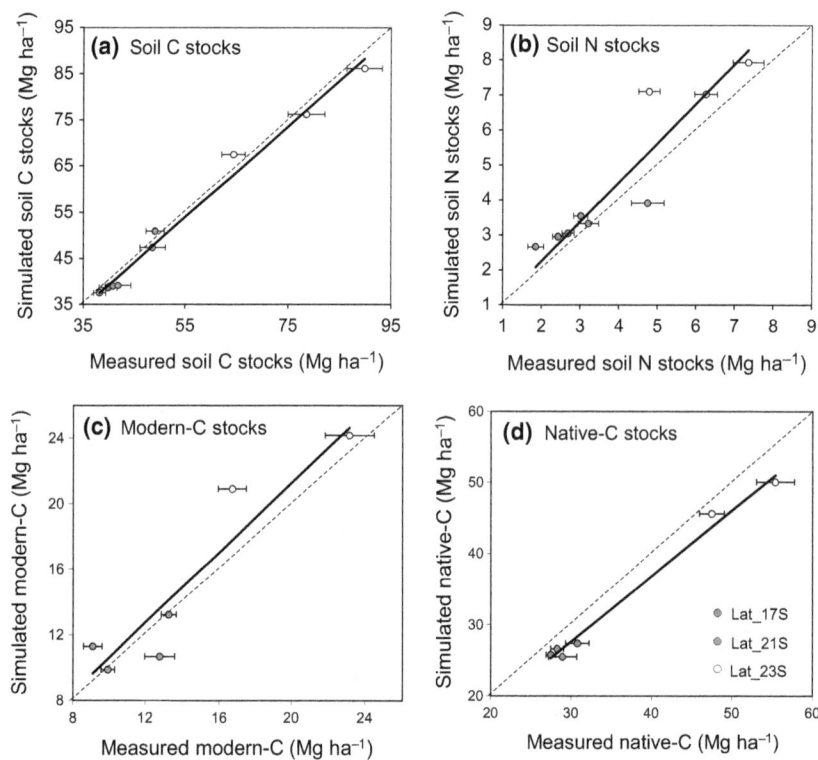

Fig. 2 Measured versus simulated soil C stocks (a), soil N stocks (b), modern-C stocks (c), and native-C stocks (d) of areas under native vegetation, pasture, and sugarcane cropping in south-central Brazil. Bars represent the standard deviation from the mean values $n = 9$.

Predicted effects of straw removal on SOC in sugarcane areas in Brazil

Straw management is a major issue affecting long-term SOC maintenance under sugarcane in Brazil (Fig. 4). The DayCent model suggested that GM would promote increased SOC, while straw removal can notably reduce SOC in sugarcane areas (Fig. 4). However, adoption of best management practices can mitigate the negative effects of straw removal, highlighting the application of organic amendments, which in our simulations showed similar results to areas under GM (Fig. 4).

The implementation of 2G technologies in Brazil will drastically alter the land demand for sugarcane

Table 2 Statistical tests applied for the validation between measured and simulated values of soil C and N stocks and C-partitioning (native-C and modern-C) of areas under native vegetation, pasture and sugarcane cropping in south-central Brazil

Statistical test	Soil C stocks	Soil N stocks	Native-C stocks	Modern-C stocks
r = Correlation Coefficient	0.98	0.91	0.99	0.94
$F = ((n-2)\,r^2)/(1-r^2)$	509.97	39.17	351.00	28.06
F-value at ($P = 0.05$)	5.59	5.59	7.71	7.71
RMSE = Root mean squared error of model	4.17%	23.73%	8.81%	15.27%
RMSE (95% Confidence Limit)	10.36%	16.17%	9.99%	12.95%
M = Mean Difference	0.95	−0.57	2.93	−0.86
t = Student's t of M	1.31	1.97	5.92	0.96
t-value (Critical at 2.5% – Two-tailed)	2.36	2.36	2.78	2.78
E = Relative Error	1.74	−14.02	8.04	−6.11
E (95% Confidence Limit). = +/−	9.63	15.17	9.23	11.72
LOFIT = Lack of Fit	418.85	74.48	555.47	252.24
F = MSLOFIT/MSE*	2.17	29.60	11.56	15.95
F (Critical at 5%)	2.19	2.19	2.65	2.65

*MS, mean squared; MSE, mean squared error.

production in the next decades. Without 2G ethanol contribution, we estimated an expansion of sugarcane planted area of 56.4% to meet the domestic ethanol demand by 2030, based on the commitments made by Brazilian government in the UNFCCC (iNDC Brazil, 2015). The contribution of 2G technologies can notably decrease the land demand for sugarcane ethanol production (Table 3). However, estimated SOC changes in a scenario where the increase in ethanol production is based on the expansion of sugarcane onto pastures areas pointed to gains in C-savings of Brazilian ethanol, while the straw removal can affect negatively the C balance by decreasing the SOC in 50 Tg between 2020 and 2050 in sugarcane areas in Brazil (Table 3).

Discussion

The DayCent model reliably reflected the main trends of SOC changes undergoing the LUC NV-PA-SG in our sites. Using the Century model, Galdos et al. (2009), Brandani et al. (2015), and Silva-Olaya et al. (2016) successfully simulated SOC changes in sugarcane areas in Brazil. Moreover, Duval et al. (2013) concluded that the DayCent model performed well for simulating SOC changes undergoing the conversion pasture–energy cane in USA.

Despite the absence of significant bias (Table 2), the model appeared to overestimate N stocks, mainly in PA and SG areas (Table 1). Conant et al. (2005) observed that the DayCent model overestimated N stocks for half of pastures evaluated in sites from USA, UK, and Canada. However, due to the rapid N transformations in a warm and humid environment, we must take into account the possibility of N losses during the sampling, transport, and initial processing of the soil samples,

which could have contributed to the discrepancies between measured and modeled soil N stocks in our study.

For the C-partitioning, some lack of fit between the measured and simulated values were observed (Table 2), mainly related with the disagreement between the measured and simulated modern-C stocks in SG areas from Lat_23S (Table 1). At this site, sugarcane was harvested with burning between 1990 and 2003 (Fig. 1). Burning events cause shifts in δ^{13}C values of C4-derived charcoal (Krull et al., 2003), which certainly interfere in the C-partitioning. Effects of pyrogenic C on estimates of SOM partitioning is not accounted for in the DayCent model estimates. Nevertheless, the RMSE values indicated that the simulated C-partitioning still fell within the 95% confidence interval for the whole dataset (Table 2).

Overall, despite the disagreements discussed above, our DayCent simulations matched the direction of the main SOM shifts undergoing the LUC NV-PA-SG for all sites evaluated, even for N stocks and C-partitioning. Smith et al. (2012) showed that widely used process-based models (including Century) simulated values in the same uncertainty range as estimates derived from field experiments in areas for biofuels production. Moreover, DayCent is the most comprehensive of the process-based models when it comes to C dynamics representing plant and soil interactions (Robertson et al., 2015).

DayCent model predicted SOC losses of 0.34 ± 0.03 Mg C ha^{-1} yr^{-1} in the transition NV-PA (Fig. 3). Assessing SOC changes associated with the LUC NV-PA in Brazil, Maia et al. (2009) and Franco et al. (2015) found losses of SOC at rates of 0.28 and 0.40 Mg ha^{-1} yr^{-1}, respectively. These SOC losses can

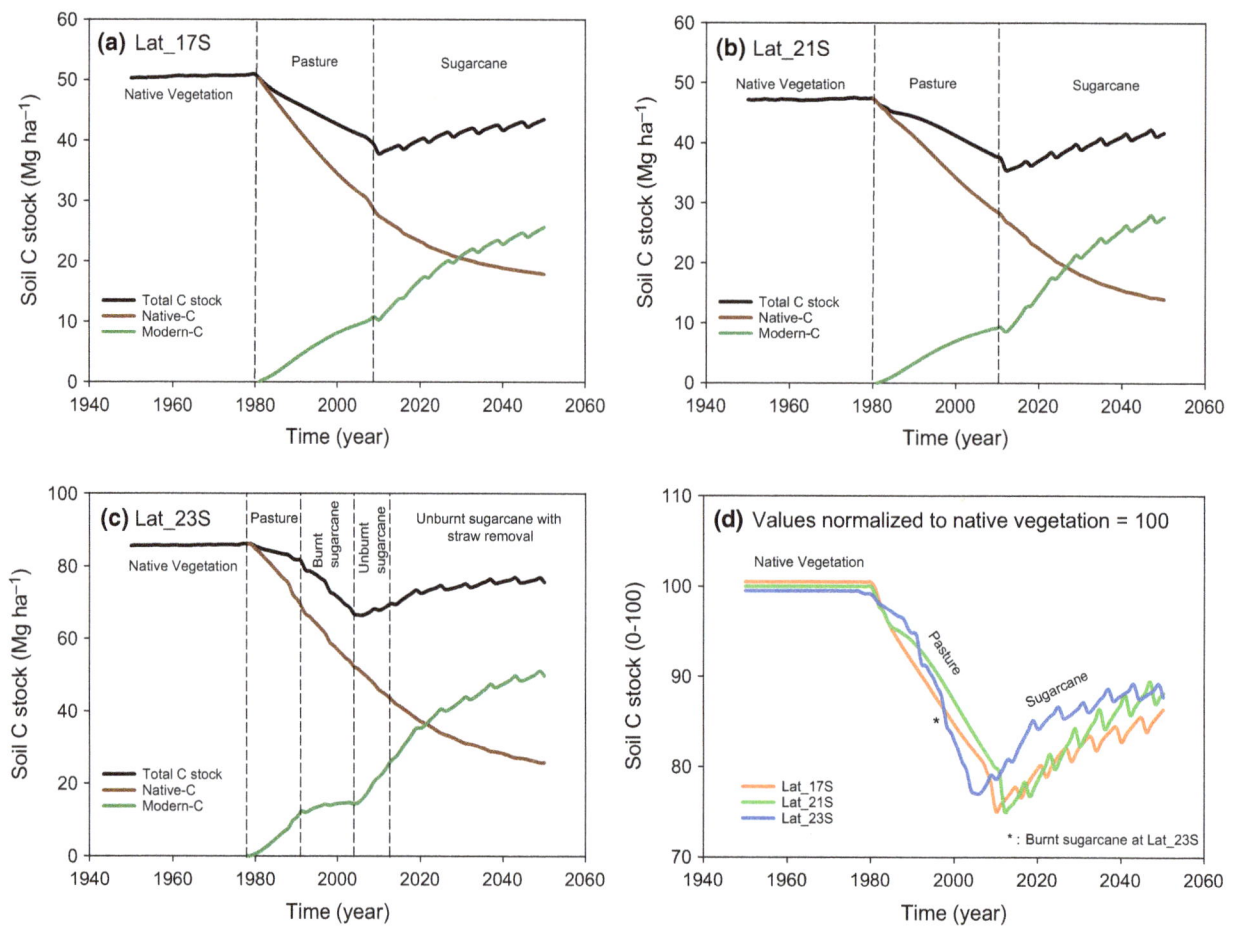

Fig. 3 Long-term simulations of soil C stocks and C-partitioning at 0–0.3 m layer of three sites – Lat_17S (a), Lat_21S (b), Lat_23S (c) – undergoing the land use transition native vegetation–pasture–sugarcane in south-central Brazil. (d): Values normalized considering the C stocks in native vegetation areas equal to 100.

be attributed to both deforestation and biomass burning effects, and subsequent processes of soil degradation in pasture areas (Maia *et al.*, 2009). After the LUC PA-SG (under GM), the simulations showed increments on SOC at a rate of 0.16 ± 0.04 Mg C ha^{-1} yr^{-1} until 2050. This result matched the previous rate (0.12 ± 0.03 Mg ha^{-1} yr^{-1}) obtained in our field-scale assessment (Oliveira *et al.*, 2016b). In USA, positive SOC changes were predicted when pastures were converted to energy cane (Duval *et al.*, 2013) or *Miscanthus* production (Dunn *et al.*, 2013). Moreover, Galdos *et al.* (2009) projected SOC gains of 0.23 Mg ha^{-1} yr^{-1} in Century simulations for SG areas under GM in Brazil.

The overall trend of increase in SOC in areas under GM is mainly related to the large input of organic material by sugarcane crop residues. In our study, the simulated C-partitioning suggested that the high input of crop residues in SG areas under GM is associated with a positive C balance, with the losses of native-C lower than the gains of modern-C, the opposite of PA areas.

In Lat_23S, when the SG was harvest with burning (1990–2003), a drop in SOC was observed (Fig. 3), as reported in other simulation studies (Galdos *et al.*, 2009; Brandani *et al.*, 2015). As a consequence, the conversion PA-SG with preharvest burning is associated with SOC losses (Mello *et al.*, 2014). However, nowadays almost all SG plantations in Brazil are green harvested (UNICA, 2015).

Sugarcane is usually replanted every sixth year. Under conventional tillage, the whole replant area is disturbed using plowing, disking, and, commonly, subsoiling. Our simulations showed that tillage operations caused a SOC loss of 1.14 ± 0.42 Mg ha^{-1} in the year right after sugarcane replanting, in agreement with previous studies in Brazil (Silva-Olaya *et al.*, 2013; Figueiredo *et al.*, 2015). The C-partitioning showed that most of the C from sugarcane (modern-C) from the previous five-year production period can be lost during the replanting period (Fig. 3). Such SOC losses are comparable with the GHG emissions from sugarcane burning

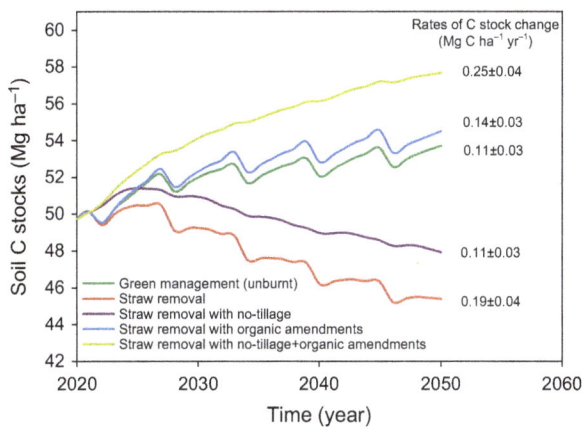

Fig. 4 Simulated soil C stocks at 0–0.3 m layer in sugarcane areas under green management, straw removal, and best management practices in Brazil. The reference for the rates of C stock change is the soil C stock in areas under green management by 2020 $n = 3$.

estimated by Bordonal *et al.* (2012). In this sense, we suggest the adoption of management systems involving 'less aggressive' tillage operations, to decrease the SOC losses in SG areas under GM in Brazil.

Nowadays, more than 60% of Brazilian pastures are in some degree of degradation (Andrade *et al.*, 2014). The replacement of degraded lands (with low soil C stocks) with high productivity energy crops may result in a positive soil C balance and additional C-savings for biofuels (Gelfand *et al.*, 2013; Gollany *et al.*, 2015). Our simulations showed that the replacement of pastures with sugarcane is associated with SOC gains, which partially offset the C debt resulting from the conversion of natural vegetation to pastures. One of the potential consequences of such LUC is the migration of livestock to other regions, increasing deforestation (Lapola *et al.*, 2010). This indirect LUC, although very controversial, is now seen to have far less impact than previously thought (Macedo *et al.*, 2015). Currently, government actions to improve pasture conditions (ABC Brazil, 2012), along with livestock production intensification, can effectively make large amounts of land available for alternative uses in Brazil. In this sense, we estimated a SOC accretion of 144 Tg if the projected increments on ethanol production were based on expansion of sugarcane into pasture areas in the next years in Brazil (Table 3).

DayCent simulations showed SOC losses of 0.19 ± 0.04 Mg ha^{-1} yr^{-1} in SG areas in Brazil with straw removal (Fig. 4). The harvest of crop residues is associated with potential environmental impacts, highlighting SOC losses (Wilhelm *et al.*, 2007; Wortmann *et al.*, 2010), as crop residues are a key component for SOC accretion (Paustian *et al.*, 2016). Modeled SOC

Table 3 Sugarcane-cultivated area and soil C changes associated with the projected increase in ethanol production to meet the domestic demand in Brazil by 2030 in different scenarios

Estimative	Value
Energy consumption in Brazil by 2030*	3529 TWh
Sugarcane contribution to the energy supply by 2030[†]	16%
Ethanol contribution in the energy supply by sugarcane[‡]	88%
Ethanol yield by 2030[§]	10 000 L ha^{-1} yr^{-1}
Ethanol yield by 2030 including 2G technology[¶]	15 000 L ha^{-1} yr^{-1}
Expected Brazilian ethanol production by 2030**	84.3 billions of L
Expected sugarcane-cultivated area by 2030[††]	8.4 Million ha
Expected sugarcane-cultivated area by 2030 (with 2G)[††]	5.6 Million ha
Soil C changes in sugarcane areas between 2020 and 2050 with the increase in ethanol production based on LUC to pastures	144 Tg C
Soil C changes in sugarcane areas between 2020 and 2050 with the increase in ethanol production based on 2G ethanol (straw removal)	−50 Tg C

*Bronzatti & Iarozinski Neto (2008).
[†]iNDC Brazil (2015).
[‡]12% to bioelectricity from sugarcane bagasse burning (Kutas, 2016).
[§]Based on the increments on ethanol yield in the last years (Goldemberg & Guardabassi, 2010).
[¶]Assuming that 2G technologies will increase the sugarcane ethanol yield about 50% (Kutas, 2016).
**Based on the sugarcane ethanol contribution in the energy supply by 2030 (iNDC Brazil, 2015).
[††]Only for ethanol production. Currently, 59.4% of the sugarcane-cultivated area is harvested to ethanol production, while the remaining (40.6%) is used by sugar industry (UNICA, 2015).

losses associated with straw removal in sweet sorghum showed that these emissions could eliminate all GHG mitigation benefits of bioethanol compared with gasoline (Wortmann *et al.*, 2010). Using DayCent simulations, Miner *et al.* (2013) concluded that all stover is needed to be left in the field to maintain SOC levels in wheat, corn and grain sorghum areas in USA. Moreover, studies from Americas (Gollany *et al.*, 2015), USA (Wilhelm *et al.*, 2007), and Australia (Zhao *et al.*, 2015) suggested that the SOC losses are the main constraint regarding the straw removal in agricultural areas for biofuels production. In this sense, the straw removal in SG areas in Brazil might be beneficial from an energy security point of view as more ethanol (or electricity) will be produced, but not necessarily will result in the higher C-savings because the potential SOC losses associated (Fig. 4).

In Brazil, the development of technologies for 2G ethanol production has been moving at a slower pace than other places for many reasons but, now, seem to be accelerating. Currently, Brazil has two commercial 2G ethanol mills in operation, three demo mills, and 20 projects in the pipeline (Kutas, 2016). Moreover, substantial investment by the private sector and government is a strong market signal that sugarcane 2G ethanol supplies are likely to increase dramatically in the next years. In addition, straw removal in SG areas is also happening to support electricity production (Walter & Ensinas, 2010), such as currently in Lat_23S site. Therefore, straw removal is likely to become a common practice in Brazilian sugarcane areas soon and management practices must be proposed in order to mitigate the negative effects of the straw removal on SOC.

The adoption of no-tillage in SG areas with straw removal can decrease the rates of SOC losses comparing with areas under straw removal only (Fig. 4). In Brazil, SOC gains have been reported in sugarcane areas under no-tillage (Segnini et al., 2013). Century simulations showed that the adoption of no-tillage reduces the losses or even result in SOC gains in corn areas with stover harvest to ethanol production in USA (Sheehan et al., 2003). However, in our study, DayCent simulations showed positive SOC changes only when another source of C (vinasse and filter cake) was added (Fig. 4). According to Century simulations, vinasse and filter cake application were predicted to increase SOC in sugarcane areas in Brazil (Brandani et al., 2015). Filter cake and vinasse are produced in large quantities by the sugar-alcohol agro-industry. Moreover, the vertical integration of the sugarcane industry in Brazil makes the distribution of these subproducts easier because of shorter distances from the refinery to the field. In this sense, filter cake and vinasse applied to the soil is a practice widely used in SG areas in Brazil (Prado et al., 2013) and, as we observed, can have a prominent role on SOC dynamics in SG areas with straw removal. Despite its benefits, the adoption of no-tillage is not common in SG areas in Brazil. However, in a scenario with straw removal and possible SOC losses, we need to consider no-tillage during sugarcane crop renovation, mainly if it is combined with other best management practices, such as vinasse and filter cake application (Fig. 4). Lastly, we must mention that soil C accretion is finite and, under the same management practices and C inputs, the soil C stocks in these areas are expected to reach a new equilibrium over the next decades.

Without the contribution of 2G ethanol, we projected that the sugarcane area in Brazil is expected to expand by 3.04 Mha by 2030. With the full implementation of 2G ethanol production in the next years, we projected an expansion on SG area of only 0.23 Mha to meet domestic ethanol demands by 2030 (Table 3). LUC projections based on feedstock demands are a quite complex task and inherently uncertain. Moreover, the possible inclusion of 2G ethanol in the Brazilian energy supply in the next years increases the uncertainty about the land required for sugarcane production. Similarly uncertain are the spatial extrapolations about SOC, as C dynamics are known to be highly dependent on environmental characteristics and local management factors. In this sense, despite the limitations discussed above, the data presented in the Table 3 aim to show the likely direction and relative magnitudes of land conversions and SOC changes related in two feasible scenarios of SG expansion in Brazil. Moreover, our projections raised concerns about the sustainability of straw removal in SG areas. The SOC changes could be greater or less than estimated here, but our research can be a starting point for development of management strategies to mitigate possible SOC losses regarding 2G ethanol production in Brazil.

Based on land availability and positive effects on C-savings of sugarcane ethanol, we believe that stakeholders involved with the governance of bioethanol expansion should consider ways to incentivize sugarcane expansion on degraded pastures in Brazil. Moreover, we are sure that 2G technology will increase notably the energy output from sugarcane, but inferences about the net mitigation potential of 2G ethanol from sugarcane will require analysis of the entire biofuel life cycle, in which possible SOC losses should be taken into account. In this sense, field studies about the environmental suitability of straw removal in sugarcane areas are mandatory before using crop residues as a source of biomass for large-scale ethanol production in Brazil. Finally, the time horizon is quite relevant when evaluating soil C dynamics in agricultural areas (e.g., time since adoption of the GM system has great impact on the potential increase in SOC in sugarcane areas). However, extensive field measurements and data collection is costly or impossible, and thus, simulation models can help researchers to expand short-term field research to longer scenarios where field measurements are difficult to conduct. Our results supported that DayCent model can complement and extend the applicability of information collected in field studies (Campbell & Paustian, 2015; Robertson et al., 2015) and may be applied to obtain credible long-term assessments of sugarcane production effects on SOC in tropical regions.

Acknowledgements

The authors gratefully thank the São Paulo Research Foundation (FAPESP) (grants # 2014/08632-9 and # 2015/14122-6) for the scholarship granted while this research was carried out, and CNPq (grants # 402992/2013-0 and # 311661/2014-9) for the financial support of the present research. We thank Mark Easter, William Parton, Melannie Hartman, and Kendrick Killian for their relevant opinions during the simulations. Anonymous reviewers are also thanked for their valuable criticisms and comments, which led to substantial improvements of this manuscript.

References

ABC Brazil (2012) Plano Setorial de Mitigação e de Adaptação às Mudanças Climáticas para a Consolidação de uma Economia de Baixa Emissão de Carbono na Agricultura, Brazilian Ministry of Agriculture, Brasília. Available at: http://www.agricultura.gov.br/arq_editor/download.pdf (accessed 5 January 2016).

Andrade RG, Teixeira AHC, Leivas JF et al. (2014) EMBRAPA: Brazilian Agriculture Observation and Monitoring System. Cerrado Pasture Degradation - Scenario 3. Available at: http://mapas.cnpm.embrapa.br/somabrasil/webgis.html (accessed 18 January 2016).

Bernoux M, Cerri CC, Neill C, De Moraes JFL (1998) The use of stable carbon isotopes for estimating soil organic matter turnover rates. Geoderma, 82, 43–58.

Bordonal RDO, Figueiredo EB, La Scala N (2012) Greenhouse gas balance due to the conversion of sugarcane areas from burned to green harvest, considering other conservationist management practices. GCB Bioenergy, 4, 846–858.

Brandani CB, Abbruzzini TF, Williams S, Easter M, Pellegrino Cerri CE, Paustian K (2015) Simulation of management and soil interactions impacting SOC dynamics in sugarcane using the CENTURY Model. GCB Bioenergy, 7, 646–657.

Bronzatti FL, Iarozinski Neto A (2008) Matrizes energéticas no Brasil: Cenário 2010-2030. In: Encontro Nacional de Engenharia de Produção, Rio de Janeiro. Available at: http://www.abepro.org.br/biblioteca/enegep2008_TN_STO_077_541_11890.pdf (accessed 19 January 2016).

Bustamante M, Nardoto G, Pinto A, Resende J, Takahashi F, Vieira L (2012) Potential impacts of climate change on biogeochemical functioning of Cerrado ecosystems. Brazilian Journal of Biology, 72, 655–671.

Campbell EE (2015) Modeling soil organic matter: theory, development, and applications in bioenergy cropping systems. PhD Thesis. Colorado State University, Fort Collins.

Campbell EE, Paustian K (2015) Current developments in soil organic matter modeling and the expansion of model applications: a review. Environmental Research Letters, 10, 123004.

Cardoso TDF, Cavalett O, Chagas MF et al. (2013) Technical and economic assessment of trash recovery in the sugarcane bioenergy production system. Scientia Agricola, 70, 353–360.

Cerri CEP, Paustian K, Bernoux M, Victoria RL, Melillo JM, Cerri CC (2004) Modeling changes in soil organic matter in Amazon forest to pasture conversion with the Century model. Global Change Biology, 10, 815–832.

Cerri CEP, Easter M, Paustian K et al. (2007) Predicted soil organic carbon stocks and changes in the Brazilian Amazon between 2000 and 2030. Agriculture, Ecosystems & Environment, 122, 58–72.

Cherubin MR, Franco ALC, Cerri CEP, Oliveira DMS, Davies CA, Cerri CC (2015) Sugarcane expansion in Brazilian tropical soils—Effects of land use change on soil chemical attributes. Agriculture, Ecosystems & Environment, 211, 173–184.

Conant RT, Paustian K, Del Grosso SJ, Parton WJ (2005) Nitrogen pools and fluxes in grassland soils sequestering carbon. Nutrient Cycling in Agroecosystems, 71, 239–248.

Del Grosso S, Parton W, Mosier A, Hartman M, Brenner J, Ojima D, Schimel D (2001) Simulated interaction of carbon dynamics and nitrogen trace gas fluxes using the DAYCENT model. In: Modeling Carbon and Nitrogen Dynamics for Soil Management (eds Schaffer M, Ma L, Hansen S), pp. 303–332. CRC Press, Boca Raton.

Del Grosso SJ, Ojima DS, Parton WJ, Stehfest E, Heistemann M, Deangelo B, Rose S (2009) Global scale DAYCENT model analysis of greenhouse gas emissions and mitigation strategies for cropped soils. Global and Planetary Change, 67, 44–50.

Dias LCP, Pimenta FM, Santos AB, Costa MH, Ladle RJ (2016) Patterns of land use, extensification, and intensification of Brazilian agriculture. Global Change Biology, 22, 2887–2903.

Dunn JB, Mueller S, Kwon H-Y, Wang MQ (2013) Land-use change and greenhouse gas emissions from corn and cellulosic ethanol. Biotechnology for Biofuels, 6, 1–13.

Duval BD, Anderson-Teixeira KJ, Davis SC, Keogh C, Long SP, Parton WJ, Delucia EH (2013) Predicting greenhouse gas emissions and soil carbon from changing pasture to an energy crop. PLoS ONE, 8, e72019.

Fargione J, Hill J, Tilman D, Polasky S, Hawthorne P (2008) Land clearing and the biofuel carbon debt. Science, 319, 1235–1238.

Figueiredo EB, Panosso AR, Reicosky DC, La Scala N (2015) Short-term CO₂-C emissions from soil prior to sugarcane (Saccharum spp.) replanting in southern Brazil. GCB Bioenergy, 7, 316–327.

Franco ALC, Cherubin MR, Pavinato PS, Cerri CEP, Six J, Davies CA, Cerri CC (2015) Soil carbon, nitrogen and phosphorus changes under sugarcane expansion in Brazil. Science of The Total Environment, 515–516, 30–38.

Galdos MV, Cerri CC, Cerri CEP, Paustian K, Antwerpen RV (2009) Simulation of Soil Carbon Dynamics under Sugarcane with the CENTURY Model. Soil Science Society of America Journal, 73, 802–811.

Galdos MV, Cerri CC, Cerri CEP, Paustian K, Van Antwerpen R (2010) Simulation of sugarcane residue decomposition and aboveground growth. Plant and Soil, 326, 243–259.

Gelfand I, Sahajpal R, Zhang X, Izaurralde RC, Gross KL, Robertson GP (2013) Sustainable bioenergy production from marginal lands in the US Midwest. Nature, 493, 514–517.

Goldemberg J, Guardabassi P (2010) The potential for first-generation ethanol production from sugarcane. Biofuels, Bioproducts and Biorefining, 4, 17–24.

Gollany HT, Titus BD, Scott DA et al. (2015) Biogeochemical research priorities for sustainable biofuel and bioenergy feedstock production in the Americas. Environmental Management, 56, 1330–1355.

Hudiburg TW, Wang W, Khanna M et al. (2016) Impacts of a 32-billion-gallon bioenergy landscape on land and fossil fuel use in the US. Nature Energy, 1, 15005.

IPEA (2016) Brazilian Gross Domestic Product. Brazilian Institute of Applied Economic Research. Available at: http://www.ipeadata.gov.br/ (accessed 28 March 2016).

Krull ES, Skjemstad JO, Graetz D, Grice K, Dunning W, Cook G, Parr JF (2003) 13C-depleted charcoal from C4 grasses and the role of occluded carbon in phytoliths. Organic Geochemistry, 34, 1337–1352.

Kutas G (2016) Application and development potential of bioethanol in Brazil –advantages and disadvantages of flexible blending. In: Fuels of the Future, Berlin. Available at: http://www.unica.com.br/files/pag=3 (accessed 10 May 2016).

Lapola DM, Schaldach R, Alcamo J, Bondeau A, Koch J, Koelking C, Priess JA (2010) Indirect land-use changes can overcome carbon savings from biofuels in Brazil. Proceedings of the National Academy of Sciences of the United States of America, 107, 3388–3393.

Lilienfein J, Wilcke W (2003) Element storage in native, agri-, and silvicultural ecosystems of the Brazilian savanna. Plant and Soil, 254, 425–442.

Macedo IDC, Nassar AM, Cowie AL et al. (2015) Greenhouse gas emissions from bioenergy. In: Bioenergy & Sustainability: Bridging the Gaps (eds Souza GM, Victoria RL, Joly CA, Verdade LM), pp. 582–616. SCOPE, São Paulo.

Maia SMF, Ogle SM, Cerri CEP, Cerri CC (2009) Effect of grassland management on soil carbon sequestration in Rondônia and Mato Grosso states, Brazil. Geoderma, 149, 84–91.

MCTI (2014) Estimativas Anuais de Emissões de Gases de Efeito Estufa no Brasil. Brazilian Ministry of Science, Technology and Innovation. Available at: http://www.mct.gov.br/upd_blob/0235/235580.pdf (accessed 17 February 2016).

Mello FFC, Cerri CEP, Davies CA et al. (2014) Payback time for soil carbon and sugar-cane ethanol. Nature Climate Change, 4, 605–609.

Miner GL, Hansen NC, Inman D, Sherrod LA, Peterson GA (2013) Constraints of No-Till Dryland Agroecosystems as Bioenergy Production Systems. Agronomy Journal, 105, 364–376.

iNDC Brazil (2015) Intended nationally determined contribution towards achieving the objective of the United Nations Framework Convention on Climate Change. Federative Republic of Brazil. Available at: http://www.itamaraty.gov.br/images/ed_desenvsust/BRAZIL-iNDC-english.pdf (accessed 21 March 2016).

Oliveira DMS, Schellekens J, Cerri CEP (2016a) Molecular characterization of soil organic matter from native vegetation–pasture–sugarcane transitions in Brazil. Science of The Total Environment, 548–549, 450–462.

Oliveira DMS, Paustian K, Davies CA, Cherubin MR, Franco ALC, Cerri CC, Cerri CEP (2016b) Soil carbon changes in areas undergoing expansion of sugarcane into pastures in south-central Brazil. Agriculture, Ecosystems & Environment, 228, 38–48.

Parton WJ, Schimel DS, Cole CV, Ojima DS (1987) Analysis of factors controlling soil organic matter levels in Great Plains grasslands. *Soil Science Society of America Journal*, **51**, 1173–1179.

Parton WJ, Hartman M, Ojima D, Schimel D (1998) DAYCENT and its land surface submodel: description and testing. *Global and Planetary Change*, **19**, 35–48.

Paustian K, Lehmann J, Ogle S, Reay D, Robertson GP, Smith P (2016) Climate-smart soils. *Nature*, **532**, 49–57.

Prado RDM, Caione G, Campos CNS (2013) Filter cake and vinasse as fertilizers contributing to conservation agriculture. *Applied and Environmental Soil Science*, **2013**, 8.

Renouf MA, Wegener MK, Nielsen LK (2008) An environmental life cycle assessment comparing Australian sugarcane with US corn and UK sugar beet as producers of sugars for fermentation. *Biomass and Bioenergy*, **32**, 1144–1155.

Robertson AD, Davies CA, Smith P, Dondini M, Mcnamara NP (2015) Modelling the carbon cycle of Miscanthus plantations: existing models and the potential for their improvement. *GCB Bioenergy*, **7**, 405–421.

Segnini A, Carvalho JLN, Bolonhezi D et al. (2013) Carbon stock and humification index of organic matter affected by sugarcane straw and soil management. *Scientia Agricola*, **70**, 321–326.

Sheehan J, Aden A, Paustian K, Killian K, Brenner J, Walsh M, Nelson R (2003) Energy and environmental aspects of using corn Stover for fuel ethanol. *Journal of Industrial Ecology*, **7**, 117–146.

Silva-Olaya AM, Cerri CEP, La Scala N Jr, Dias CTS, Cerri CC (2013) Carbon dioxide emissions under different soil tillage systems in mechanically harvested sugarcane. *Environmental Research Letters*, **8**, 015014.

Silva-Olaya AM, Cerri CEP, Williams S, Cerri CC, Davies CA, Paustian K (2016) Modelling SOC response to land use change and management practices in sugarcane cultivation in South-Central Brazil. *Plant and Soil*, **00**, 1–16.

Smith P, Smith JU, Powlson DS et al. (1997) Evaluation and comparison of soil organic matter models: a comparison of the performance of nine soil organic matter models using datasets from seven long-term experiments. *Geoderma*, **81**, 153–225.

Smith WN, Grant BB, Campbell CA, Mcconkey BG, Desjardins RL, Kröbel R, Malhi SS (2012) Crop residue removal effects on soil carbon: measured and inter-model comparisons. *Agriculture, Ecosystems & Environment*, **161**, 27–38.

Souza AP, Grandis A, Leite DCC, Buckeridge MS (2014) Sugarcane as a bioenergy source: history, performance, and perspectives for second-generation bioethanol. *BioEnergy Research*, **7**, 24–35.

UNICA (2015) Coletiva de imprensa: Estimativa safra 2015/2016. Brazilian Sugarcane Industry Association. Available at: http://www.unica.com.br/documentos/apresentacoes/unica/ (accessed 26 January 2016).

USDA (2014) *Keys to Soil Taxonomy*. USDA—Natural Resources Conservation Service, Washington.

Walter A, Ensinas AV (2010) Combined production of second-generation biofuels and electricity from sugarcane residues. *Energy*, **35**, 874–879.

Wei Y, Liu S, Huntzinger DN et al. (2014) NACP MsTMIP: Global and North American Driver Data for Multi-Model Intercomparison. ORNL Distributed Active Archive Center. Available at: http://daac.ornl.gov (accessed 23 December 2015).

Wilhelm WW, Johnson JMF, Karlen DL, Lightle DT (2007) Corn stover to sustain soil organic carbon further constrains biomass supply. *Agronomy Journal*, **99**, 1665–1667.

Wortmann CS, Liska AJ, Ferguson RB, Lyon DJ, Klein RN, Dweikat I (2010) Dryland performance of sweet sorghum and grain crops for biofuel in Nebraska. *Agronomy Journal*, **102**, 319–326.

Zhao G, Bryan BA, King D, Luo Z, Wang E, Yu Q (2015) Sustainable limits to crop residue harvest for bioenergy: maintaining soil carbon in Australia's agricultural lands. *GCB Bioenergy*, **7**, 479–487.

5

Hemp (*Cannabis sativa* L.) leaf photosynthesis in relation to nitrogen content and temperature: implications for hemp as a bio-economically sustainable crop

KAILEI TANG[1,2] (iD), PAUL C. STRUIK[1], STEFANO AMADUCCI[2], TJEERD-JAN STOMPH[1] and XINYOU YIN[1]

[1]Centre for Crop Systems Analysis, Department of Plant Sciences, Wageningen University & Research, PO Box 430, Wageningen, 6700 AK, The Netherlands, [2]Department of Sustainable Crop Production, Università Cattolica del Sacro Cuore, via Emilia Parmense 84, Piacenza, 29122, Italy

Abstract

Hemp (*Cannabis sativa* L.) may be a suitable crop for the bio-economy as it requires low inputs while producing a high and valuable biomass yield. With the aim of understanding the physiological basis of hemp's high resource-use efficiency and yield potential, photosynthesis was analysed on leaves exposed to a range of nitrogen and temperature levels. Light-saturated net photosynthesis rate (A_{max}) increased with an increase in leaf nitrogen up to 31.2 ± 1.9 µmol m^{-2} s^{-1} at 25 °C. The A_{max} initially increased with an increase in leaf temperature (T_L), levelled off at 25–35 °C and decreased when T_L became higher than 35 °C. Based on a C$_3$ leaf photosynthesis model, we estimated mesophyll conductance (g_m), efficiency of converting incident irradiance into linear electron transport under limiting light (κ_{2LL}), linear electron transport capacity (J_{max}), Rubisco carboxylation capacity (V_{cmax}), triose phosphate utilization capacity (T_p) and day respiration (R_d), using data obtained from gas exchange and chlorophyll fluorescence measurements at different leaf positions and various levels of incident irradiance, CO$_2$ and O$_2$. The effects of leaf nitrogen and temperature on photosynthesis parameters were consistent at different leaf positions and among different growth environments except for κ_{2LL}, which was higher for plants grown in the glasshouse than for those grown outdoors. Model analysis showed that compared with cotton and kenaf, hemp has higher photosynthetic capacity when leaf nitrogen is <2.0 g N m^{-2}. The high photosynthetic capacity measured in this study, especially at low nitrogen level, provides additional evidence that hemp can be grown as a sustainable bioenergy crop over a wide range of climatic and agronomic conditions.

Keywords: Hemp (*Cannabis sativa* L.), model, nitrogen, photosynthesis, sustainable crop, temperature

Introduction

The multiple societal challenges such as climate change, natural resource scarcity and environmental pollution have fuelled interest in bio-economy (Jordan *et al.*, 2007). Previous comprehensive research programmes indicated that hemp (*Cannabis sativa* L.) fits well in the concept of bio-economy (Mccormick & Kautto, 2013; Amaducci *et al.*, 2015). Hemp has the potential to produce up to 27 Mg ha^{-1} biomass yield (Tang *et al.*, 2016) at relatively low inputs (Struik *et al.*, 2000; Amaducci *et al.*, 2002) and has a positive impact on the environment (Bouloc & Van der Werf, 2013; Barth & Carus, 2015). Its stem contains high-quality cellulose (De Meijer

& Van der Werf, 1994), the seeds contain high-quality oil (Oomah *et al.*, 2002), and the inflorescence contains valuable resins (Bertoli *et al.*, 2010). From speciality pulp and paper to nutritional food, medicine and cosmetics, there are as many as 50 000 uses claimed for hemp products derived from its stem, seed and inflorescence (Carus *et al.*, 2013; Carus & Sarmento, 2016). Recent research demonstrated that hemp is also a suitable feedstock for bioenergy production (Rice, 2008; Kreuger *et al.*, 2011; Prade *et al.*, 2011).

Although once an important crop for the production of textiles and ropes, hemp has not been subjected to the intensive research that has driven great improvements in major crops in the last 50 years (Amaducci *et al.*, 2015; Salentijn *et al.*, 2015) due to the continuous decrease in hemp acreage after the Second World War and its slow revival in the last couple of decades

Correspondence: Xinyou Yin
e-mail: xinyou.yin@wur.nl

(Wirtshafter, 2004; Allegret, 2013). To advance research needed to consolidate and expand the market of hemp renewable materials, within the frame of the EU funded project Multihemp (www.multihemp.eu), it was proposed to develop a process-based hemp growth model similar to the successful models for major staple crops (Bouman et al., 2007). With the aim of understanding the physiological basis of hemp's high resource-use efficiency and yield potential using a modelling approach, this study focuses on analysing leaf photosynthesis of hemp as a primary source of biomass production.

Very few studies report on leaf photosynthesis of hemp. De Meijer et al. (1995) reported a light-saturated rate of leaf photosynthesis for hemp of 30 kg CO_2 ha^{-1} h^{-1} (equivalent to 19 µmol m^{-2} s^{-1}) under field conditions. Chandra et al. (2008, 2011a,b, 2015) showed the response of leaf photosynthesis of hemp to irradiance intensity, CO_2 concentration and temperature by measuring gas exchange of leaves from glasshouse-grown plants. Marija et al. (2011) found that nitrogen fertilization significantly affected different aspects of photosynthetic photochemistry, as shown by chlorophyll a fluorescence analysis. To the best of our knowledge, a comprehensive analysis of the relation between leaf nitrogen status and photosynthesis rate is not yet available for hemp.

Leaf photosynthesis rate depends on both nitrogen nutrition status and environmental conditions (Sinclair & Horie, 1989). Thanks to a thorough understanding of the biochemical mechanisms of leaf photosynthesis, the response of leaf photosynthesis to irradiance intensity and CO_2 concentration can be modelled (Farquhar et al., 1980; Yin et al., 2006; Von Caemmerer et al., 2009). Such a model dissects net leaf photosynthesis into mesophyll conductance (g_m), linear electron transport capacity (J_{max}), Rubisco carboxylation capacity (V_{cmax}), triose phosphate utilization capacity (T_p) and day respiration (R_d). The effects of leaf nitrogen status and temperature on leaf photosynthesis are considered through their effects on these photosynthetic parameters (Hikosaka et al., 2016). Experimental protocols for parameterizing the biochemical photosynthesis model have been well documented (Sharkey et al., 2007; Yin et al., 2009; Bellasio et al., 2015), and the model has been successfully embedded as a submodel in process-based crop growth models for upscaling to canopy photosynthesis and crop production (Yin & Struik, 2009), such as the GECROS crop model (Yin & Van Laar, 2005). Therefore, parameterizing the photosynthesis model for hemp is an excellent opportunity to understand its photosynthetic resource-use efficiency, as well as to provide essential information for modelling hemp growth.

The first objective of this study was to analyse leaf photosynthesis of hemp as affected by irradiance intensity, CO_2 concentration, temperature and nitrogen status. Secondly, this study aimed to parameterize a widely used C_3 leaf photosynthesis model (Farquhar et al., 1980; Yin et al., 2006) for hemp. In the final section, the photosynthetic capacity of hemp is compared with that of two other bio-economic crops, cotton (Gossypium hirsutum L.) and kenaf (Hibiscus cannabinus L.), using a modelling method. Cotton and kenaf were chosen because they are bio-economically important crops and, in particular, kenaf is considered as an alternative for hemp in tropical and subtropical climates (Lips & van Dam, 2013; Patanè & Cosentino, 2013; Alexopoulou et al., 2015).

Materials and methods

Plant growth and data collection

Three independent experiments were carried out at the research facilities of the Università Cattolica del Sacro Cuore (45.0°N, 9.8°E, 60 m asl; Piacenza, Italy). Seeds of hemp (cv. Futura 75) were received from the Fédération National des Producteurs de Chanvre, Le Mans, France. The plants were grown outdoors in 2013 and 2014 and in a glasshouse in 2015.

An experiment on the effect of nitrogen on leaf photosynthetic capacity (N-trial)

Seeds were sown in 18 containers ($40 \times 40 \times 30$ cm^3) placed outdoors on 9 May 2014. Each container was filled with 23 kg of soil (dry weight) that contained 0.22% total nitrogen and had a clay–silt–sand ratio of 30:43:27. After seedling emergence, the plants were hand-thinned to 18 plants per container and three levels of urea fertilization were applied (0, 1.0 and 2.0 g N per container, respectively). There were six containers for each fertilization level. Other nutrients (e.g. phosphate and potassium) were assumed not limiting factors according to historic experience in the field from which the soil was collected. The same applies to the other two trials. During plant growth, all containers were positioned randomly and tightly in one block surrounded by a green shading net (transmitting 3% of the light). The net height was adjusted daily according to the increment of plant height. The plants were well watered during the entire experiment. The daily temperature and global radiation during the growth period are presented in Fig. S1.

Photosynthetic measurements were started on 46 days after sowing (the 6th–8th pair of leaves had appeared) in a growth chamber with the temperature set at 25 °C. The container was moved into the growth chamber 2 hrs before measurements. On one representative plant in each container, the middle leaflets of the youngest, fully expanded top leaf and of the middle leaf (i.e. two nodes below the top leaf) were measured. Simultaneous gas exchange (GE) and chlorophyll fluorescence (CF) measurements were implemented in situ using a portable open gas exchange system with a 1.7-cm^2 clamp-on leaf chamber (CIRAS-2, PP Systems international, Inc., Amesbury, MA, USA) combining with FMS2 (Hansatech Instruments Ltd, King's Lynn, Norfolk, UK). The system set-up of the combined CIRAS-2 and FMS2 for performing simultaneous GE and CF measurements was implemented according to the instructions provided by PP Systems International, Inc., USA. Light response curve of net photosynthesis rate (A) (A-I_{inc}) and its CO_2 response curve (A-C_a) were assessed for each leaf under ambient O_2 (i.e. 21%) conditions. The A-I_{inc} curves were assessed by decreasing incident light intensity (I_{inc}) as 2000, 1500, 1000, 500, 300, 200, 150, 100, 60 and 30 µmol m^{-2} s^{-1}, while keeping leaf chamber CO_2 concentration (C_a) at 400 µmol mol^{-1}. At the end of assessing the A-I_{inc} curve, the light source was turned off for 15 min to measure leaf respiration in darkness (R_{dk}). The A-C_a curves were assessed by changing C_a as 400, 250, 150, 80, 70, 60, 50, 400, 400, 600, 800, 1000 and 1500 µmol mol^{-1}, while keeping I_{inc} at 1000 µmol m^{-2} s^{-1}. Leaf temperature (T_L) and vapour pressure of supplying air during measurements were set constant at 25 °C and 2 kPa, respectively. The response curves were started when the leaf had adapted to the condition at the first I_{inc} or C_a level for 30 min. Data were recorded programmatically with 2-min interval for A-I_{inc} curves and 3-min interval for A-C_a curves. Premeasurements indicated these time intervals were sufficiently long for A to reach a steady state. Three plants were measured for each fertilization level.

To obtain a calibration factor that can properly convert fluorescence-based PSII efficiency into linear electron transport rate, parts of A-I_{inc} and A-C_a curves were also assessed under 2% O_2. This condition was realized by supplying the CIRAS-2 with a humidified mixture of 2% O_2 and 98% N_2. To avoid O_2 leakage, the air-in pump in the CIRAS-2 was replaced by a sealed one according to the manufacturer's instruction. The curves for 2% O_2 were assessed in accordance with the ones for ambient O_2, but the A-I_{inc} curves were only assessed at I_{inc} ≤150 µmol m^{-2} s^{-1} and the A-C_a curves were only assessed at C_a ≥600 µmol mol^{-1}. These particular I_{inc} and C_a conditions are required for obtaining the calibration factor (Yin et al., 2009), that is to ensure that A is limited by electron transport.

When the photosynthetic measurements were completed, SPAD, a proxy for chlorophyll concentration, was measured using a SPAD-502 (Minolta, Japan). Leaf area was determined from scans using IMAGEJ (version 1.49; https://imagej.nih.gov/). Dry weight was measured after drying at 75 °C until constant weight. Total leaf nitrogen concentration was analysed using a CN analyser (Vario Max CN Analyzer; Elementar Americas, Inc., Hanau, Germany). Specific leaf nitrogen (SLN; g N m^{-2}) was calculated for each measured leaf using the leaf dry weight, leaf area and nitrogen concentration. CO_2 leakage of the CIRAS-2 leaf chamber was assessed by performing A-C_a curves on three heat-killed leaves. Based on these measurements, values of A and the intercellular CO_2 concentration (C_i) of A-C_a curves were recalculated using the CIRAS-2 built-in formulae.

An experiment on the effect of temperature on leaf photosynthetic capacity (T-trial)

Seeds were sown in six pots (10 × 10 × 15 cm^3) placed in a glasshouse on 12 February 2015. Each pot contained 1 kg of soil

that had identical properties with the ones in the N-trial. The temperature in the glasshouse was maintained at approximately 25 °C. A LED lamp (270 Watt, Shenzhen GTL Lighting Co., Ltd, China) mounted 50 cm above the canopy for 16 hrs each day gave the light level in glasshouse of approximately 600 µmol m^{-2} s^{-1}. After emergence, the plants were hand-thinned to two plants per pot, and urea fertilization was applied (0.3 g N per pot). The plants were well watered during growth.

Starting on 46 days after sowing, GE measurements were conducted in a temperature-controllable chamber. On one plant in each pot, the middle leaflet of the youngest, fully expanded top leaf was measured. The A-I_{inc} and A-C_a curves were assessed subsequently at T_L 15, 20, 25, 30, 35 and 40 °C. The levels of I_{inc} and C_a were set in accordance with the N-trial under ambient O_2. During the measurements, the temperature in the growth chamber was controlled close to the targeting T_L and the vapour pressure of supplying air was set at 1.5 kPa for all temperature levels except for 15 °C, when it was set at 1.0 kPa to avoid water condensation. Three plants were measured. SPAD, SLN and gas leakage were analysed using the procedures described for the N-trial.

An experiment on leaf photosynthesis in response to fluctuating temperature under different leaf nitrogen levels (TN-trial)

Seeds were sown in 18 containers (60 × 20 × 18 cm^3) placed outdoor on 5 August 2013. Each container was filled with 10 kg of soil that contained 0.11% of total nitrogen and had a clay–silt–sand ratio of 15:22:63. After seedling emergence, the plants were hand-thinned to 10 plants per container and three levels of urea fertilization were applied (0, 0.78 and 1.95 g N per container, respectively). Each fertilization level had six containers. The plants were well watered during growth. Because of very late sowing, a halogen lamp (54 Watt) that was mounted at 50 cm from the top of canopy was turned on for 16 hrs per day to prevent plants from flowering. The daily temperature and radiation during the growth period are presented in Fig. S1.

Starting on 50 days after sowing (the 8th – 10th pair of leaves had appeared), GE measurements were conducted outdoors on three representative plants for each nitrogen level. A-I_{inc} and A-C_a curves were assessed on the middle leaflet of the youngest, fully expanded leaf. The levels of light for the A-I_{inc} curves were identical to those in the N-trial under ambient O_2, while the A-C_a curves were assessed by increasing C_a as: 50, 60, 70, 80, 150, 250, 400, 650, 1000 and 1500 µmol mol^{-1} while keeping I_{inc} at 1000 µmol m^{-2} s^{-1}. During measurement, T_L and vapour pressure were not controlled and, therefore, varied depending on ambient conditions. A response curve was started when the leaf had adapted to the leaf chamber for 15 min at the first I_{inc}/C_a level. Data were recorded manually when the real-time net photosynthesis (A) had apparently reached steady state (~ 3 min for A-I_{inc} and ~ 5 min for A-C_a). SPAD, SLN and gas leakage were analysed using the procedures described for the N-trial.

Model analysis

The photosynthesis model of Farquhar *et al.* (1980) coupled with CO_2 diffusion model, as described in Yin & Struik (2009), was used in this study.

Modelling net leaf photosynthesis rate at the carboxylation sites of Rubisco

The net leaf photosynthesis rate (A, μmol m^{-2} s^{-1}) was modelled as the minimum of the Rubisco-limited rate (A_c), the electron transport-limited rate (A_j) and the triose phosphate utilization-limited rate (A_p):

$$A = \min(A_c, A_j, A_p) \tag{1}$$

A_c is described, following the Michaelis–Menten kinetics, as:

$$A_c = \frac{(C_c - \Gamma^*)V_{cmax}}{C_c + K_{mc}(1 + O/K_{mo})} - R_d \tag{2}$$

where C_c (μmol mol^{-1}) and O (mmol mol^{-1}) are the CO_2 and O_2 levels at the carboxylation sites of Rubisco; V_{cmax} (μmol m^{-2} s^{-1}) is the maximum rate of carboxylation; K_{mc} (μmol mol^{-1}) and K_{mo} (mmol mol^{-1}) are Michaelis–Menten constants of Rubisco for CO_2 and O_2, respectively; R_d (μmol m^{-2} s^{-1}) is the day respiration (respiratory CO_2 release other than by photorespiration); and Γ^*(μmol mol^{-1}) is the CO_2 compensation point in the absence of R_d.

A_j is described as:

$$A_j = \frac{(C_c - \Gamma^*)J}{4C_c + 8\Gamma^*} - R_d \tag{3}$$

where J (μmol m^{-2} s^{-1}) is the potential linear e$^-$ transport rate that is used for CO_2 fixation and photorespiration, and it is described as:

$$J = \frac{\kappa_{2LL}I_{inc} + J_{max} - \sqrt{(\kappa_{2LL}I_{inc} + J_{max})^2 - 4\theta J_{max}\kappa_{2LL}I_{inc}}}{2\theta} \tag{4}$$

where J_{max} (μmol m^{-2} s^{-1}) is the maximum value of J under saturated light; I_{inc} is the incident light (μmol m^{-2} s^{-1}); κ_{2LL} (mol mol^{-1}) is the conversion efficiency of incident light into J at strictly limiting light; and θ (dimensionless) is convexity factor for the response of J to I_{inc}.

A_p is described as:

$$A_p = 3T_p - R_d \tag{5}$$

where T_p (μmol m^{-2} s^{-1}) is the rate of triose phosphate export from the chloroplast.

The T_L response of R_d, T_p and kinetic properties of Rubisco (involving V_{cmax}, K_{mc}, K_{mo} and Γ^*) are described using an Arrhenius function normalized with respect to their values at 25 °C (Eqn 6) while the response of J_{max} is described using a peaked Arrhenius function (Eqn 7):

$$X = X_{25}\exp\left[\frac{E_x(T_L - 25)}{298R(T_L + 273)}\right] \tag{6}$$

$$X = X_{25}\exp\left[\frac{E_x(T_L - 25)}{298R(T_L + 273)}\right]\left[\frac{1 + \exp\left(\frac{298S_x - D_x}{298R}\right)}{1 + \exp\left(\frac{(T_L + 273)S_x - D_x}{R(T_L + 273)}\right)}\right] \tag{7}$$

where X_{25} is the value of each parameter at 25 °C (i.e. R_d, V_{cmax}, K_{mc}, K_{mo}, Γ^* and J_{max}). E_x and D_x are the energies of activation and deactivation (i.e. E_{Rd}, E_{Vcmax}, E_{Kmc}, E_{Kmo}, E_{Tp}, E_{Γ^*}, E_{Jmax} and D_{Jmax}, all in J mol^{-1}); S_x is the entropy term (S_{Jmax} in J K^{-1} mol^{-1}); and R is the universal gas constant (=8.314 J K^{-1} mol^{-1}).

Modelling mesophyll conductance for CO_2

The CO_2 concentration at intercellular space (C_i) was taken from gas exchange measurement whereas the estimation of C_c relies on proper estimation of mesophyll conductance (g_m). g_m, calculated by the variable J method (Harley *et al.*, 1992a), appeared to vary with CO_2 and irradiance levels (see section Result). Whether or not g_m varies with CO_2 and irradiance levels is debatable (Flexas *et al.*, 2007, 2012). We used the model of Yin *et al.* (2009) that is able to deal with both constant and variable g_m models, and have a similar form as Eqn (8):

$$g_m = g_{m0} + \frac{\delta(A + R_d)}{C_c - \Gamma^*} \tag{8}$$

where g_{m0} (mol m^{-2} s^{-1}) is the minimum g_m if irradiance approaches zero; parameter δ (dimensionless) in this model defines the $C_c : C_i$ ratio at saturating light as $(C_c - \Gamma^*)/(C_i - \Gamma^*) = 1/(1 + 1/\delta)$. Any positive value of δ predicts a variable g_m pattern in response to C_i and I_{inc}, and a higher δ implies higher g_m and therefore a higher $C_c : C_i$ ratio. If $\delta = 0$, Eqn (8) predicts an independence of g_m on C_i and I_{inc} (i.e. $g_m = g_{m0}$), equivalent to the constant-g_m model.

Model parameterization and validation

The data collected in the N-trial was used to assess the effect of leaf nitrogen on the values of model parameters at 25 °C. The data collected in the T-trial were used to assess the effect of leaf temperature on the values of (peaked) Arrhenius model parameters. The parameterized model was validated against the data collected in the TN-trial. In the model, Rubisco kinetic property-related parameters (i.e. K_{mc}, K_{mo} and Γ^*) and θ, convexity factor for the response of J to I_{inc}, are conserved among C$_3$ species (Von Caemmerer *et al.*, 2009). Thus, the value of θ was set to 0.7 (Ögren & Evans, 1993); the values of K_{mc}, K_{mo} and Γ^* at 25 °C were set to 272 μmol mol^{-1}, 165 mmol mol^{-1} and 37.5 μmol mol^{-1} (at 21% O_2), respectively (Bernacchi *et al.*, 2002). The energies of activation E_{Kmc}, E_{Kmo} and E_{Γ^*} were adapted from the values of Bernacchi *et al.* (2002) as $E_{Kmc} = 80990$ J mol^{-1}, $E_{Kmo} = 23720$ J mol^{-1} and $E_{\Gamma^*} = 24460$ J mol^{-1}.

Model parameterization with data collected in the N-trial: nitrogen effect

The stepwise parameterizing procedures described by Yin *et al.* (2009) were adapted in this study. Specifically:

Step 1: Estimating electron transport parameters (J_{max} and κ_{2LL}) and R_d

According to Yin *et al.* (2009), the observed A_j under non-photorespiratory conditions can be expressed using Eqn (9):

$$A_j = \frac{s I_{inc} \Phi_2}{4} - R_d \qquad (9)$$

$$s = \beta \rho_2 \left(1 - \frac{f_{pseudo(b)}}{1 - f_{cyc}}\right) \qquad (9a)$$

where s is a lumped parameter; Φ_2 is PSII operating efficiency, usually assessed from the chlorophyll fluorescence measurements, indicating quantum efficiency of PSII e⁻ flow on PSII-absorbed light basis; β is leaf absorptance; ρ_2 is proportion of absorbed I_{inc} partitioned to PSII; and f_{cyc} and $f_{pseudo(b)}$ are the fraction of cyclic and basal pseudocyclic electron transport, respectively. Thus, a simple linear regression can be performed for the observed A against $(I_{inc}\Phi_2/4)$ using data of the e⁻ transport-limited range under nonphotorespiratory conditions (measurements conducted at 2% O_2). The slope of the regression yields an estimate of the calibration factor s, and the intercept gives an estimate of R_d under 2% O_2 condition. The estimated s allowed the conversion of CF-based PSII operating efficiency into the actual rate of linear electron transport as:

$$J = s I_{inc} \Phi_2 \qquad (10)$$

Thus, J_{max} and κ_{2LL} can be estimated from fitting Eqn (4) to the values of J.

The same linear regression for the observed A against $(I_{inc}\Phi_2/4)$ using data of the e⁻ transport-limited range may be applied as well to photorespiratory conditions (i.e. ambient O_2) for estimating R_d although the slight variation in C_i with I_{inc} can have bearing under these conditions (Yin *et al.*, 2009, 2011).

Step 2: Parameterization of the g_m model and V_{cmax} and T_p

Combining Eqn (8) with Eqn (2) and Eqn (3), and replacing C_c with $(C_i - A/g_m)$ yields (Yin *et al.*, 2009):

$$A_c \text{ or } A_j = \frac{-b - \sqrt{b^2 - 4ac}}{2a} \qquad (11)$$

where

$$a = x_2 + \Gamma^* + \delta(C_i + x_2)$$

$$b = -\Big\{(x_2 + \Gamma^*)(x_1 - R_d) + (C_i + x_2)[g_{m0}(x_2 + \Gamma^*) + \delta(x_1 - R_d)] + \delta[x_1(C_i - \Gamma^*) - R_d(C_i + x_2)]\Big\}$$

$$c = [g_{m0}(x_2 + \Gamma^*) + \delta(x_1 - R_d)][x_1(C_i - \Gamma^*) - R_d(C_i + x_2)]$$

$$\text{with } x_1 = \begin{cases} V_{cmax} & \text{for } A_c \\ \frac{J}{4} & \text{for } A_j \end{cases}$$

$$\text{and } x_2 = \begin{cases} K_{mc}\left(1 + \frac{O}{K_{mo}}\right) & \text{for } A_c \\ 2\Gamma^* & \text{for } A_j \end{cases}$$

Thus, V_{cmax}, T_p and δ (or g_{m0}) can be estimated simultaneously by fitting Eqn (1), Eqn (4), Eqn (5) and Eqn (11) to A-I_{inc} and A-C_i using pre-estimated J_{max}, κ_{2LL} and R_d as input.

As it is uncertain if g_m varies with CO_2 and irradiance levels, g_m was first assessed according to the variable J method (Harley *et al.*, 1992a):

$$g_m = \frac{A}{C_i - \frac{\Gamma^*[J + 8(A + R_d)]}{J - 4(A + R_d)}} \qquad (12)$$

where A and C_i were taken from gas exchange measurements and J was calculated by Eqn (10). If g_m does vary in response to changing C_i and I_{inc}, we could fit only δ by fixing g_{m0} to 0 (Yin *et al.*, 2009). In such a case, g_m can be calculated as:

$$g_m = \frac{A + \delta(A + R_d)}{C_i - \Gamma^*}. \qquad (13)$$

Model parameterization with data collected in the T-trial: temperature effect

By assuming the value of δ is independent of leaf temperature, the values of J_{max}, κ_{2LL}, V_{cmax} and T_p at each leaf temperature were solved from Eqn (1), Eqn (4), Eqn (5) and Eqn (11) by simultaneously fitting A-I_{inc} and A-C_i curves. Subsequently, the parameter values at different T_L were fitted to either Eqn (6) for estimating E_{Rd}, E_{Vcmax}, E_{Tp}, or Eqn (7) for estimating E_{Jmax}, D_{Jmax} and S_{Jmax}.

Model validation

The parameterized model was validated against the data obtained in the TN-trial. The model parameters R_d, J_{max}, V_{cmax} and T_p at 25 °C were derived from their linear relationships with SLN (see section Result), and the effect of T_L on the values of these parameters was quantified through Eqn (6) or Eqn (7) with the estimated E_{Rd}, E_{Vcmax}, E_{Tp}, E_{Jmax}, D_{Jmax} and S_{Jmax}.

Comparison of hemp leaf photosynthetic competence with that of cotton and kenaf

To illustrate the leaf photosynthetic competence of hemp in comparison with cotton and kenaf, A-C_i, A-I_{inc}, A-T_L and A-SLN curves were constructed for hemp using the validated model while those of cotton and kenaf were constructed using the FvCB models and corresponding parameters reported in Harley *et al.* (1992b) for cotton (*cv.* Coker 315) and in Archontoulis *et al.* (2011) for kenaf (*cv.* Everglades 41).

Statistics

Simple linear regression was performed using Microsoft Excel. Nonlinear fitting was carried out using the GAUSS method in PROC NLIN of SAS (SAS Institute Inc., Cary, NC, USA). If parameters were proven independent from leaf nitrogen or temperature, the dummy variables method was used to estimate one common value (Yin *et al.*, 2009). The goodness of fit was assessed by calculating the coefficient of determination (r^2) and the relative root mean square (rRMSE). The effect of leaf position on parameter values was tested by performing ANOVA test considering leaf nitrogen as covariance.

Results

Results of the N-trial: nitrogen-dependent photosynthetic capacity

Measurements to assess the effect of leaf nitrogen on leaf photosynthetic capacity of hemp (N-trial) were conducted on leaves having an average SLN of 0.87 g N m^{-2}, 1.25 g N m^{-2} and 1.75 g N m^{-2} at the top of the canopy, or 0.65 g N m^{-2}, 0.78 g N m^{-2} and 1.22 g N m^{-2} at the middle of the canopy, for the three N treatments, respectively. Examples of A-I_{inc} and A-C_i curves at different SLN levels are shown in Fig. 1. The R_{dk} (μmol m^{-2} s^{-1}; leaf respiration in the dark) and light-saturated net photosynthesis rate (A_{max}; measured at 2000 μmol m^{-2} s^{-1}) increased linearly with increasing SLN, and these linear relationships did not differ between the top and middle leaves (Fig. 2).

Using the data of electron transport-limited range under nonphotorespiratory conditions (i.e. at 2% O_2, C_a ≥600 μmol mol^{-1} in the A-C_a curve and I_{inc} ≤150 μmol m^{-2} s^{-1} in the A-I_{inc} curve), parameter s was estimated as the slope of a linear regression of A against ($I_{inc}\Phi_2/4$). The value of s was independent of SLN and canopy position ($P > 0.05$; see Fig. S2a). Thus, a common s (0.33 ± 0.01) was estimated from pooled data. κ_{2LL} and J_{max} were estimated from fitting Eqn (4) to the data on calculated J from Eqn (10). A preliminary estimation indicated that κ_{2LL} was unlikely to change with SLN and canopy position ($P > 0.01$; Fig. S2b). Thus, a common κ_{2LL} (0.21 ± 0.004 mol mol^{-1}) was estimated together with J_{max} using the dummy variable method. The J_{max} ranged from 116.1 μmol m^{-2} s^{-1} to 316.4 μmol m^{-2} s^{-1} and increased linearly with an increase in SLN at the rate of 132.9 μmol s^{-1} (g N)$^{-1}$ (Fig. 3a). The relationship between J_{max} and SLN was independent of canopy position ($P > 0.05$).

The estimated R_d values at 21% O_2 were roughly in line with the ones at 2% O_2 (see Fig. S3). Although the latter were on average 25% lower, a test of covariance indicated that R_d did not differ significantly between the different O_2 levels ($P = 0.17$). At 21% O_2, R_d ranged from 0.29 μmol m^{-2} s^{-1} to 1.61 μmol m^{-2} s^{-1}, increasing linearly with SLN at a rate of 0.85 μmol s^{-1} (g N)$^{-1}$ (Fig. 3b). The R_d-SLN relationship did not differ much between the middle and top leaves ($P > 0.05$).

The g_m calculated using the variable J method, Eqn (12), indicated that it varied with changing I_{inc} and C_i (Fig. 4a, b). A preliminary analysis indicated that the value of g_{m0} in Eqn (8) was close to zero. By fixing g_{m0} to zero, a common value of δ (2.12 ± 0.09) was estimated together with V_{cmax} and T_p using the dummy variable method. With the estimated δ, Eqn (13) estimates that g_m changes with I_{inc} and C_i in a similar trend as observed for the g_m calculated using Eqn (12); the latter, however, was 38% lower (Fig. 4a, b), probably as a result that the variable J method assumes the limitation on photosynthesis by electron transport over the full range of A-I_{inc} and A-C_i curves (Yin et al., 2009). The estimated g_m with Eqn (13) increases with an increase in SLN (Fig. 4c). The estimated V_{cmax} ranged from 53.7 μmol m^{-2} s^{-1} to 163.2 μmol m^{-2} s^{-1} and increased linearly with an increase in SLN at the rate of 76.2 μmol s^{-1} (g N)$^{-1}$ (Fig. 3c). The estimated T_p ranged from 6.9 μmol m^{-2} s^{-1} to 11.5 μmol m^{-2} s^{-1} and increased linearly with an increase in SLN at the rate of 4.2 μmol s^{-1} (g N)$^{-1}$ (Fig. 3d). The effects of SLN on V_{cmax} and T_p were independent of leaf position ($P > 0.05$). With the estimated R_d, κ_{2LL}, J_{max}, δ, V_{cmax} and T_p, the r^2 and $rRMSE$ of the model description of the measured A in the N-trial were 0.99 and 18.5%, respectively.

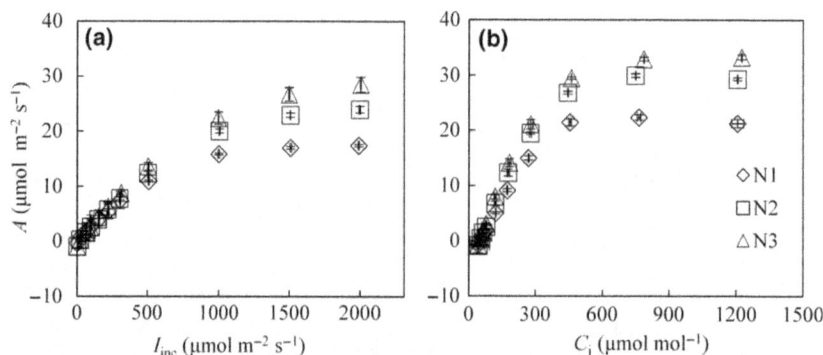

Fig. 1 The net leaf photosynthesis (A) in response to incident irradiance (I_{inc}; Panel a) and intercellular CO_2 concentration (C_i; Panel b) under different leaf nitrogen levels. Data presented were measured at 21% O_2 on the top leaves in the N-trial. N1, N2 and N3 correspond to nitrogen treatments, resulting in average specific leaf nitrogen values of 0.87 g N m^{-2}, 1.25 g N m^{-2} and 1.75 g N m^{-2}, respectively. The bars indicate standard errors of the mean ($n = 3$).

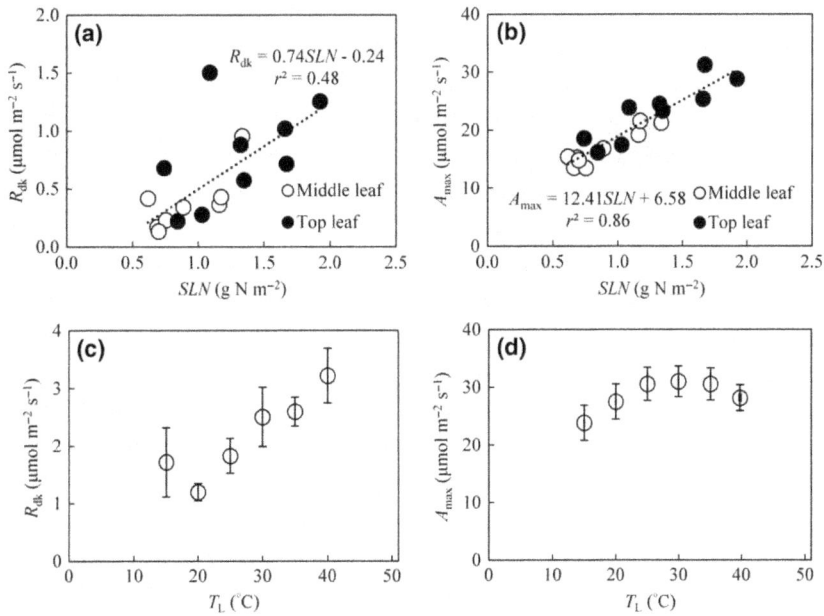

Fig. 2 The response of leaf respiration in dark (R_{dk}, panels a and c) and maximum light-saturated net photosynthesis rate (A_{max}; panels b and d) to specific leaf nitrogen (SLN; panels a and b) and leaf temperature (T_L; panels c and d). R_{dk} was measured after adapting leaves in dark for 15 min after measuring the A - I_{inc} curve. A_{max} was measured at 2000 µmol m^{-2} s^{-1} for incident light intensity and 400 µmol mol^{-1} for ambient CO_2 concentration. The data presented in Panel a and Panel b were obtained in the N-trial while those in Panel c and Panel d were obtained in the T-trial. The bars in panels c and d indicate standard errors of the mean ($n = 3$).

Results of T-trial: temperature-dependent photosynthetic capacity

The R_{dk} increased continuously from 0.9 µmol m^{-2} s^{-1} to 4.1 µmol m^{-2} s^{-1} at increasing T_L from 15 to 40 °C while the A_{max} initially increased with increasing T_L, levelled off at 25–35 °C and decreased when T_L became higher than 35 °C (Fig. 2c, d).

The estimated R_d increased continuously with an increase in T_L, ranging from 0.3 µmol m^{-2} s^{-1} until 3.2 µmol m^{-2} s^{-1} (Fig. 5a). The κ_{2LL}, J_{max}, V_{cmax} and T_p were estimated simultaneously by assuming $\delta = 2.12$ (estimated in N-trial) at each T_L. With the constant δ, the model predicted that g_m changed with an increase in T_L following a similar trend as A_{max} (cf. Figs 2d and 4d). A preliminary analysis indicated that κ_{2LL} was conserved at different levels of T_L ($P > 0.05$; see Fig. S2c) but significantly higher than the value estimated in the N-trial (i.e. $\kappa_{2LL} = 0.21 \pm 0.004$ mol mol^{-1}). Thus, a common κ_{2LL} (0.37 ± 0.01 mol mol^{-1}) was estimated together with J_{max}, V_{cmax} and T_p using the dummy variable method. The J_{max}, V_{cmax} and T_p at 25 °C were comparable with those derived from the N-trial (Fig. 3). The value of T_p increased consistently with an increase in T_L from 15 to 30 °C (Fig. 5d). When T_L was higher than 30 °C, the curve fitting failed to assess T_p properly because the triose phosphate utilization is not limited at

such high temperatures (Sage & Kubien, 2007; Busch & Sage, 2016). Therefore, T_p limitation was excluded to estimate J_{max} and V_{cmax} at 35 and 40 °C. The V_{cmax} increased continuously at increasing T_L from 15 to 40 °C while the value of J_{max} peaked at 30–35 °C (Fig. 5b, c).

By fitting the R_d-T_L, V_{cmax}-T_L and T_p-T_L to Eqn (6), the activation energies E_{Rd}, E_{Vcmax} and E_{Tp} were estimated at 21634.8 ± 4085.5 J mol^{-1}, 63042.7 ± 1562.2 J mol^{-1} and 34417.8 ± 5297.7 J mol^{-1}, respectively. By fitting J_{max}-T_L to Eqn (7), the values of E_{Jmax}, D_{Jmax} and S_{Jmax} were estimated at 67292.1 ± 35985.5 J mol^{-1}, 114701.0 ± 28709.6 J mol^{-1} and 375.6 ± 82.3 J K^{-1} mol^{-1}, respectively. With the estimated parameters, the model described well the response of A to changing I_{inc} and C_i at different T_L ($r^2 = 0.94$ and $rRMSE = 24.1\%$).

Model validation

The measurements in the TN-trial were conducted on leaves with SLN ranging from 0.63 g N m^{-2} to 1.44 g N m^{-2}. During the measurement, the T_L ranged from 21 to 33 °C, and VPD ranged from 0.61 kPa to 2.61 kPa.

The parameterized model was validated against the data obtained in the TN-trial. The measured A was overestimated with either the κ_{2LL} derived in the N-trial

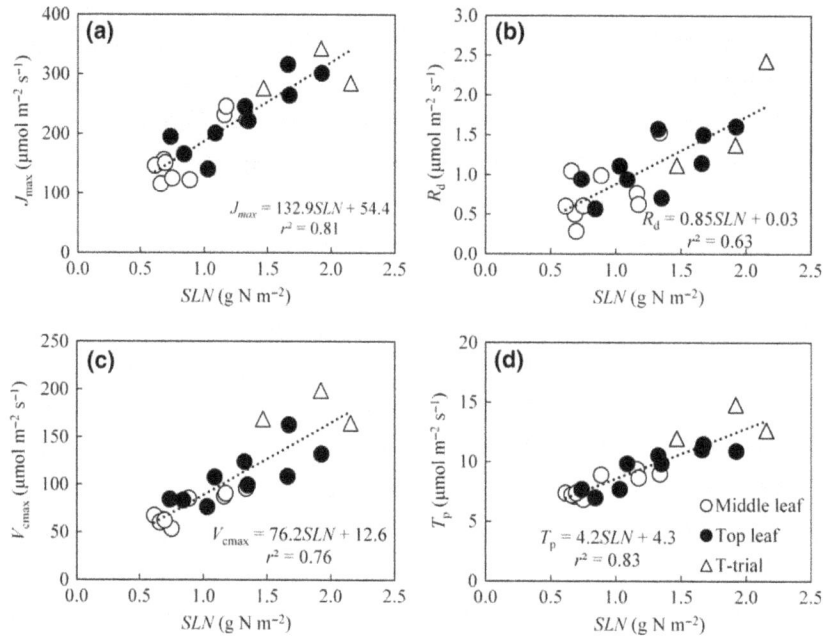

Fig. 3 Dependence of maximum potential linear e⁻ transport rate (J_{max}; Panel a), day respiration (R_d; Panel b), maximum rate of carboxylation (V_{cmax}; Panel c) and the rate of triose phosphate export from the chloroplast (T_p: Panel d) on specific leaf nitrogen (SLN). Values indicated as circles (O and ● denote leaves at the middle and top of canopy, respectively) were derived from the data collected in the N-trial; values indicated as triangles (Δ) were derived from the data collected in the T-trial at a leaf temperature of 25 °C.

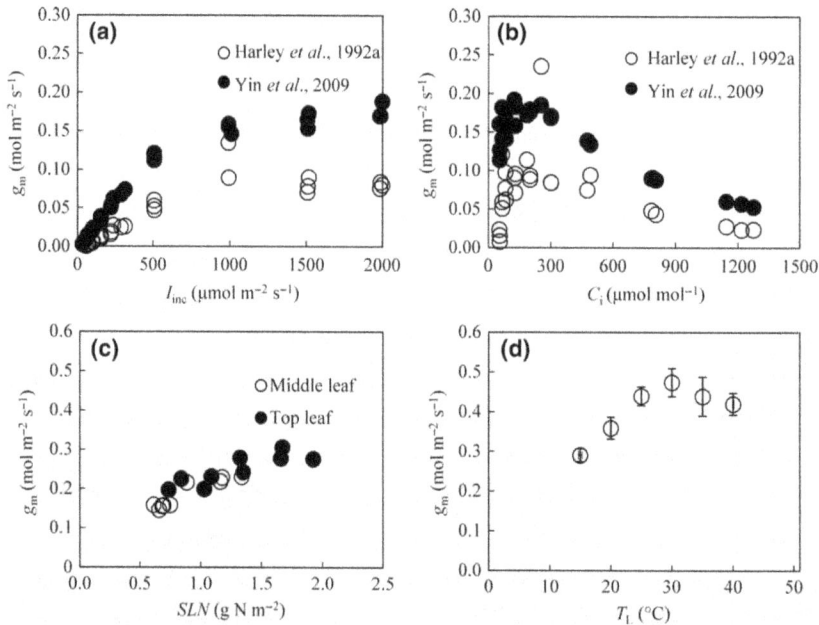

Fig. 4 Illustration of mesophyll conductance (g_m) in relation to changing incident irradiance (I_{inc}: Panel a), intercellular CO_2 concentration (C_i: Panel b), specific leaf nitrogen (SLN; Panel c) and leaf temperature (T_L: Panel d). In panels a and b, the data presented were obtained from the leaves at the middle of the canopy in the treatment without nitrogen fertilization in the N-trial; the open (O) and closed (●) circles were calculated using the variable J method of Harley et al. (1992a) (see Eqn 12 in the text) and the method of Yin et al. (2009) (see Eqn 13 in the text), respectively. In Panel c, the data presented were obtained at I_{inc} = 1000 μmol m⁻² s⁻¹ and C_a = 400 μmol mol⁻¹ in the N-trial; the open (O) and closed (●) circles represent data obtained from leaves from the middle and the top of the canopy, respectively. In Panel d, the data presented were obtained at I_{inc} = 1000 μmol m⁻² s⁻¹ and C_a = 400 μmol mol⁻¹ in the T-trial; the bars indicate standard errors of the mean (n = 3). Note the differences in scale along the y-axes.

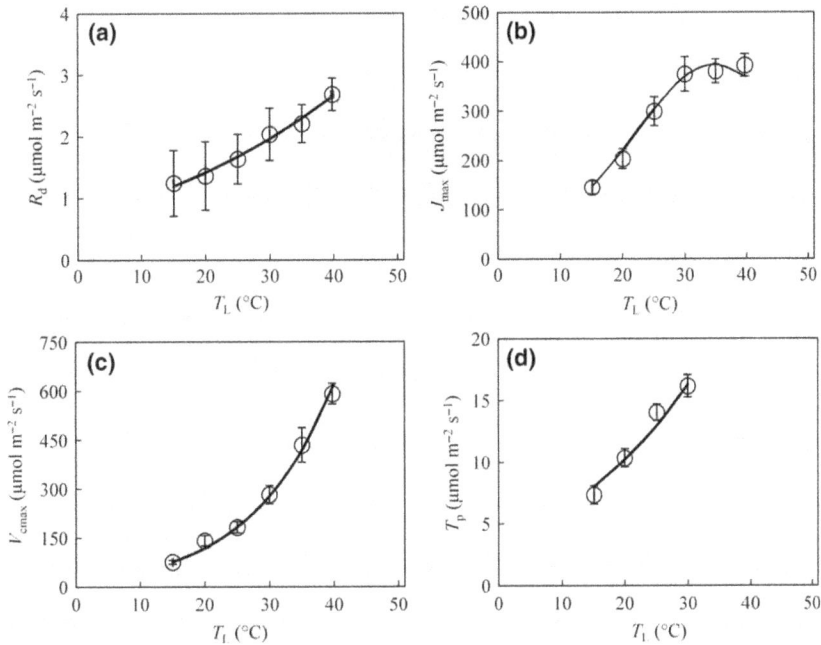

Fig. 5 Response of day respiration (R_d; Panel a), maximum potential linear e⁻ transport rate (J_{max}; Panel b), maximum rate of carboxylation (V_{cmax}; Panel c) and the rate of triose phosphate export from the chloroplast (T_p; Panel d) to leaf temperature (T_L). The solid lines denote the predicted relations according to Eqn (6) or Eqn (7) with values presented in Table 1. The bars indicate standard errors of the mean ($n = 3$).

(κ_{2LL} = 0.21 mol mol⁻¹) or in the T-trial (κ_{2LL} = 0.37 mol mol⁻¹) (Fig. 6a, b). The *rRMSE* reduced significantly with decreasing value of κ_{2LL} until 0.13 mol mol⁻¹ (Fig. 6c). Assuming κ_{2LL} = 0.13 mol mol⁻¹ for the TN-trial, the r^2 and *rRMSE* were 0.94 and 26%, respectively; the error of model prediction distributed evenly across measured *SLN* and T_L (see Fig. S4).

Leaf photosynthetic competence of hemp in comparison with kenaf and cotton

Comparison of leaf photosynthetic competence of hemp with kenaf and cotton is presented in Fig. 7. The values of the main parameters are summarized in Table 1. In this illustration, we considered the uncertainty in estimated values of parameters (i.e. R_d, J_{max}, V_{cmax} and T_p) for their linear relationships with *SLN* and nonlinear relationships with T_L (presented as the shaded area). The modelled values of A for hemp are shown using lower and upper bounds of 95% confidence interval of these parameter values. Given that there was a large variation in the value of κ_{2LL} among different growth environments and each estimate of κ_{2LL} had a very small standard error (Table 1), the lower bounds were combined with κ_{2LL} of 0.21 mol mol⁻¹ (derived from N-trial) while the upper bounds were combined with κ_{2LL} of 0.37 mol mol⁻¹ (derived from T-trial).

For the response to C_i, these three crops had similar A at the current atmosphere CO_2 level (Fig. 7a). In case of a further increase in CO_2 level in the future, kenaf may become more productive than hemp. For both crops, there was a large uncertainty in the responses of A to I_{inc} and T_L (Fig. 7b, c) because these curves are affected by the value of κ_{2LL}. When using κ_{2LL} of 0.37 mol mol⁻¹, a value close to that of healthy C_3 leaves (presented as dashed black lines), the calculated A for hemp was similar to that for kenaf across different I_{inc} levels, but was slightly higher than for cotton at intermediate I_{inc}. Reducing κ_{2LL} to 0.21 mol mol⁻¹ (presented as solid black lines) resulted in a reduction of A under light limiting condition and in a reduction of the optimal temperature. For the response to leaf nitrogen, the leaf photosynthetic competence of hemp, including its 95% confidence interval, was consistently higher than that of cotton and kenaf at $SLN < 2.0$ g N m⁻², which is close to the maximum *SLN* measured in this study (Fig. 7d).

Discussion

Hemp is considered an ideal annual crop for the bioeconomy as it has the potential to produce a high multipurpose biomass yield while requiring little inputs (Finnan & Burke, 2013; Tang *et al.*, 2016). However, very limited information is available on the physiological basis of hemp resource-use efficiency. With the aim of

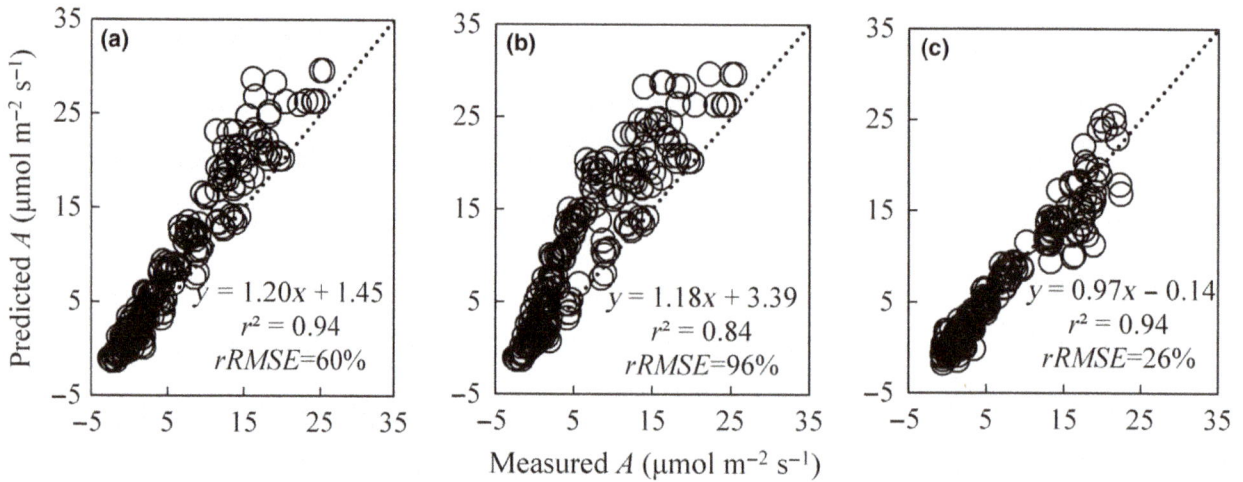

Fig. 6 Results of model validation against the data measured net photosynthesis rate (A) in the TN-trial. The dotted lines represent the 1:1 line. The predicted A values in panels a, b and c were with a value of $\kappa_{2LL} = 0.21$ mol mol^{-1} (derived from the N-trial), $\kappa_{2LL} = 0.37$ mol mol^{-1} (derived from the T-trial) and $\kappa_{2LL} = 0.13$ mol mol^{-1} (obtained by minimizing prediction error of A), respectively.

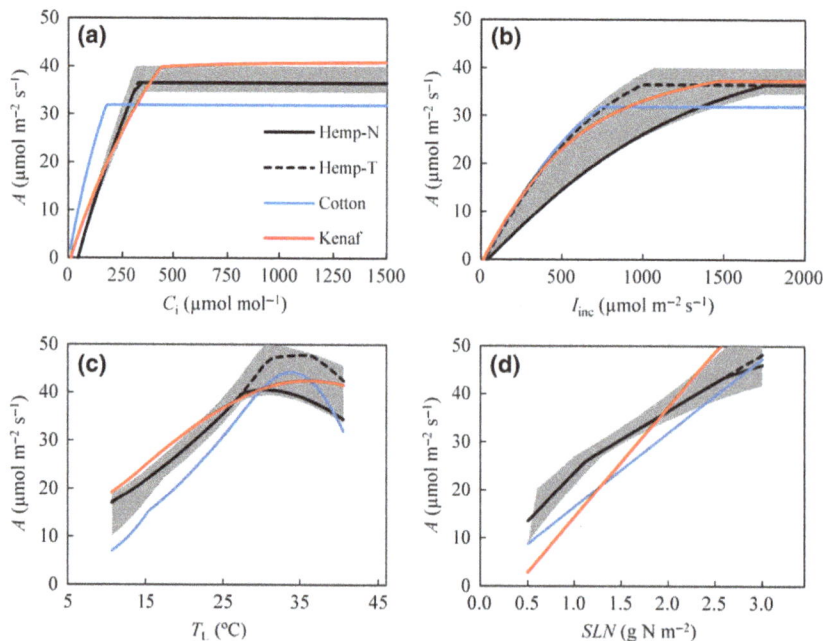

Fig. 7 Simulation of leaf photosynthetic capacity (A) of hemp (black lines), kenaf (red line) and cotton (blue line) in response to intercellular CO_2 concentration (C_i, Panel a), incident light intensity (I_{inc}, Panel b), leaf temperature (T_L, Panel c) and leaf nitrogen (SLN, Panel d). The hemp leaf photosynthesis presented by a continuous line was simulated with $\kappa_{2LL} = 0.21$ mol mol^{-1} (derived from the N-trial) while the dashed line was simulated with $\kappa_{2LL} = 0.37$ mol mol^{-1} (derived from the T-trial). The shaded area presents 95% confidence interval of hemp leaf photosynthesis. The photosynthesis rates of cotton were simulated using the model and values described in Harley *et al.* (1992b) while for kenaf the model and values came from Archontoulis *et al.* (2011). Except when used as the independent variable, the variables were set constant as $C_a = 400$ μmol mol^{-1}, $I_{inc} = 2000$ μmol m^{-2} s^{-1}, $SLN = 2.0$ g N m^{-2} and $T_L = 25$ °C.

understanding the response of leaf photosynthesis capacity of hemp to leaf nitrogen status and environmental factors and setting the basis for a hemp growth model, this study presents the results of extensive hemp leaf photosynthetic measurements and parameterization of a widely used photosynthesis model.

Table 1 List of model parameters (± standard errors if available) of hemp, cotton and kenaf

Parameter		Unit	Hemp	Cotton†	Kenaf§
Respiration					
R_d-SLN	Slope	$\mu mol\ s^{-1}\ (g\ N)^{-1}$	0.85 ± 0.15	0‡	0.80
	Intercept	$\mu mol\ m^{-2}\ s^{-1}$	0.03 ± 0.19	0.82‡	−0.37
E_{Rd}		$J\ mol^{-1}$	21634 ± 4085	84450	83440
e⁻ transport parameters					
J_{max}-SLN	Slope	$\mu mol\ s^{-1}\ (g\ N)^{-1}$	132.9 ± 14.6	98.1	122.1
	Intercept	$\mu mol\ m^{-2}\ s^{-1}$	54.4 ± 18.8	−4.6	−47.6
E_{Jmax}		$J\ mol^{-1}$	67292 ± 35986	79500	28149
D_{Jmax}		$J\ mol^{-1}$	114701 ± 28710	201000	474614‖
S_{Jmax}		$J\ K^{-1}\ mol^{-1}$	375 ± 82	650	1482‖
κ_{2LL}		$mol\ mol^{-1}$	0.21 ± 0.004 (N-trial)	0.24*	0.28
			0.37 ± 0.01 (T-trial)		
θ		-	0.70*	0.83*	0.63
Rubisco parameters					
V_{cmax}-SLN	Slope	$\mu mol\ s^{-1}\ (g\ N)^{-1}$	76.2 ± 9.8	60.0¶	66.7¶
	Intercept	$\mu mol\ m^{-2}\ s^{-1}$	12.6 ± 12.5	−9.6¶	26.0¶
E_{vcmax}		$J\ mol^{-1}$	63024 ± 1562	116300	61812
TPU parameters					
T_p-SLN	Slope	$\mu mol\ s^{-1}\ (g\ N)^{-1}$	4.2 ± 0.4	5.1¶	NA
	Intercept	$\mu mol\ m^{-2}\ s^{-1}$	4.3 ± 0.6	0.6¶	NA
E_{Tp}		$J\ mol^{-1}$	34417 ± 5298	53100	NA
g_m parameters					
δ		-	2.12 ± 0.09	NA	NA
g_{m0}		$mol\ m^{-2}\ s^{-1}$	0*	NA	NA

NA: not estimated or not available.

*Parameter values are fixed beforehand.

†Parameter values are derived from Harley *et al.* (1992b) with plants grown at an ambient [CO_2] of 35 Pa; the parameter values of temperature response are converted to fit Eqn (6) or Eqn (7) in the text; the value of θ is converted to fit Eqn (4) in the text.

‡R_d was held constant at different nitrogen levels and equal to 0.82 $\mu mol\ m^{-2}\ s^{-1}$.

§Parameter values are derived from Archontoulis *et al.* (2011). In their paper, the value of E_{Rd} is a function of SLN. The value presented here is derived at $SLN = 2.0$ g N m^{-2}. Slopes of R_d-SLN are calculated from simulation of R_d against SLN using original model.

¶Note that the absolute value of these parameters may be lower than the presented one when g_m is considered;

‖The optimum temperature J_{max} was not observed, so its J_{max} was fitted to the Arrhenius Eqn (6); thereby, D_{Jmax} and S_{Jmax} were not estimated. The presented value gave equal temperature sensitivities, but it was rejected by the authors due to a high standard error of the estimate.

Parameterization of the leaf photosynthesis model for hemp

Theoretically, the method to estimate R_d (day respiration) works best for the NPR (nonphotorespiratory) condition (Yin *et al.*, 2011). The estimated R_d in this study did not differ significantly between PR (photorespiratory) and NPR conditions ($P > 0.05$). This result suggests that estimating R_d from Eqn (9) is practicable even under PR condition (Yin *et al.*, 2009, 2011). Note that assessing the true R_d is somewhat difficult and the estimated R_d differs according to methodologies. A comparison of the method used in this study with other

ones to estimate R_d is discussed in Yin *et al.* (2011). The estimated R_d values were on average 20% lower than R_{dk} values (respiration in the dark) in line with other reports (Brooks & Farquhar, 1985; Yin *et al.*, 2009, 2011). An *in vivo* metabolic study (Tcherkez *et al.*, 2005) indicated that the main inhibited steps were the entrance of hexose molecules into the glycolytic pathway and the Krebs cycle. Nevertheless, detailed mechanism of this difference still needs further research (Tcherkez *et al.*, 2012).

Both R_d and R_{dk} increased monotonically with an increase in SLN and T_L (Figs 2, 3 and 5) within the tested ranges. The result agrees with those of Yin *et al.*

(2009, 2011), but does not support those in Harley *et al.* (1992b) for cotton, where a constant R_d was considered at changing nitrogen and temperature. For hemp, Chandra *et al.* (2008, 2011a) reported that R_{dk} levelled off or slightly decreased with an increase in temperature from 30 to 40 °C. This was not confirmed in the present study, although the highest R_{dk} measured at 25 °C in our study is comparable with the value observed in Chandra *et al.* (2008, 2011a). The reason for such discrepancy of R_{dk} in response to T_L is not clear. It is probably due to an artefact of different protocols or due to changes in thermal sensitivity of respiration at different growth environments and plant status (e.g. drought, nutrient availability and sugar concentration) (Atkin *et al.*, 2005; Katja *et al.*, 2012). If an increase of respiration with increasing SLN and T_L is proven for hemp, it could counteract, at least partly, the positive effects of SLN and T_L on A (net photosynthesis rate) when considering at daily basis.

Based on the findings that the maximum quantum yields (the initial slopes of the response of CO_2 uptake to photon absorption) were conserved across age classes within species or across the mature photosynthetic organs of different species (Long *et al.*, 1993), κ_{2LL} was often fixed as a constant across different growth environments and species in studies of plant photosynthesis (Harley *et al.*, 1992b; Medlyn *et al.*, 2002). However, very different values have been assumed in different studies without clear explanation, ranging from 0.18 mol mol^{-1} until 0.39 mol mol^{-1} (Harley *et al.*, 1992b; Wullschleger, 1993; Medlyn *et al.*, 2002; Yamori *et al.*, 2010). The estimated κ_{2LL} in the present study did not change with SLN and with T_L, but it was not constant across growth environments (0.21 mol mol^{-1} for the N-trial; 0.37 mol mol^{-1} for the T-trial and 0.13 mol mol^{-1} resulted in the best prediction of measurements in the TN-trial), in line with Archontoulis *et al.* (2011) who observed that cardoon (*Cynara cardunculus*) had a higher κ_{2LL} in the cold season than in the warm season. The reason for the variation in κ_{2LL} in different environments is still not fully understood. We speculate that the low κ_{2LL} in the N-trial and the TN-trial in comparison with the κ_{2LL} in the T-trial is a consequence of photoinhibition that occurs naturally in field plants grown in West Europe when the temperature is low and the sky is clear (Long *et al.*, 1994). The plants of the N-trial and the TN-trial were grown outdoors, with fluctuations in temperature and irradiance; particularly, the plants in the TN-trial experienced a sudden drop of temperature five days before measuring (Fig. S1). These conditions could have resulted in severe photoinhibition (Long *et al.*, 1983; Powles *et al.*, 1983) causing a reduction in Φ_{2LL} (PSII quantum use efficiency under strictly limiting light) and an increase in the fraction of alternative electron transport (i.e. $\frac{f_{pseudo(b)}}{1-f_{cyc}}$; cf. Eqn 9a) (Curwiel & Van Rensen, 1993; Murata *et al.*, 2012), hence a low κ_{2LL}. In contrast, the plants of the T-trial were grown in the glasshouse where both light intensity and temperature were controlled at a condition free of photoinhibition. Thus, the value of κ_{2LL} (0.37 mol mol^{-1}) was high and close to the range for healthy C_3 leaves (between 0.32 mol mol^{-1} and 0.35 mol mol^{-1}) (Hikosaka *et al.*, 2016 and their references). Moreover, the variation in κ_{2LL} could be partly attributed to the change in β (leaf absorbance; cf. Eqn 9a) as a result of environmental acclimation (Archontoulis *et al.*, 2011). A higher β in the T-trial than in the N-trial and the NT-trial is reflected by the higher $SPAD$ values when considered at the same SLN (Fig. S5). Given that the value of κ_{2LL} varied significantly across different environments and that it affected significantly the prediction of photosynthesis when electron transport was limited (i.e. A_j) (Fig. 7), caution is needed when modelling photosynthesis rate using a value of κ_{2LL} derived from different environments, particularly if these include both glasshouse and open field conditions. To improve modelling of crop growth in field conditions, further study should be conducted to investigate the mechanisms underlying variation in κ_{2LL} during the whole growth season.

The relationships J_{max}-SLN, V_{cmax}-SLN and T_p-SLN were consistent across canopy positions and growth environments whereas linear regression of these relationships resulted in negative intersections at the x-axis (Fig. 3), in line with Akita *et al.* (2012) but different from Archontoulis *et al.* (2011) and Braune *et al.* (2009) where the intersection of linear extrapolating resulted in a minimum SLN required for photosynthesis (SLN_b). Given that it is not physiologically possible to have a negative SLN_b, the results in this study indicate that the relationships J_{max}-SLN, V_{cmax}-SLN and T_p-SLN for hemp may not be perfectly linear. Further study would be needed to elucidate the relationship between these parameters and SLN at SLN levels approaching zero.

It is well recognized that g_m is not infinite (Bernacchi *et al.*, 2002). Using both the variable J method and the modelling method, our analysis for hemp (Fig. 4) supports that g_m varies with changing C_i and I_{inc} (Flexas *et al.*, 2007, 2012; Yin *et al.*, 2009), which is in contrast with the assumption that g_m is independent of C_i and I_{inc} (Bernacchi *et al.*, 2002). This highlights an important uncertainty in the present understanding of CO_2 diffusion processes in leaves. The g_m obtained from Eqn (13) with a constant δ changed in line with A (cf. Figs 2 and 4), confirming the assumption of Piel *et al.* (2002) and Ethier *et al.* (2006) that g_m is correlated with A. The value of δ (2.12) is lower than that of wheat (2.54) (Yin *et al.*, 2009) but higher than that of rice (0.45~1.57) (Gu *et al.*, 2012).

Does hemp have high photosynthetic competence?

The observed A_{max} was observed to be levelled off at 25–35 °C (Fig. 2d) that is comparable with the 27 °C reported in Cosentino *et al.* (2012) and the 30 °C reported in Chandra *et al.* (2011a) for hemp leaf photosynthesis. The wide range of optimal temperature for leaf photosynthesis confirms the fact that hemp has been cultivated from the tropic (Tang *et al.*, 2012) to the polar circle (Pahkala *et al.*, 2008).

The highest A_{max} (light-saturated net photosynthesis rate) at 25 °C was measured at 31.2 ± 1.9 µmol m^{-2} s^{-1} (Fig. 2b). This value is higher than the highest value reported for hemp in De Meijer *et al.* (1995) and (Chandra *et al.*, 2008, 2011a), which were 19.0 µmol m^{-2} s^{-1} and 24.0 µmol m^{-2} s^{-1}, respectively. The highest A_{max} in this study is comparable with that of other C_3 bioenergy crops. Archontoulis *et al.* (2011) reported that the highest A_{max} of kenaf, sunflower (*Helianthus annuus* L.) and cardoon ranged between 30 µmol m^{-2} s^{-1} and 35 µmol m^{-2} s^{-1} under optimum temperature.

As direct comparison of A_{max} among crops is difficult due to the variation in experimental protocols and plant status, we constructed A-C_i, A-I_{inc}, A-T_L and A-SLN curves for hemp, cotton and kenaf with the same values of variables (i.e. C_i, I_{inc}, T_L and SLN) (Fig. 7). The comparison highlighted that hemp has higher leaf photosynthesis rate than cotton and kenaf at a low nitrogen condition (i.e. SLN < 2.0 g N m^{-2}). This was presumably because hemp has a relatively low SLN_b. Analysis of newly senesced hemp leaves resulted in a nitrogen content of 0.25 ± 0.01 g N m^{-2}. This value is at the low range of SLN_b among C_3 crops and weeds (average value = 0.31 ± 0.03 g N m^{-2}) and is considerably lower than the estimation for kenaf (0.39 ± 0.13 g N m^{-2}) (Archontoulis *et al.*, 2011).

The high photosynthesis rate of hemp at low nitrogen condition is in line with its observed high productivity at low nitrogen input (Struik *et al.*, 2000; Finnan & Burke, 2013) and puts hemp ahead of cotton and kenaf from a perspective of bio-economy. However, our model approach has limitations. Firstly, the comparison was based on parameters derived from different studies conducted in different environments. Secondly, even though the FvCB model is biochemically based and the relationships J_{max}-SLN, V_{cmax}-SLN and T_p-SLN were consistent in this study across canopy positions and growth environments (Fig. 3), increasing evidences show that the model parameters may change when plant acclimates to growing environments. For example, Harley *et al.* (1992b) reported that the slope of V_{cmax}-SLN decreased with an increase in CO_2 concentration in the growth environment. The present study also indicated that the value of κ_{2LL} may differ among growth environments. Thirdly, variation in photosynthetic competence among cultivars has been reported

for hemp (Chandra *et al.*, 2011b). As only one cultivar was studied, it is not clear whether the advantage of photosynthetic competence of hemp is persistent across cultivars. Therefore, to consolidate the potential of hemp as a bioeconomic sustainable crop, further study is needed to compare hemp leaf photosynthetic competence with those of cotton, kenaf and other bioenergy crops in the same growing environment with multiple cultivars.

Acknowledgements

The research leading to these results has received funding from the European Union's Seventh Framework Programme for research, technological development and demonstration under grant agreement no 311849.

References

Akita R, Kamiyama C, Hikosaka K (2012) Polygonum sachalinense alters the balance between capacities of regeneration and carboxylation of ribulose-1,5-bisphosphate in response to growth CO_2 increment but not the nitrogen allocation within the photosynthetic apparatus. *Physiologia Plantarum*, **146**, 404–412.

Alexopoulou E, Li D, Papatheohari Y *et al.* (2015) How kenaf (*Hibiscus cannabinus* L.) can achieve high yields in Europe and China. *Industrial Crops and Products*, **68**, 131–140.

Allegret S (2013) The history of hemp. In: *Hemp: Industrial Production and Uses* (eds Allegret S, Bouloc P, Arnaud L), pp. 4–26. CPi Group (UK) Ltd, Croydon, UK.

Amaducci S, Errani M, Venturi G (2002) Response of hemp to plant population and nitrogen fertilisation. *Italian Journal of Agronomy*, **6**, 103–111.

Amaducci S, Scordia D, Liu FH *et al.* (2015) Key cultivation techniques for hemp in Europe and China. *Industrial Crops and Products*, **68**, 2–16.

Archontoulis SV, Yin X, Vos J *et al.* (2011) Leaf photosynthesis and respiration of three bioenergy crops in relation to temperature and leaf nitrogen: how conserved are biochemical model parameters among crop species? *Journal of Experimental Botany*, **63**, 895–911.

Atkin OK, Bruhn D, Hurry VM *et al.* (2005) The hot and the cold: unravelling the variable response of plant respiration to temperature. *Functional Plant Biology*, **32**, 87–105.

Barth M, Carus M (2015) *Carbon Footprint and Sustainability of Different Natural Fibres for Biocomposites and Insulation Material*, Hürth, Germany, nova-Institute. Available at: http://bio-based.eu/ecology/ (accessed 24 January 2017).

Bellasio C, Beerling DJ, Griffiths H (2015) An Excel tool for deriving key photosynthetic parameters from combined gas exchange and chlorophyll fluorescence: theory and practice. *Plant, Cell and Environment*, **69**, 80–97.

Bernacchi CJ, Portis AR, Nakano H *et al.* (2002) Temperature response of mesophyll conductance. Implications for the determination of Rubisco enzyme kinetics and for limitations to photosynthesis in vivo. *Plant Physiology*, **130**, 1992–1998.

Bertoli A, Tozzi S, Pistelli L *et al.* (2010) Fibre hemp inflorescences: from crop-residues to essential oil production. *Industrial Crops and Products*, **32**, 329–337.

Bouloc P, Van der Werf HMG (2013) The role of hemp in sustainable development. In: *Hemp: Industrial Production and Uses* (eds Bouloc P, Allegret S, Arnaud L), pp. 278–289. CPi Group (UK) Ltd, Croydon, UK.

Bouman BAM, Feng L, Tuong TP *et al.* (2007) Exploring options to grow rice using less water in northern China using a modelling approach: II. Quantifying yield, water balance components, and water productivity. *Agricultural Water Management*, **88**, 23–33.

Braune H, Müller J, Diepenbrock W (2009) Integrating effects of leaf nitrogen, age, rank, and growth temperature into the photosynthesis-stomatal conductance model LEAFC3-N parameterised for barley (*Hordeum vulgare* L.). *Ecological Modelling*, **220**, 1599–1612.

Brooks A, Farquhar GD (1985) Effect of temperature on the CO_2/O_2 specificity of ribulose-1,5-bisphosphate carboxylase/oxygenase and the rate of respiration in the light. *Planta*, **165**, 397–406.

Busch FA, Sage RF (2016) The sensitivity of photosynthesis to O_2 and CO_2 concentration identifies strong Rubisco control above the thermal optimum. *New Phytologist*, **213**, 1036–1051.

Carus M, Sarmento L (2016) *The European Hemp Industry: Cultivation, processing and applications for fibres, shivs and seeds.* pp 1-9, European Industrial Hemp Association (EIHA), Hürth, Germany. Available at: http://eiha.org/media/2016/05/16-05-17-European-Hemp-Industry-2013.pdf (accessed 24 January 2017).

Carus M, Karst S, Kauffmann A et al. (2013) The European Hemp Industry: Cultivation, processing and applications for fibres, shivs and seeds. European Industrial Hemp Association (EIHA), Hürth, Germany. Available at : http://eiha.org/media/2014/10/13-06-European-Hemp-Industry.pdf (accessed 24 January 2017).

Chandra S, Lata H, Khan IA et al. (2008) Photosynthetic response of Cannabis sativa L. to variations in photosynthetic photon flux densities, temperature and CO2 conditions. Physiology and Molecular Biology of Plants, 14, 299–306.

Chandra S, Lata H, Khan IA et al. (2011a) Temperature response of photosynthesis in different drug and fiber varieties of Cannabis sativa L. Physiology and Molecular Biology of Plants, 17, 297–303.

Chandra S, Lata H, Khan IA et al. (2011b) Photosynthetic response of Cannabis sativa L., an important medicinal plant, to elevated levels of CO2. Physiology and Molecular Biology of Plants, 17, 291–295.

Chandra S, Lata H, Mehmedic Z et al. (2015) Light dependence of photosynthesis and water vapor exchange characteristics in different high Δ9-THC yielding varieties of Cannabis sativa L. Journal of Applied Research on Medicinal and Aromatic Plants, 2, 39–47.

Cosentino SL, Testa G, Scordia D et al. (2012) Sowing time and prediction of flowering of different hemp (Cannabis sativa L.) genotypes in southern Europe. Industrial Crops and Products, 37, 20–33.

Curwiel VB, Van Rensen JJS (1993) Influence of photoinhibition on electron transport and photophosphorylation of isolated chloroplasts. Physiologia Plantarum, 89, 97–102.

De Meijer EPM, Van der Werf HMG (1994) Evaluation of current methods to estimate pulp yield of hemp. Industrial Crops and Products, 2, 111–120.

De Meijer WJM, Van Der Werf HMG, Mathijssen EWJM, Van Den Brink PWM (1995) Constraints on dry matter production in fibre hemp (Cannabis sativa L.). European Journal of Agronomy, 4, 109–117.

Ethier G, Livingston N, Harrison D, et al. (2006) Low stomatal and internal conductance to CO2 versus Rubisco deactivation as determinants of the photosynthetic decline of ageing evergreen leaves. Plant, Cell and Environment, 29, 2168–2184.

Farquhar GD, Von Caemmerer S, Berry JA (1980) A biochemical model of photosynthetic CO2 assimilation in leaves of C3 species. Planta, 149, 78–90.

Finnan J, Burke B (2013) Nitrogen fertilization to optimize the green gas balance of hemp crops grown for biomass. GCB Bioenergy, 5, 701–712.

Flexas J, Diaz-Espejo A, Galmes J et al. (2007) Rapid variations of mesophyll conductance in response to changes in CO2 concentration around leaves. Plant, Cell and Environment, 30, 1284–1298.

Flexas J, Barbour MM, Brendel O et al. (2012) Mesophyll diffusion conductance to CO2: an unappreciated central player in photosynthesis. Plant Science, 193, 70–84.

Gu JF, Yin XY, Stomph TJ et al. (2012) Physiological basis of genetic variation in leaf photosynthesis among rice (Oryza sativa L.) introgression lines under drought and well-watered conditions. Journal of Experimental Botany, 63, 5137–5153.

Harley PC, Loreto F, Di Marco G et al. (1992a) Theoretical considerations when estimating the mesophyll conductance to CO2 flux by analysis of the response of photosynthesis to CO2. Plant Physiology, 98, 1429–1436.

Harley PC, Thomas RB, Reynolds JF et al. (1992b) Modelling photosynthesis of cotton grown in elevated CO2. Plant, Cell and Environment, 15, 271–282.

Hikosaka K, Noguchi K, Terashima I (2016) Modeling leaf gas exchange. In:Canopy Photosynthesis: From Basics to Applications (eds Hikosaka K, Niinemets Ü, Anten NPR), pp. 61–100. Springer, London, UK.

Jordan N, Boody G, Broussard W et al. (2007) Sustainable development of the agricultural bio-economy. Science, 316, 1570–1571.

Katja H, Irina B, Hiie I et al. (2012) Temperature responses of dark respiration in relation to leaf sugar concentration. Physiologia Plantarum, 144, 320–334.

Kreuger E, Prade T, Escobar F et al. (2011) Anaerobic digestion of industrial hemp–Effect of harvest time on methane energy yield per hectare. Biomass and Bioenergy, 35, 893–900.

Lips SJJ, van Dam JEG (2013) Kenaf fibre crop for bioeconomic industrial development. In: Kenaf: A Multi-Purpose Crop for Several Industrial Applications (eds Monti A, Alexopoulou E), pp. 105–143. Springer, London, UK.

Long S, East T, Baker N (1983) Chilling damage to photosynthesis in young Zea mays I. Effects of light and temperature variation on photosynthetic CO2 assimilation. Journal of Experimental Botany, 34, 177–188.

Long S, Postl WF, Bolhár-Nordenkampf HR (1993) Quantum yields for uptake of carbon dioxide in C3 vascular plants of contrasting habitats and taxonomic groupings. Planta, 189, 226–234.

Long S, Humphries S, Falkowski PG (1994) Photoinhibition of photosynthesis in nature. Annual Review of Plant Biology, 45, 633–662.

Marija M, Māra V, Veneranda S (2011) Changes of photosynthesis-related parameters and productivity of Cannabis sativa under different nitrogen supply. Environmental and Experimental Biology, 9, 61–69.

Mccormick K, Kautto N (2013) The bioeconomy in Europe: an overview. Sustainability, 5, 2589–2608.

Medlyn BE, Dreyer E, Ellsworth D et al. (2002) Temperature response of parameters of a biochemically based model of photosynthesis. II. A review of experimental data. Plant, Cell and Environment, 25, 1167–1179.

Murata N, Allakhverdiev SI, Nishiyama Y (2012) The mechanism of photoinhibition in vivo: re-evaluation of the roles of catalase, α-tocopherol, non-photochemical quenching, and electron transport. Biochimica et Biophysica Acta (BBA) - Bioenergetics, 1817, 1127–1133.

Ögren E, Evans J (1993) Photosynthetic light-response curves. Planta, 189, 182–190.

Oomah BD, Busson M, Godfrey DV, Drover JCG (2002) Characteristics of hemp (Cannabis sativa L.) seed oil. Food Chemistry, 76, 33–43.

Pahkala K, Pahkala E, Syrjala H (2008) Northern limits to fiber hemp production in Europe. Journal of Industrial Hemp, 13, 104–116.

Patanè C, Cosentino SL (2013) Yield, water use and radiation use efficiencies of kenaf (Hibiscus cannabinus L.) under reduced water and nitrogen soil availability in a semi-arid Mediterranean area. European Journal of Agronomy, 46, 53–62.

Piel C, Frak E, Le Roux X et al. (2002) Effect of local irradiance on CO2 transfer conductance of mesophyll in walnut. Journal of Experimental Botany, 53, 2423–2430.

Powles SB, Berry JA, Bjorkman O (1983) Interaction between light and chilling temperature on the inhibition of photosynthesis in chilling-sensitive plants. Plant, Cell and Environment, 6, 117–123.

Prade T, Svensson S-E, Andersson A et al. (2011) Biomass and energy yield of industrial hemp grown for biogas and solid fuel. Biomass and Bioenergy, 35, 3040–3049.

Rice B (2008) Hemp as a feedstock for biomass-to-energy conversion. Journal of Industrial Hemp, 13, 145–156.

Sage RF, Kubien DS (2007) The temperature response of C3 and C4 photosynthesis. Plant, Cell and Environment, 30, 1086–1106.

Salentijn EMJ, Zhang Q, Amaducci S et al. (2015) New developments in fiber hemp (Cannabis sativa L.) breeding. Industrial Crops and Products, 68, 32–41.

Sharkey TD, Bernacchi CJ, Farquhar GD et al. (2007) Fitting photosynthetic carbon dioxide response curves for C3 leaves. Plant, Cell and Environment, 30, 1035–1040.

Sinclair TR, Horie T (1989) Leaf nitrogen, photosynthesis, and crop radiation use efficiency: a review. Crop Science, 29, 90–98.

Struik PC, Amaducci S, Bullard MJ et al. (2000) Agronomy of fibre hemp (Cannabis sativa L.) in Europe. Industrial Crops and Products, 11, 107–118.

Tang Z, Hu X, Sun T et al. (2012) Adaptability of different hemp varieties (lines) in Xishuangbanna prefecture. Journal of Southern Agriculture, 43, 160–163. (Chinese with English abstract).

Tang K, Struik PC, Yin X et al. (2016) Comparing hemp (Cannabis sativa L.) cultivars for dual-purpose production under contrasting environments. Industrial Crops and Products, 87, 33–44.

Tcherkez G, Cornic G, Bligny R et al. (2005) In vivo respiratory metabolism of illuminated leaves. Plant Physiology, 138, 1596–1606.

Tcherkez G, Boex-Fontvieille E, Mahé A et al. (2012) Respiratory carbon fluxes in leaves. Current Opinion in Plant Biology, 15, 308–314.

Von Caemmerer S, Farquhar G, Berry J (2009) Biochemical model of C3 photosynthesis. In: Photosynthesis In Silico (eds Laisk A, Nedbal L, Govindjee), pp. 209–230. Springer, London, UK.

Wirtshafter DE (2004) Ten years of a modern hemp industry. Journal of Industrial Hemp, 9, 9–14.

Wullschleger SD (1993) Biochemical limitations to carbon assimilation in C3 plants: a retrospective analysis of the A/Ci curves from 109 species. Journal of Experimental Botany, 44, 907–920.

Yamori W, Evans JR, Von Caemmerer S (2010) Effects of growth and measurement light intensities on temperature dependence of CO2 assimilation rate in tobacco leaves. Plant, Cell and Environment, 33, 332–343.

Yin XY, Struik PC (2009) C3 and C4 photosynthesis models: an overview from the perspective of crop modelling. NJAS - Wageningen Journal of Life Sciences, 57, 27–38.

Yin XY, Van Laar HH (2005) Crop Systems Dynamics: An Ecophysiological Simulation Model for Genotype-by-Environment Interactions. Wageningen, Wageningen Academic.

Yin XY, Harbinson J, Struik PC (2006) Mathematical review of literature to assess alternative electron transports and interphotosystem excitation partitioning of steady-state C-3 photosynthesis under limiting light. Plant Cell and Environment, 29, 1771–1782.

Yin XY, Struik PC, Romero P et al. (2009) Using combined measurements of gas exchange and chlorophyll fluorescence to estimate parameters of a biochemical C3 photosynthesis model: a critical appraisal and a new integrated approach applied to leaves in a wheat (Triticum aestivum L.) canopy. Plant, Cell and Environment, 32, 448–464.

Yin XY, Sun ZP, Struik PC et al. (2011) Evaluating a new method to estimate the rate of leaf respiration in the light by analysis of combined gas exchange and chlorophyll fluorescence measurements. Journal of Experimental Botany, 62, 3489–3499.

6

Alternative operation models for using a feed-in terminal as a part of the forest chip supply system for a CHP plant

KARI VÄÄTÄINEN[1] (iD), ROBERT PRINZ[1], JUKKA MALINEN[2], JUHA LAITILA[1] and LAURI SIKANEN[1]

[1]*Natural Resources Institute Finland (LUKE), Joensuu, Finland*, [2]*University of Eastern Finland (UEF), Joensuu, Finland*

Abstract

The fuel supply of forest chips has to adapt to the annual fluctuations of power and heat generation. This creates inefficiency and unbalances the capacity utilization of the fuel supply fleet in the direct fuel supplies from roadside storages to power and heat generation. Terminals can offer an alternative approach for the fleet management of fuel supplies in terms of smoothing the unbalanced fleet use towards more even year-round operations. The aim of the study was to compare the supply costs of a conventional direct forest chip supply to an alternative fuel supply with the use of a feed-in terminal using the discrete-event simulation method. The influences of the terminal location, terminal investment cost, outbound terminal transport method, terminal truck utilization and quality changes of terminal-stored forest chips for the fuel supply cost were studied in the case environment. By introducing a feed-in terminal and a shuttle truck for the transports of terminal-stored forest chips, the total supply cost was 1.4% higher than the direct fuel supply scenario. In terminal scenarios, the supply costs increased 1–2% if the cost of the terminal investment increased 30%, the distance to the terminal increased from 5 to 30 km or the total annual use of a terminal truck decreased 1500 h. Moreover, a 1 per cent point per month increase in the dry matter loss of terminal-stored chips increased the total supply cost 1%. The study revealed that with the relatively low additional cost, the feed-in terminal can be introduced to the conventional forest chip supply. Cost compensation can be gained through the higher annual use of a fuel supply fleet and more secured fuel supply to power plants by decreasing the need for supplement fuel, which can be more expensive at a time of the highest fuel demand.

Keywords: discrete-event simulation, feed-in terminal, forest chips, fuel supply system, logistics, supply costs

Introduction

Material-storing terminals are widely used in various material and goods supply logistics. Terminals and intermediate storages are required for the smooth and time-dependent running of the material supply from the initial point of supply to the final end user (Stampfer & Kanzian, 2006; Kanzian *et al.*, 2009; Kons *et al.*, 2014; Wolfsmayr & Rauch, 2014; Virkkunen *et al.*, 2015). Respectively, biomass supply logistics for energy generation, especially in large scale, require fuel terminals in the supply system (Virkkunen *et al.*, 2015). The use of terminals in the forest fuel supply has notably increased in Nordic countries (Kärhä, 2011; Palander & Voutilainen, 2013; Routa *et al.*, 2013; Kons *et al.*, 2014; Virkkunen *et al.*, 2015; Strandström, 2016).

The operations of a fuel supply fleet in traditional fuel supply systems have to adapt to the annual fluctuations of power and heat production. This increases the inefficiency and lowers the capacity utilization of machines, thus negatively influencing the cost efficiency of the fuel supply. For permanent workers, stable and year-round working opportunities are difficult to provide in the forest chip supply business. Difficulties also occur in the recruitment of additional workforce for the relatively short time periods during the high heating season. The use of terminals can act as a balancing element enabling year-round working opportunities and year-round machine use with benefits for both the fuel supply entrepreneurs and their operating personnel (Väätäinen *et al.*, 2014; Raitila & Korpinen, 2016).

Understandably, a terminal as a part of the forest fuel supply structure generates additional costs compared to a direct fuel supply system from roadsides to a power plant. Depending on the size and the function of a terminal, basic investment and construction costs for its establishment can be notable (Virkkunen *et al.*, 2016). Additional transportation, material handling and fuel quality maintenance in terminal produce extra costs (Karttunen *et al.*, 2013; Virkkunen *et al.*, 2016).

Correspondence: Kari Väätäinen
e-mail: kari.vaatainen@luke.fi

In Finland during 2015, 29% of forest-based fuel was transported to, stored and comminuted in terminals before further transport to energy use (Strandström, 2016). Currently, statistics on the amounts of roadside chipped material stored in terminals are not available. However, small amounts of energy biomass as comminuted material are stored for some months in terminals.

Storing demonstrations and tests have been accomplished mainly for comminuted forest biomass as in larger heaps, to follow up regarding changes in quality and material loss over monitored time spans (Björklund, 1982; Nurmi, 1990; Jirjis, 2005; Laitila & Nuutinen, 2015; Raitila & Korpinen, 2016). It is well known from previous studies that as comminuted biomass is stored, microbial activity will occur resulting in heat generation and dry matter losses (Kubler, 1987; Nurmi, 1990, 1999; Jirjis, 2005). Fines of <3 mm in length represent a health hazard because they reduce air circulation during storage, supporting bacteria proliferation with an increased risk of combustion (Jirjis, 2005). It is also known that chips made of fresh wood generate more heat and suffer greater dry material loss (DML) than if they are made of seasoned material (Björklund, 1982; Kubler, 1987; Routa et al., 2015).

The reasons for introducing a terminal to a fuel supply system can vary. Terminals offer safety to fulfil the fuel demand during divergent fuel demand and supply occasions, often when the demand becomes unexpectedly high, or when the fuel supply fleet is unable to operate normally due to challenges in the operational environment (Malinen et al., 2014). External terminals are also needed for securing fuel deliveries in cases when the size of a power plant's buffer is limited or the number of direct deliveries from roadside storages during weekends remain low (Virkkunen et al., 2015). Bigger terminals allow better quality monitoring on fuel, and therefore, quality-based control and supply of required fuel quality could be executed. In addition, terminals offer a new approach for the fleet management of fuel supplies in terms of smoothing unbalanced fleet use towards more even year-round operations. During the low season, fuel supply units can conduct fuel transports to fill the terminals, enabling terminals to take part in fuel supply to plants during the high season to increase the year-round utilization of fuel supply units (Raitila & Korpinen, 2016).

Biomass terminals for heat and energy generation can be categorized by their main functions as a part of the fuel supply, by their size in hectares or by the capacity of stored biomass in energy or volume (Kons 2015, Virkkunen et al., 2015). To classify terminals with the functions of terminals as part of the fuel supply, the transshipment terminals are small areas, where raw and uncomminuted material is transported to and comminuted before further transports to heat or energy production. Feed-in terminals usually contain more infrastructures, such as a paved terminal area and a scaling place. The main function of the feed-in terminal is to balance the fuel supply for power and heat generation. So-called satellite terminals have their own terminal machinery and a large storing capacity with several fuel assortments. Satellite terminals are for year-round operations and serve distant and large customers (Virkkunen et al., 2015). In Sweden, with a 74% share of all forest biomass terminals, small terminals with <2 ha area are a majority (Kons et al., 2014), and in Finland 1–3 ha or 10 to 100 GWh per a terminals dominate the total number of terminals (Raitila & Virkkunen, 2016).

Thus far, only a few studies related to forest biomass supply systems for energy use have considered the influence of having a terminal as a part of the fuel supply on total supply costs (e.g. Karttunen et al., 2013; Korpinen & Aalto, 2016; Virkkunen et al., 2016). This study tackles the theme of the balanced use of a supply fleet and a more secured fuel supply to a power plant by comparing the conventional supply system to the terminal-based supply, where the feed-in terminal has been introduced into the conventional supply system.

The aim of the study was to compare the costs of the conventional direct forest chip supply to an alternative fuel supply with the use of a feed-in terminal using discrete-event simulations. The operational environment was a typical forest chip supply environment in Eastern Finland with 517 GWh of forest chips used in a combined heat and power plant (CHP plant). In terminal transport functions, the truck transport methods of the supplier's chip truck and designated shuttle truck from the feed-in terminal to the CHP plant were studied. Furthermore, the influences of the terminal location, terminal investment cost, shuttle truck utilization and quality changes in the terminal-stored forest chips were examined.

Materials and methods

Modelling the system environment

To model the system environment and to conduct the simulations of determined scenarios, WITNESS simulation software applying the discrete-event simulation method integrated with Excel-based parameter input was used. The forest chip supply environment was framed to begin from the roadside storages of forest biomass and to end at the power plant. The simulation environment corresponded to the typical forest chip supply conditions located in Eastern Finland in the Joensuu district. In addition, the model's parameter data closely resembled the business as usual (BAU) situation for the CHP plant's forest chip supply. However, some simplifications and parameter adjustments in the system environment were carried out, expressed as followed.

The model construction was carried out by following the BAU case in the forest chip supply to the CHP plant in the city of Joensuu. The simulation model consisted of four forest chip suppliers operating with one truck-mounted chipper and two chip trucks. As a simplification in the model, each supplier had a similar operational environment in terms of operation area in size, characteristics of roadside storages and operation model of the forest chip supply. In addition, the fuel suppliers operated with a one shift weekly working schedule having Sundays off from work (Fig. 1). The sizes of the roadside storages of logging residues corresponded to the real situation of spruce-dominated final fellings located in North-Karelia, Eastern Finland (Windisch *et al.*, 2015). The storage location, storage size in solid-m³ and the number of storages in a defined storage cluster were determined by theoretical distributions presented in Table 1.

Within the simulations, the fuel supply from roadsides was based on a distance orientation by having a mean transport distance of 60–64 km (Table 1). During the low heating season, the chip supply was oriented to the longest distance classes, whereas during the high heating season to the shortest distance classes. During the moderate heating seasons of autumn and spring, the mid-distance classes were stressed most. In terminal scenarios, forest chip transports from the roadside storages to

Table 1 Modelling the characteristics of roadside storages by theoretical distributions in simulations. Distance classes were defined in straight line distances. Tnormal is the truncated normal distribution; SD is the standard deviation

	Mean	SD	Min	Max	Distribution
Roadside storage size, solid-m³	150	100	40	600	Tnormal
Number of storages in cluster	2	2	1	4	Tnormal
Location and proportion of storages					
- 2 km–20 km, 10%			2	20	Uniform
- 20 km–40 km, 25%			20.01	40	Uniform
- 40 km–60 km, 30%			40.01	60	Uniform
- 60 km–80 km, 25%			60.01	80	Uniform
- 80 km–100 km, 10%			80.01	100	Uniform

the terminal were directed to long distances during the low heating season.

The material for the chipping was logging residues (tops and branches) from spruce-dominated final felling. The *net calorific value* of forest chips was set as 19.2 MJ kg⁻¹ corresponding to the representative value from Finnish studies

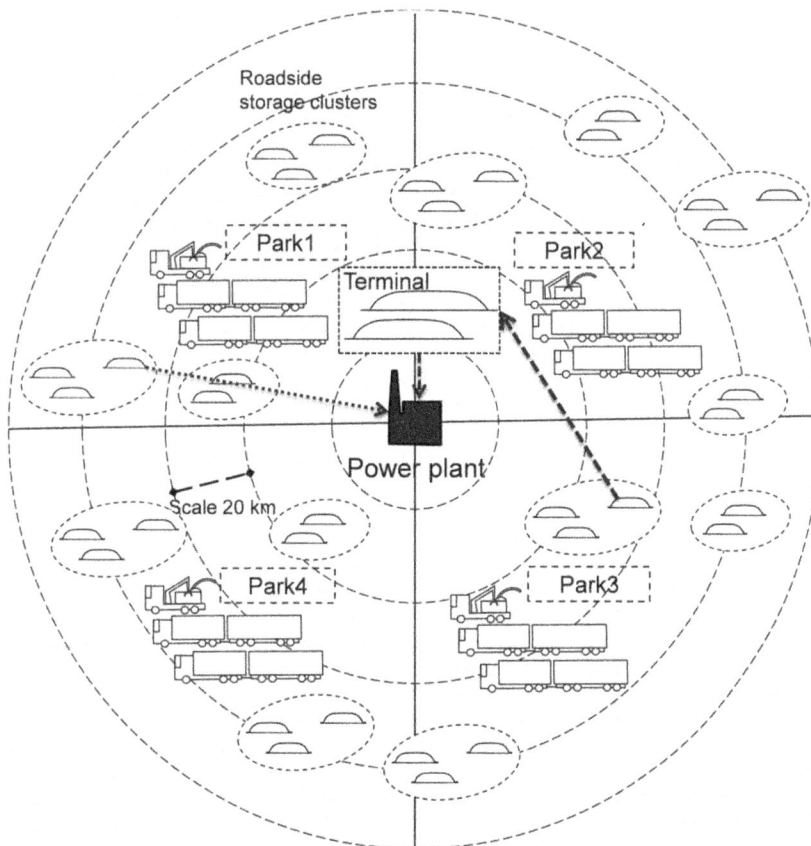

Fig. 1 General picture of the simulation environment of the forest chip supply system.

(Alakangas, 2005). The dry matter weight of forest chips was set as 445 kg per solid-m³ (Hakkila, 1985). The average moisture content of 47% was applied, varying from a minimum of 25% to a maximum of 65%, which corresponded to the moisture content of forest chips at the fuel receiving station in a large-scale CHP plant (Hakkila, 2004; Alakangas, 2005). To apply the annual moisture variation of forest chips into the model, monthly correction factors were used (Hakkila, 2004) (Table 2). The truncated normal distribution (mean, SD, min and max; 47 + [correction factor], 7, 25, 65) was used to determine the moisture content of roadside storages.

The monthly demand of forest chips used in simulations was derived from the local CHP plant in the city of Joensuu (Table 2). The daily variation of forest chip demand was introduced using the truncated normal distribution, where the mean value was the average daily demand of a particular month, the SD was 20% from the average, the minimum was 80% and the maximum value was 120% from the average demand. At the fuel reception, fixed parameter values were used for both driving and scaling-unloading, which were 0.05 and 0.3 h, respectively. In addition, the model recorded queuing times for each truck if queuing occurred. The capacity of the buffer for forest chips at the CHP plant was set to 6.0 GWh.

Each fuel supplier had an individual operating sector covering the whole supply area (Fig. 1). The operating radius from the business premises of each contractor was set to 55 km, wherein the roadside storages were generated. Both the distance to the plant and the storage cluster distribution controlled the generation of a new roadside storage location. The exact location was defined using a random number generator (−100 km, 100 km for x-coord and y-coord) to pinpoint the coordinates from the permissible distribution class by having 0.01 × 0.01 km grid level accuracy. The power plant's location was at the origin (0, 0). To include the winding of roads depending on the density and the structure of the road network, a winding factor of 1.3 was used to convert straight line distances to road distances.

In BAU scenarios, where the feed-in terminal was not used, fuel suppliers had scheduled holiday periods during the low heating season. The fuel supply fleet of two suppliers was not in operation in the first two holiday months of May and June, and the fleet of the other two suppliers was in lay-down the following two months, June and August. In the terminal scenarios, the fuel supply fleet of all suppliers was able to be operated year round by having one shift system and substitute truck drivers operating during permanent drivers' holidays.

According to the test simulation runs of one year, the terminal was emptied before the end of April. Consequently, the start of the scenario runs was defined as the beginning of May. For all terminal scenarios, initial seed buffer was set to zero for the terminal storage, thus improving the comparison between scenarios of direct supply and terminal-based supply.

Both the daily fuel demand and the alarm levels of the power plant's buffer controlled the transports of forest chips from roadside storages to the power plant or to the terminal and from the terminal to the power plant. The lower alarm level of the plant's buffer was set to 40% and the higher level at 90% of the buffer capacity. In terminal scenarios, during the low heating season when the buffer exceeded 90% of the capacity, fuel transports were directed to terminals, whereas if the buffer size decreased to <40% of the capacity, transports from the terminal to the CHP plant were ignited (Table 3). In BAU cases, only the high alarm level of 90% was used. If the buffer exceeded the higher alarm level, forest chip transports were stopped for the rest of the day after the unfinished load cycle of each chip truck. In situations in which the buffer was emptied, supplementary fuel (such as peat, sawdust, bark or forest chips carried out by a supplementary fuel supplier) with the fixed supply cost was introduced to the model to feed into the plant and fulfil the gap of demand and supply of forest chips.

Chippers had a 9-h normative shift schedule all weekdays except Sunday, whereas chip trucks were operated on 8.5-h shift schedule, respectively. Mobile chippers were modelled to arrive to roadside storage earlier than chip trucks to prepare the setup before the arrival of the first chip truck. If the scheduled work shift was left less than one hour, chip trucks were directed to the supplier's business premises from the plant or from the terminal after unloading. In cases in which the scheduled work shift ended at the time when the chip truck was arriving to the roadside storage or chipping was ongoing, the chipping was carried out until the chip truck was loaded. Then, chip truck was directed as loaded to the business premises of the contractor. At the start of next working day, the loaded truck was directed either to the plant or the terminal depending on the scenario and the simulation stage. During the high heating season from December to February, 8.5- to 9-h work shifts were scheduled on Sundays as well. The logics of chippers' and chip trucks' daily operation are presented as flow charts in Figs 2 and 3.

The chip trucks in use were conventional container-based truck and trailer units with seven axels allowing a 64-tonne max vehicle weight in Finnish transportation. The total weight

Table 2 Moisture correction factor for correcting the mean value (47%) of moisture for each month (Hakkila, 2004) and monthly forest fuel demand of the CHP plant in the simulation in GWh

	January	February	March	April	May	June	July	August	September	October	November	December
Correction for the mean value of moisture, per cent point	+1	+4	+4	+4	+4	0	−8	−6	−3	−3	+5	+1
Forest fuel demand, GWh	79.9	70.5	56.4	32.9	18.8	14.4	9.4	18.8	37.6	47	59.2	70

Table 3 Control of fuel transports to the power plant in the simulation scenarios. RS is roadside; PP is power plant

Main scenarios	Buffer size at power plant	Controlling rules
BAU scenarios	Between 0% and 90%	Transports from RS to PP or from truck park to PP
	Filled over 90% level	Transports of forest chips stopped after unfinished loads
	Emptied to zero	Use of supplementary fuel
Terminal scenarios	Between 40% and 90%	Transports from RS to PP or from truck park to PP
	Filled over 90% level	Transports from RS to terminal or from truck park to terminal
	Emptied under 40% level	Transports from terminal to PP and RS to PP
	Emptied to zero	Use of supplementary fuel

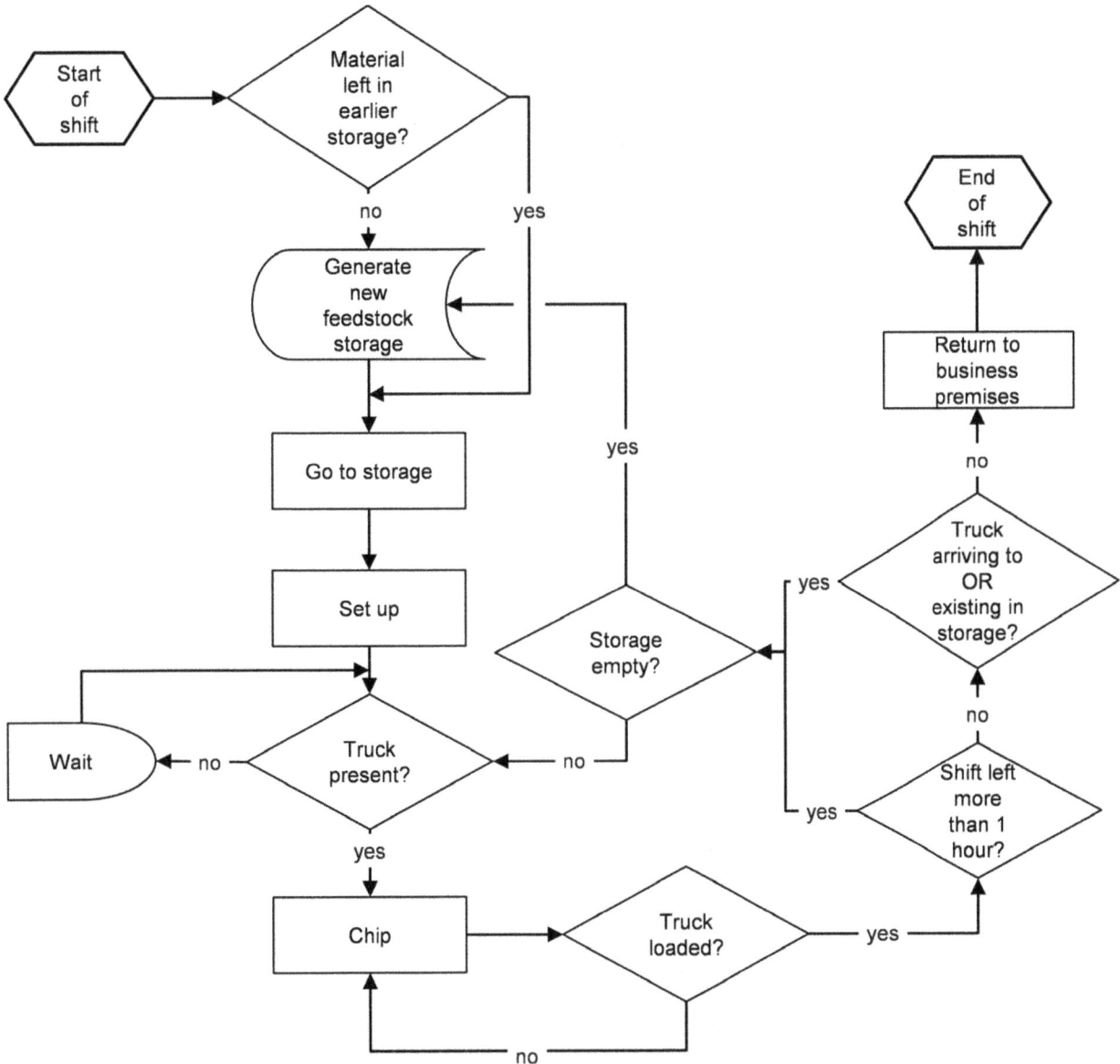

Fig. 2 Flow chart of the logic the mobile chipper follows in the simulation model.

of empty chip truck units was 23 tonnes, thus resulting in a 41-tonne maximum payload. The frame volume of containers was 131.6 frame-m³ and load density was 0.38 solid-m³ per frame-m³ resulting in 50 solid-m³. Preventing the overweight of trucks, the loading was controlled by calculating the fresh weight of solid-m³ of forest chips. If the moisture of forest chips

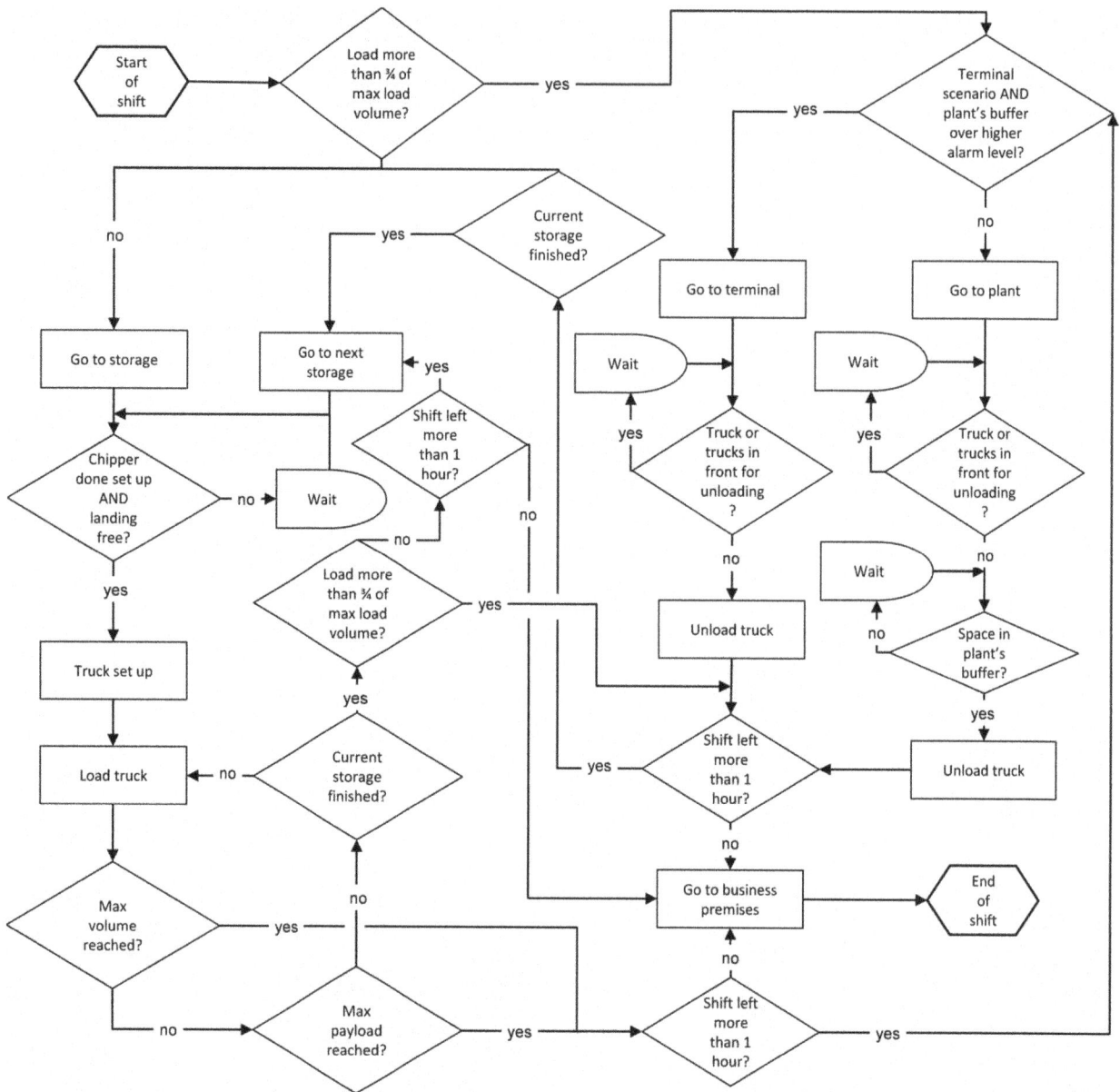

Fig. 3 Flow chart of the logic the chip truck follows for the terminal scenarios in the simulation model.

in storage reached 45.7%, the vehicle weight restricted over-loading, whereas if moisture was under the breakeven value, the frame volume of the chip truck restricted overloading. The driving speed of trucks was determined by the speed functions of timber truck and trailer units introduced by Nurminen & Heinonen (2007).

The mobile chipper was a conventional truck-mounted drum chipper typically used in chipping at roadside storages. Normal distribution was used to determine the chipping productivity in each of the roadside storages. The mean productivity was 50 solid-m³ per effective hour, and the standard deviation was 5. Negative exponential distribution was used for generating breakdowns. The mean value for the time between failures was 9.5 h and for the duration of breakdowns was 0.5 h. A

fixed setup time of 0.25 h occurred every time the chipper arrived to the roadside storage.

Terminal characteristics

From the results of terminal scenario runs, the maximum amount of stored chips at the terminal was used to determine the area required for storing forest chips (Table 4) utilizing the area requirement factor 1.2 solid-m³ m⁻² (Impola & Tiihonen, 2011; Virkkunen et al., 2015). In addition, for the total space requirement of terminals, 0.7 ha was added to the calculated area requirement, thus allowing for possible greater use of terminals with additional infra. Other cost parameters for defining the investment cost of terminal, following the values used in

Table 4 Characteristics and cost factors of the feed-in terminal used in the cost calculations of the simulation scenarios 1B and 1C

Terminal area					
Space requirement for forest chips	1.2	Solid-m^3 m^{-2}	Purchase price of land	5000	€ ha^{-1}
Maximum storage size in scenarios			Gravel paved cost	30	€ m^{-2}
- 1B. Suppliers' chip trucks (simulation value)	45 876	Solid-m^3	Residual price	5000	€ ha^{-1}
- 1C. Shuttle truck (simulation value)	45 146	Solid-m^3	Interest	4	%
Land area	4.5	ha	Investment total	1 372 500	€
Depreciation time	15	a	TOTAL per year	117 900	€
Wheeled loader					
Terminal working hours	1500	h	Hourly costs, idle	40	€ h^{-1}
Total working hours	3000	h	Hourly costs, operating	60	€ h^{-1}

Virkkunen *et al.*'s (2015) study, are introduced in Table 3. The feed-in terminal included a wheeled loader for terminal operations of building 5- to 7-m high heaps with forest chips, the loading of chip trucks, keeping the area free from snow and defending against the burning of chips in heaps (Table 4). The operator of the wheeled loader was the person responsible for the serving and maintenance at the terminal.

A shuttle truck was used in the terminal scenario, where outbound terminal transports were carried out with a higher capacity truck and trailer unit. The shuttle truck was a 9-axle truck-trailer unit having a maximum weight of 76 tonnes with a 157.9 loose-m^3 or 60 solid-m^3 load capacity. All loads were fully loaded to either the maximum weight or frame volume. In the shuttle scenario, the shuttle truck operated between the feed-in terminal and CHP plant. The chip transport was activated when the CHP plant's buffer decreased under a lower alarm level of 40% of the buffer capacity. During the high heating season, the shuttle truck operated one 8-h shift per day through the week if the fuel demand of the plant required a complementary fuel supply from the terminal (Table 2). During other seasons except the high heating season, the truck was used in mixed cargo transports. The amount of other transport operations was determined to be 1500 operating hours. In addition, cost sensitivity analysis with 0 and 750 h of other uses of the shuttle truck was conducted. The initial terminal location was 5 km from the CHP plant to the north.

In the initial terminal scenarios, the dry matter loss of stored material was expected to be 1% per month higher in chipped material than if stored as unchipped material at roadside storage. The dry matter loss of terminal-stored material was defined as fixed regardless of the arrival time. Fuel chip arrival and departure at the terminal was based on the first in-first out procedure. Respectively, the moisture content in initial scenarios was expected to stay at the same moisture level upon arrival at the terminal.

Simulation scenarios

The BAU 1A1 scenario resembled the typical fuel supply scenario of the conventional forest chip supply from roadside storages to CHP plant in Eastern Finland. All main parameters were kept as presented earlier. The main difference compared to the terminal scenarios was more fluctuated forest chip supply from roadside storages stressed to high heating season and 2 months sequenced lay-down time of the supply fleet for each

entrepreneur. In addition to the BAU 1A1 scenario, scenario 1A2 with a 2-shift system during the high heating season was introduced to fulfil the initial forest chip demand of the power plant. All scenario runs are presented in Table 5.

Two main options were simulated in the terminal scenarios. Scenario 1B consisted of the utilization of forest chip suppliers' own chip trucks for transporting fuel from the terminal to the power plant. In 1B, while the outbound terminal transports were needed, one of each contractor's chip trucks was called for terminal transports and operated in terminal transports until the lower alarm level of the power plant's buffer was fulfilled. Another chip truck of each contractor operated normally transporting chips from roadside storages to the plant. In scenario 1C, a separate shuttle truck was used for terminal transports to the power plant. In all scenario simulations, the distances had to match the distribution of distances from roadsides to the power plant to have valid correspondence among scenarios.

Four distance scenarios with distances of 5, 10, 20 and 30 km from the power plant to the terminal were used to explore the influence of the terminal location both on supply operations and costs. The locations of terminal were in line from the power plant towards the north (Fig. 1).

The third main scenario comparison included an evaluation of the impact of the dry matter loss of chipped material in storing compared to the unchipped material storing at roadsides. In 3C1, no additional dry matter loss compared to the BAU scenario occurred. In 3C2, dry matter loss of 1% per month was defined for terminal-stored material as chips, whereas 3C3 had 2% per month dry matter loss, respectively. In the fourth main scenario, the influence of terminal-stored forest chip moisture in fuel supply costs was studied. In the scenarios of 4C4 and 4C5, the moisture increased 6 and 10 per cent points, both having dry matter loss of 2% per month (Table 5).

For each scenario simulation, seven replications with varying seed numbers for determining unique random number streams were applied. Average values, standard deviations and 95% confidence levels were calculated for the main output parameters of each simulation scenario. In addition to the terminal scenario runs, sensitivity analyses for the operating hours of the shuttle truck and the investment level of the terminal were investigated. The shuttle truck's operating hours for uses other than terminal use varied 0, 750 and 1500 h, whereas the terminal's investment cost had 30% lower and 30% higher values than the initial investment cost.

Table 5 Simulation scenarios used in the study (pp, percentage point)

Main scenario	Abbreviation	Terminal in use	Definitions	Equals to
1: Business as usual (BAU) vs. terminal	1A1	No	BAU - one shift; 2 + 2 months off-shift at summer	
	1A2	No	BAU - 1 to 2 shifts; 2 + 2 months off-shift at summer	
	1B	Yes	Suppliers' chip trucks used for terminal transports	
	1C	Yes	Separate shuttle truck for terminal transports	
2: Terminal location	2B1	Yes	Terminal 5 km from CHP plant, Suppliers' trucks	1B
	2B2	Yes	Terminal 10 km from CHP plant, Suppliers' trucks	
	2B3	Yes	Terminal 20 km from CHP plant, Suppliers' trucks	
	2B4	Yes	Terminal 30 km from CHP plant, Suppliers' trucks	
	2C1	Yes	Terminal 5 km from CHP plant, Shuttle truck	1C
	2C2	Yes	Terminal 10 km from CHP plant, Shuttle truck	
	2C3	Yes	Terminal 20 km from CHP plant, Shuttle truck	
	2C4	Yes	Terminal 30 km from CHP plant, Shuttle truck	
3: Dry matter loss (DML)	3C1	Yes	Moisture and DML equal with BAU	
	3C2	Yes	Moisture equal, DML 1 pp per month higher than BAU	1C
	3C3	Yes	Moisture equal, DML 2 pp per month higher than BAU	
4: Moisture change	4C1	Yes	Moisture 3 pp lower, DML 1 pp per month higher than BAU	
	4C2	Yes	Moisture 3 pp higher, DML 1 pp per month higher than BAU	
	4C3	Yes	Moisture 6 pp higher, DML 1 pp per month higher than BAU	
	4C4	Yes	Moisture 6 pp higher, DML 2 pp per month higher than BAU	
	4C5	Yes	Moisture 10 pp higher, DML 2 pp per month higher than BAU	

Cost calculation

The costs of forest chip supply operations were calculated with typical cost accounting tables for machines using Excel spreadsheets. The main cost parameters accounting for the mobile chipper, the chip truck and the shuttle truck are presented in Tables 6 and 7. All cost values exclude value-added tax. The supply costs were calculated after simulations using the average values of performance and time data. The cost of the supplementary fuel supply as delivered to the plant was set to 8 € per MWh. To more clearly visualize the differences in the supply costs between scenarios, the purchase prices of fuels (forest chips and supplementary fuel) were not included. In this respect, supply costs included the costs of chipping, truck transports and terminal costs (investment and operations).

Results

Scenarios with higher amounts of forest chip supply than the BAU scenario 1A1 resulted in longer transport distances (Fig. 4). In terminal scenarios, forest chip transports to longer distances were stressed starting from the 20–40 km distance class.

Within the BAU scenario, four supply units of one mobile chipper and two chip trucks were not sufficient to meet the power plant's annual demand while operating with one shift (Fig. 5). From the total fuel demand, 19.3% as supplement fuel was used in addition to the base supply of forest chips. To meet the demand for the forest chip supply, the external shift during the high heating season was needed for the supply fleet (scenario

1A2). In both terminal scenarios, the forest chip supply was merely enough to fulfil the demand of the power plant; the use of supplement fuel was 6.3% and 3.4% from the total demand in terminal scenarios 1B and 1C. The share of fuel delivery via the terminal was 18.0% and 17.6% in scenarios 1B and 1C.

In the 1A2 scenario with extra shift during high heating season, the average work shift length was 21% higher than the BAU scenario (Table 8). Respectively, in scenario 1B the average work shift length was 8% longer than in scenario 1A1 due to the time used for terminal transports from long distances and the timing of operations at the end of the scheduled shift. Due to the full loads of the outbound terminal transports, the mean load size of 64-tonne truck-trailer unit was slightly higher in scenario 1B than in other scenarios. In addition, in scenario 1C the shuttle truck was able to transport maximum loads as in volume due to the moisture levels of forest chips and the high transport capacity in the mass/frame volume ratio. The scenario-average load size was 90–92% from the load capacity between scenarios.

The effective chipping time was notably lower in 1B compared to other scenarios having 38.3% from the total time use (Fig. 6). In 1B, the lowest share in chipping and the highest in waiting was a result of the chip truck control for transporting forest chips from terminals during the high heating season. When comparing the work element shares of chip trucks, the greatest differences

Table 6 Main performance and cost factors for the mobile chipper

Performance			Capital cost		
Productivity, effective time	50	Solid-m³ per h_0	Investment price	600 000	€
Setup time	0.25	h	Depreciation time	7	y
Time between failures (mean)	9.5	h	Depreciation rate	17	%
Duration of breakdowns (mean)	0.5	h	Interest rate	4	%
Employer cost			**Variable cost**		
Salary	16	€	Fuel price	0.9	€ L^{-1}
Indirect salary cost	68	%	Fuel consumption		
Daily allowance	18	€ per d	-chipping	124	L h^{-1}
			-driving	55	L h^{-1}
Overheads and risk margin			-other	14	L h^{-1}
Insurance	10 000	€ per y	Repairs and maintenance	0.8	€ m^{-3}
Administration and overheads	7500	€ per y	Oil and lubricants	0.08	€ m^{-3}
Risk margin	5	%	Blades	13	€ h^{-1}

Table 7 Main performance and cost factors for chip suppliers' chip trucks and the shuttle truck

Performance	Chip truck	Shuttle truck		Capital cost	Chip truck	Shuttle truck	
Load space	131.6	157.9	Loose-m³	Truck	144 200	175 000	€
Setup time at roadside storage	0.25	-	h	Trailer and equipments	142 800	180 000	€
Speeds (loaded, unloaded)	From Nurminen & Heinonen (2007)			Tyre, truck; trailer	725; 450	725; 450	€ per piece
Loading with front loader	0.4	0.4	h	Truck tyres, total	5800	8700	€
Unloading (terminal and plant)	0.3	0.3	h	Trailer tyres, total	7200	8100	€
				Depreciation time	7	7	y
Employer cost				Depreciation rate			
Salary	16	16	€	- truck	20	20	%
Indirect salary cost	68	68	%	- trailer	25	25	%
				Interest rate	4	4	%
Overheads and risk margin							
Insurance	5500	6000	€ per a	**Variable cost**			
Traffic fees	2500	2500	€ per a	Fuel price	1.0	1.0	€ L^{-1}
Administration and maintenance	6500	7000	€ per a	Fuel consumption	55	66	L per 100 km
Uncompensated driving	2000	2000	€ per a	Repairs and maintenance	14 000	14 000	€ per a
Risk margin	5	5	%	Oil and lubricants	2000	3000	€ per a
				Tyres (coating)	200	200	€ per piece

involved elements of waiting time. Within scenarios 1A1 and 1A2, chip trucks' waiting times were longer at roadsides, whereas in terminal scenarios 1B and 1C, chip trucks had a longer waiting time at the power plant (Fig. 6). At the beginning of shifts, the number of arrivals to the CHP plant was high, resulting the longest queuing times, while filled up trucks from the earlier shift arrived to unload at the plant.

Forest fuel transports were analysed at the weekly level in the 1A1, 1A2, 1B and 1C scenario simulations (Fig. 7). In scenario 1A1, the fuel transports from four suppliers were able to fulfil the plant's demand until week 43, whereas scenario 1A2 fulfilled the demand until week 1. In both scenarios, the initial 4 months starting from week 18 were operated by two suppliers fulfilling the fuel demand. Between the time of calendar week 47 and 8, the amount of weekly transports was rising due to the change in the selection order of roadside storages emphasizing shorter distances.

In both terminal scenarios, 1B and 1C, all four fuel suppliers were operating during the summer being able both to fulfil the plant's fuel demand and to supply the fuel to the terminal (Fig. 7). In total, the transported forest chip amount at the terminal was 92 950 MWh (45 876 solid-m³) and 90 978 MWh (45 146 solid-m³) in scenarios 1B and 1C, respectively.

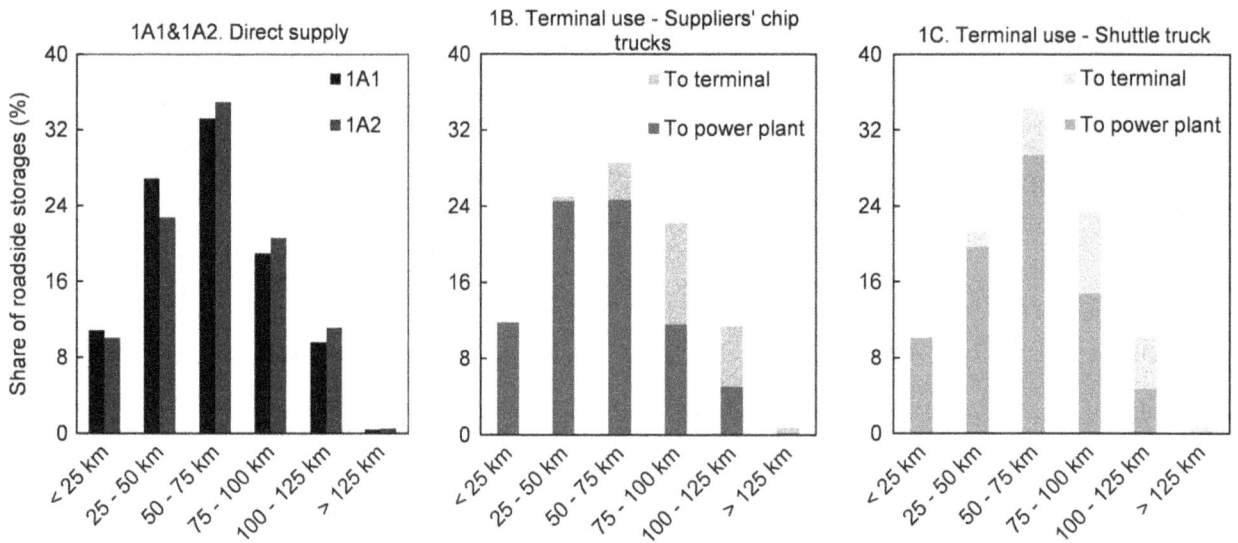

Fig. 4 Resulting distance distributions of roadside storages in the 1A1, 1A2, 1B and 1C simulation scenarios. Distances are road distances. The mean distance in each scenario was 60.8, 63.5, 62.3 and 63.8 km in scenarios 1A1, 1A2, 1B and 1C, respectively.

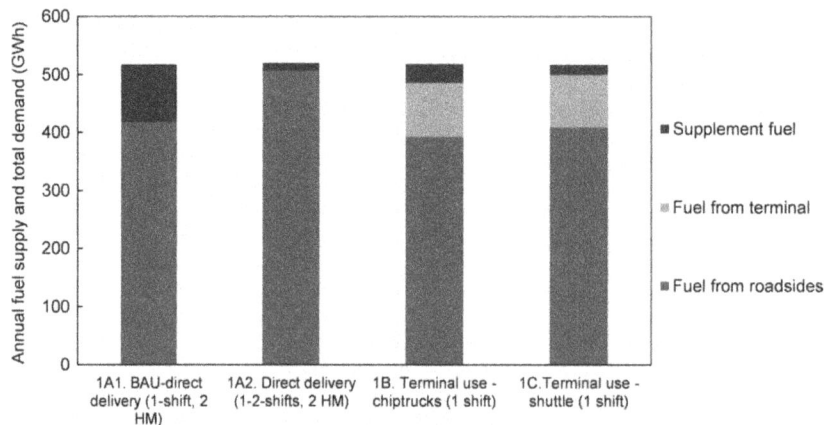

Fig. 5 The annual supply of forest chips and the use of supplement fuel to meet the initial forest chip demand of the CHP plant in the four simulation scenarios. (HM = holiday month).

In scenario 1B, the more the chip transports were carried out from the terminal, the less the chips were transported from the roadsides. That was a result of the supply control, where one of the each supplier's two chip trucks was determined to carry out the outbound terminal transports to meet the increased fuel demand at the CHP plant (Fig. 7). In scenario 1C, separate shuttle truck transports did not influence the forest chip suppliers' operations, thus increasing the total supply compared to scenario 1B.

While comparing the annual supply costs, the least costly scenario was 1A2, with an extra shift during the high season resulting in 7.1% lower supply costs compared to the BAU scenario (1A2) while an assumption of an 8 € per MWh supplement fuel cost was used

(Fig. 8). From the terminal scenarios, the cheapest option with the lowest supply cost was 1C, with separate terminal shuttle use having a 1.4% higher annual supply cost than the BAU scenario 1A1. Respectively, terminal scenario 1B had a 3.1% higher supply cost. The share of terminal costs from the total supply costs was 4.7% and 4.8% in terminal scenarios 1B and 1C, whereas the separated costs of outbound terminal transports (terminal to power plant) were 2.3% and 1.5%, respectively.

In all scenarios, chipping had the highest operational cost, varying from 3.67 to 3.91 € per MWh, whereas the transport costs from roadsides varied from 3.45 to 3.74 € per MWh (Fig. 9). While taking the total forest chip supply costs into account in the terminal scenarios, the share of the terminal cost including terminal transports

Table 8 Operational statistics of mobile chipper and chip truck units in the simulation scenarios of 1A1, 1A2, 1B and 1C. The confidence interval is presented as a per cent-share from the mean value (CI 95%)

	1A1. BAU-direct delivery (1-shift)		1A2. delivery (1-2-shifts)		Direct	1B. Terminal use – chip trucks (1 shift)		1C. Terminal use – shuttle (1 shift)		1C. *Shuttle data*	
	Mean	CI-%	Mean	CI-%		Mean	CI-%	Mean	CI-%	Mean	CI-%
Working days, d per a	273	-	273	-		326	-	326	-	*111*	-
Mobile chippers											
Shift length, h	8.8	0.3	10.6	0.7		9.5	0.4	9.1	0.6		
Annual production, solid-m³ per a	50 107	1.0	60 731	0.9		58 517	0.5	60 148	0.7		
Productivity, h_0 per solid-m³	49.52	0.4	49.52	0.5		49.54	0.3	49.48	0.3		
Road side storages, No.	289	1.3	349	0.7		335	0.5	347	0.5		
Total driving, km per a	26 751	2.1	28 281	1.9		32 392	2.5	32 121	2.0		
Daily driving, km per d	98.0	2.1	103.6	1.9		99.4	2.5	98.5	2.0		
Chip trucks and *Shuttle*											
Shift length, h	8.6	0.3	10.4	0.3		9.2	0.1	8.9	0.2	*8.4*	*0.3*
Production from roadsides, solid-m³	25 054	0.8	30 365	0.6		29 258	0.3	30 074	0.4		
Production from terminal, solid-m³						5735	0.5			*45 146*	*1.4*
Load size, solid-m³	45.1	0.6	45.4	0.3		45.9	0.3	45.3	0.2	*59.5*	*0.3*
Load size, MWh	94.7	0.7	95.1	0.3		96.8	0.3	94.9	0.3	*119.9*	*0.4*
Total driving, km per a	80 280	1.2	95 048	0.8		99 238	0.4	99 641	0.5	*8252*	*1.4*
Driving per load, km per load	144.7	1.7	142.1	1.1		130.3	0.6	151.1	0.8	*10.9*	*0.2*

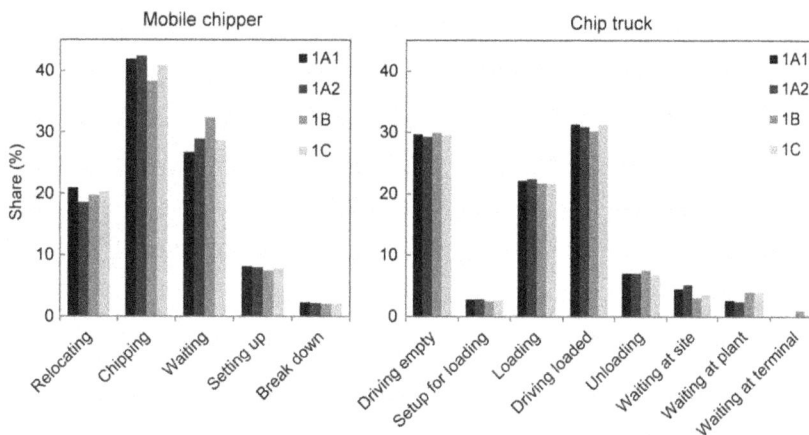

Fig. 6 Distribution of work elements for the mobile chipper and chip truck in scenarios 1A1, 1A2, 1B and 1C.

to power plant was 7.5% and 6.5% in scenarios 1B and 1C. A sensitivity analysis of the variation of terminal investment in 30% less and a higher cost resulted in a 0.9% difference in the total supply cost both in scenarios 1B and 1C.

In all distances, the terminal scenarios of using suppliers' own chip trucks for terminal transports resulted in 1.0–2.3% higher total supply costs compared to using a separate shuttle truck for terminal transports (Fig. 10). For distances of 5–30 km between the terminal and the plant, the total supply cost varied from 7.95–8.17 € per

MWh in scenarios of suppliers' own trucks, whereas in terminal shuttle scenarios, the supply cost variation was 7.83–7.98 € per MWh. The cost difference between scenarios B and C increased based on the distance between the terminal and the plant. While using suppliers' own chip trucks for terminal-plant transports at longer distances, mobile chippers had more idle time, and thus, chipping costs increased.

To obtain cost-effective trucking costs for the shuttle truck, terminal transports necessitated at least a moderate other use for the shuttle truck unit during the low

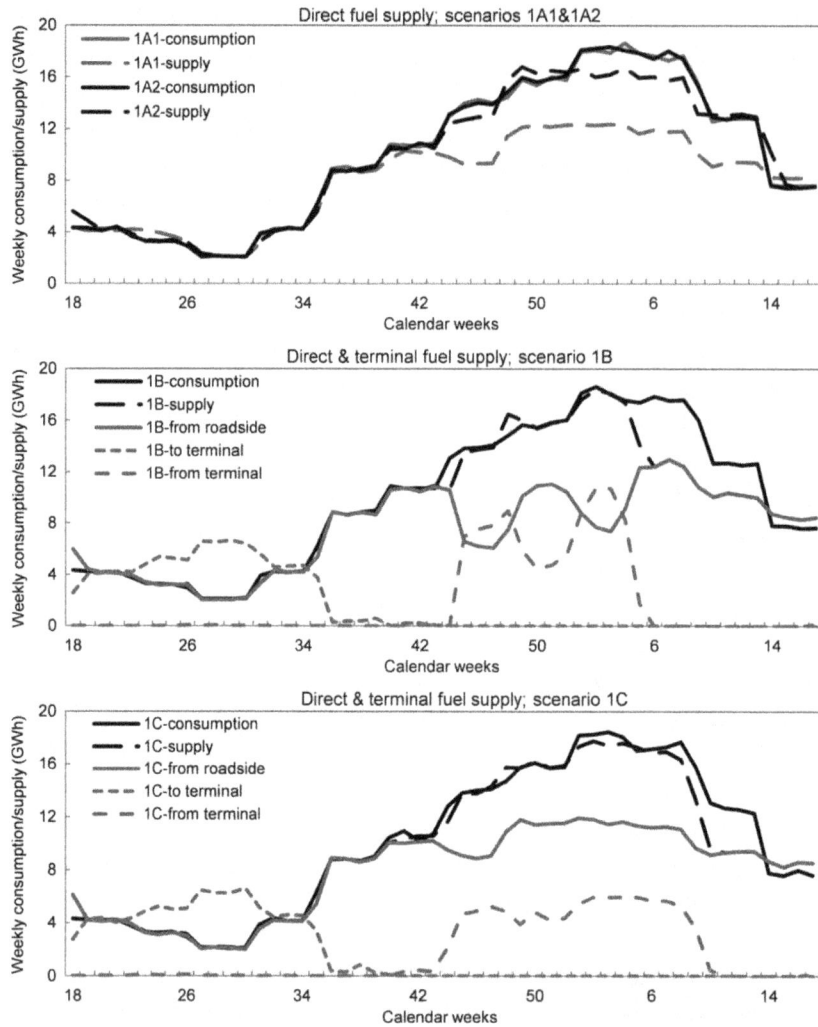

Fig. 7 Weekly level statistics for the forest chip demand and transports during 1 year. The compared scenarios were 1A1, 1A2, 1B and 1C.

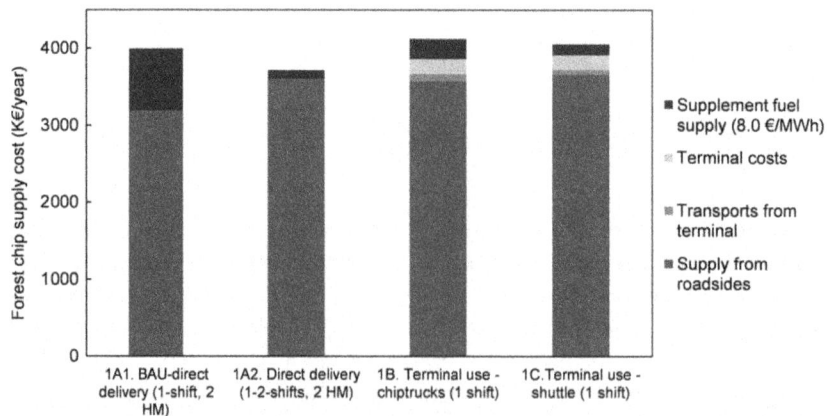

Fig. 8 Annual fuel supply costs of the power plant in four simulation scenarios. Terminal costs include investment and operational costs of the terminal.

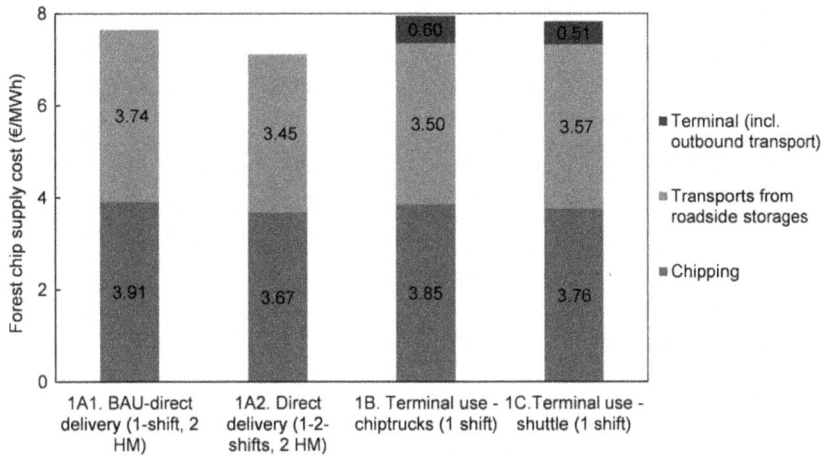

Fig. 9 Supply costs of forest chips in each simulation scenario.

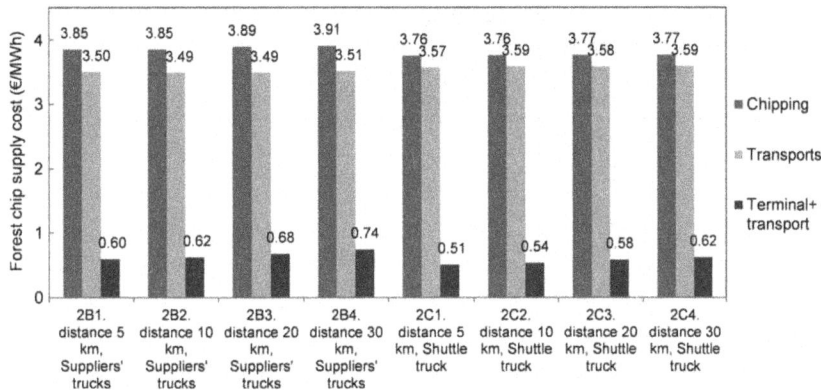

Fig. 10 Supply cost structure of forest chips based on the distance between the terminal and the power plant (5, 10, 20 and 30 km). Transports represent direct transports to the power plant and to the terminal. Suppliers' own chip trucks and a separate shuttle truck were used in the terminal scenarios.

heating season (Table 9). Compared to the simulation scenario of 1500 h of other use, at a 5-km distance, the pure terminal use (0 h of other use) option had 99.2% and 750 h of the other use option had 22.0% higher trucking costs. At a 30-km distance, the respective values were 36.8% and 12.0% for 0 h and 750 h other use of trucks. The same comparison for the total fuel supply costs resulted in 1.5% and 0.3% higher costs at 5 km and 1.1% and 0.4% higher costs at 30 km, respectively.

The cost influence of the changes of terminal-stored forest chips in DML and moisture content was identified in the main scenarios of 3 and 4 (Table 10). If the moisture content of terminal-stored chips was expected to remain the same as the moisture content upon arrival, and, dry matter loss would increase more than in roadside storing as uncomminuted, the cost increase in the total forest chip supply was 1.1% and 2.0%, in 1% per month DML and 2% per month DML scenarios. A change in the dry matter loss had a clear impact on the

energy content of terminal-stored material as well as on the transport efficiency by reducing the load size in MWh. If the moisture content had a 6 per cent point increase during terminal storing, the supply cost increase was 0.5% while DML remained the same and 0.8% if DML increased from 1% per month to 2% per month.

Discussion

Discrete-event simulation was applied in analysing the operational differences of a terminal-aided forest chip supply compared to a conventional direct chip supply in terms of supply costs, the machine utilization of the supply fleet and the supplied energy amount. Weekly work shift arrangements, lay-down seasons, variations in the power plant's daily fuel demand, distance-oriented storage selection, fuel supply control by the power plant's buffer level, as well as detailed machine

Table 9 Influence of the terminal shuttle truck's other use in trucking costs and the total supply cost of forest chips at 5 km and 30 km terminal distances from the power plant

Other use, h	Terminal use, h	Distance from power plant, km	Terminal transport cost related to terminal fuel supply, € per MWh	Terminal transport cost related to total fuel supply, € per MWh	Total fuel supply cost, € per MWh
0	933	5	1.33	0.24	7.95
750	933	5	0.82	0.15	7.86
1500*	933	5	0.67	0.12	7.83
0	1532	30	1.87	0.33	8.07
750	1532	30	1.53	0.27	8.01
1500*	1532	30	1.36	0.24	7.98

*Used in simulation scenarios.

Table 10 The effect of dry matter loss (DML in % per month) and moisture content change in terminal-stored forest chips on the total supply costs, terminal costs and transported energy amount of the shuttle truck (MC, moisture content; pp, percentage point)

	Unit costs for total supply of forest chips				Statistics for terminal and shuttle truck		
	Chipping, € per MWh	Transports – chip trucks, € per MWh	Terminal costs, € per MWh*	Supply costs, € per MWh	Terminal costs, € per MWh†	Total, MWh	Load size, MWh
1A1. BAU	3.91	3.74		7.65			
3C1. MC and DML equal with BAU	3.71	3.53	0.503	7.75	2.64	96 147	125.4
3C2. MC equal, DML 1% per month	3.76	3.57	0.507	7.83	2.79	90 978	119.9
3C3. MC equal, DML 2% per month	3.79	3.60	0.511	7.90	2.96	85 482	113.4
4C1. MC 3 pp lower, DML 1% per month	3.74	3.55	0.506	7.80	2.76	91 901	121.8
4C2. MC 3 pp higher, DML 1% per month	3.76	3.57	0.510	7.84	2.84	89 479	116.5
4C3. MC 6 pp higher, DML 1% per month	3.77	3.59	0.515	7.87	2.90	88 167	111.4
4C4. MC 6 pp higher, DML 2% per month	3.82	3.63	0.516	7.97	3.08	82 531	108.1
4C5. MC 10 pp higher, DML 2% per month	3.83	3.64	0.523	8.00	3.17	80 857	100.5

*Includes terminal investment, terminal operation and terminal transports from terminal to power plant. Costs for total supply.
†Includes terminal investment, terminal operation and terminal transports from terminal to power plant. Costs for terminal supply.

interactions were compiled within the model. This enabled the reliable investigation of the influence of the terminal on the whole forest fuel supply of a CHP plant. While comparing the supply costs to other studies, it is important to note that neither the purchase price of the fuel material, harvesting and extraction costs to roadside storage, fuel supplier entrepreneurs' profit nor the costs of fuel supply management were taken into account in the supply costs.

Windisch *et al.* (2015) studied the influence of the smart selection of roadside storages for the supply of forest chips to power plants in selection terms of

moisture content, the location of roadside storages and the size of roadside storages. However, they did not take into account terminals in the simulations. Eriksson (2016) has also studied supply chains without terminal options by simulations, whereas Karttunen *et al.* (2013) and Korpinen & Aalto (2016) have included terminals into forest fuel supply system analyses. Karttunen *et al.* (2013) conducted a simulation study of intermodal container supply of forest chips including big satellite terminal and train transports for outbound transports.

Terminal options were rather cost-competitive alternatives to the BAU chip supply operated by one shift system. In the study case of a 517 GWh forest chip supply per annum, by introducing a feed-in terminal and a shuttle truck for transports of terminal-stored chips, the total supply cost was 1.4% higher than the BAU fuel supply. In the terminal scenarios, around 20% of the total forest chip use was circulated through the terminal. If operators were available for operating an additional shift during the high fuel demands, the direct chip supply would have been 7% cheaper than the BAU chip supply. Applying a terminal to the fuel supply can be a feasible option, especially in supply situations where an additional and professional workforce is not possible to recruit during the high heating season, which is currently a significant problem in Finland (Väätäinen et al., 2014).

The influences of terminal investment costs with a variation of ±30%, the terminal location variation with distances 5–30 km from the CHP plant and the variation in the annual utilization of the shuttle truck (0, 750, 1500 h of other use) did not have a notable difference in the total supply cost, as changes were 1–2% in the annual supply costs for each case compared to base scenarios of 1B and 1C. Moreover, possible deterioration of the fuel quality in terms of increased dry matter loss or moisture content was detected for the terminal-stored material. As a directional estimate, a 1% per month increase in the DML of terminal-stored chips increased the total supply cost 1%, whereas a 6% increase in the moisture content resulted in a 0.5–1.0% increase depending on the strength of the concurrent change of DML.

The supply cost of the supplementary fuel was defined to be 4.4% higher than the cost of forest chips in the direct supply in the BAU scenario, which could be realistic, especially during the season of the highest fuel demand with typically higher costs of fuel supply. To create successful terminal scenarios with lower costs than the BAU scenario, the breakeven points for the supply cost of the supplementary fuel were a) 8.7 € per MWh for the terminal scenario with the shuttle truck (1C) and b) 9.8 € per MWh for the terminal scenario with the use of chip suppliers' trucks (1B). Furthermore, in terminal scenarios and the direct supply scenario of 1A2, the higher forest chip supply calls for a larger forest chip supply area. This was taken into account in the distance allocation of roadside storages (see Fig. 4).

The productivity of the chipper with 50 solid-m^3 per H$_0$ for logging residues corresponded to the results of Kärhä et al. (2011a,b) and Mola-Yudego et al. (2015) for the corresponding engine power of chippers. The machine utilization percentage of mobile chippers was found to be low compared to earlier studies (Asikainen, 1995; Spinelli & Visser, 2009; Eliasson et al., 2012). The results were quite similar to the findings of Windisch

et al. (2015) by having arguments of including all time elements in daily operation in the simulations. The shares of the relocations and idling are relatively large, as it is in reality in similar forest chip supply units. Relatively high chipping productivity resulted in a high share of idling for chippers, thus questioning the feasibility to use two chip trucks instead of three. The use of two trucks and one mobile chipper is most common in Finland, as it was the de facto for the suppliers operating in that area.

Currently, half (49%) of the truck and trailer units for energy chip transports are <65 tonnes (Venäläinen & Poikela, 2016); however, a trend to increase the total weight is ongoing in Finland (Venäläinen & Poikela, 2016). According to Korpilahti (2015), the transport costs of forest chips for 60- and 68-tonne truck-trailer units were in line with the costs of this study; for 64-tonne trucks in 60–63 km distances, unit costs followed the costs of 60 tonners applied from Korpilahti (2015).

The defined setup at the feed-in terminal was solely based on the storage of comminuted forest biomass. The terminal structure did not include a weighing station or other additional features, such as the chipping of uncomminuted material, which are more common to terminals over 5 hectares (Korpinen & Aalto, 2016). A wheeled loader operator was expected to be available for terminal operations instantly without any delays. In addition, the overheating of chip heaps and risk of fire requires the monitoring of the terminal area manually or by utilizing sensors, thus making it obligatory to have at least one person to monitor and carry out the extinguishing of possible burning zones.

In scenario comparisons, the values for DML in the terminal storing of chipped material had 1–2 per cent unit higher DML compared to the values of uncomminuted logging residues at roadside storages. The latest results for ensuring the loss of dry matter of logging residue chips in terminal storing are 2.7% and 2.2% per month, on average, having storing times of 8 months and 4–8 months, respectively (Raitila & Sikanen, 2016). On the other hand, alternative storing in the form of uncomminuted logging residue material in windrows at roadsides have generated similar DML as well (Jirjis & Lehtikangas, 1993; Nurmi, 1999; Routa et al., 2015). Therefore, the selected DML values for the scenarios can be stated to be rather conservative for the terminal storing conditions of forest chips.

The use of a feed-in terminal as a part of the supply system is a promising option for direct supply, which unbalances the use of workforce and machinery, according to this study. Decreasing fluctuation of the workload also intensifies the use of machines and human resources. Balancing the seasonal fluctuation was stated as one of the most significant factors in developing the

forest biomass transports in the study of Väätäinen *et al.*
(2014). Terminals increase their importance in situations
when road connections and accessibility to roadside
storages drop down. The cost savings and the secured
supply would have been higher for terminal scenarios if
bad road conditions would have been taken into
account in the simulations. There are various possibilities for utilizing terminals in terms of the size, transported material and the operations included in the
terminals.

There are still many improvement possibilities in
biomass supply to energy generation. For the future,
research remarks including a terminal secured supply
model could allow various analysis and scenario comparisons. How is the fuel supply fulfilled in cases of
fuel demand variations between years? What is impact
on the supply when the parameters of the roadside
storages change in terms of storage size, fuel type,
moisture level and location, for example? What is the
influence of different sizes of trucks and mobile chippers on supply performance? How do different operations models affect the supply cost and performance?
What is the effect of road trafficability and bad road
seasons on the fuel supply in scenarios of direct and
terminal-supported supplies? What is the role of biomass terminals for the drying of fuel by introducing
different methods for enhancing the drying such as
covering the stored biomass or having larger terminal
area?

Acknowledgements

The authors gratefully acknowledge financial support from the
Strategic Research Consortium Project FORBIO (3500010200),
BEST Programme and Wood Procurement Logistics -Project.
We thank Dr. Johannes Windisch for valuable comments on
earlier simulation model drafts. Thanks are also due to the Lan-
ner team for their technical support.

References

Alakangas E (2005) Properties of wood fuels used in Finland. Technical Research
    Centre of Finland, VTT Processes, Project report PRO2/P2030/05. 90 p. + 10 p
Asikainen A (1995) Discrete-event simulation of mechanized wood-harvesting sys-
    tems. Academic dissertation. Research notes 38. Faculty of Forestry, University of
    Joensuu, PP. 86.
Björklund L (1982) Lagring av bränsleflis I fraktionen 25–30 mm [Storage of fuel-
    wood chips in fraction 25–30 mm]. *Sveriges Lantbruksuniversitet. Institutionen för
    Virkeslära. Uppsatser*, **115**, 26. (In Swedish with English summary).
Eliasson L, Granlund P, von Hofsten H, Björheden R (2012) Studie av en last-
    bilsmonterad kross – CBI 5800 study of a truck-mounted CBI 5800 grinder. *Arbets-
    srapport Från Skogforsk nr*, **775**, 16. [in Swedish with English abstract].
Eriksson A (2016) Improving the efficiency of forest fuel supply chains. Diss. (sam-
    manfattning/summary) Uppsala: Sveriges lantbruksuniv. *Acta Universitatis Agri-
    culturae Sueciae*, **101**, 1652–6880.
Hakkila P (1985) Pienpuun ja metsätähteen korjuu energiakäyttöön. (Harvesting of
    small-wood and forest residues for energy use). In: Metsäenergian mahdollisu-
    udet Suomessa, PERA-projektin väliraportti. Folia Forestalia 624. Finnish Forest
    Research Institute. PP. 76 (In Finnish with English summary).

Hakkila P (2004) Developing technology for large-scale production of forest chips.
    Wood energy technology programme 1999–2003, Technology programme report
    6/2004. Final report. PP. 99.
Impola R, Tiihonen I (2011) Terminaalikäsikirja (Terminal guidebook). VTT-R-08634-
    11. Jyväskylä: VTT Technical research centre of Finland. PP. 38. (In Finnish).
Jirjis R (2005) Effects of particle size and pile height on storage and fuel quality of
    comminuted *Salix viminalis*. *Biomass and Bioenergy*, **28**, 193–201.
Jirjis R, Lehtikangas P (1993) Branslekvalitet och substansforluster vid valtlagring av
    hyggesrester (Fuel quality and dry matter losses during storing of logging resi-
    dues in windrows). Sveriges Lantbruksuniversitet. Institutionen for virkeslara.
    Rapport nr. 236. Sweden.
Kanzian C, Holzleitner F, Stampfer K, Ashton S (2009) Regional energy wood logis-
    tics – optimizing local fuel supply. *Silva Fennica*, **43**, 113–128.
Kärhä K (2011) Industrial supply chains and production machinery of forest chips in
    Finland. *Biomass and Bioenergy*, **35**, 3404–3413.
Kärhä K, Hautala A, Mutikainen A (2011a) Jenz HEM 581 DQ hakkuutähteiden ja
    pienpuun tienvarsihaketuksessa. Metsätehon tuloskalvosarja 5/2011. (In Finnish).
Kärhä K, Hautala A, Mutikainen A (2011b) Heinola 1310 ES hakkuutähteiden ja
    pienpuun tienvarsihaketuksessa. Metsätehon tuloskalvosarja 9/2011. (In Finnish).
Karttunen K, Lättilä L, Korpinen O-J, Ranta T (2013) Cost-efficiency of intermodal
    container supply chain for forest chips. *Silva Fennica*, **47**, 24.
Kons K (2015) Forest biomass terminal properties and activities. Licentiate Thesis.
    Swedish University of Acricultural Sciences, Umeå.
Kons K, Bergström D, Eriksson U, Athanassiadis D, Nordfjell T (2014) Characteris-
    tics of Swedish forest biomass terminals for energy. *International Journal of Forest
    Engineering*, **25**, 238–246.
Korpilahti A (2015) Bigger vehicles to improve forest energy transport. Metsätehon
    tuloskalvosarja 2/2015. 33 slides.
Korpinen O-J, Aalto M (2016) Assessing the performance of biomass terminals and
    supply systems with simulation approaches. In: *Energy Biomass Supply Chain Con-
    cepts Including Terminals* (eds Sikanen L, Korpinen O-J, Tornberg J, Saarentaus T,
    Leppänen K, Jahkonen M), pp. 18–25. Luke reports. Luonnonvarakeskus (Luke),
    Helsinki.
Kubler H (1987) Heat generation process as cause of spontaneous ignition in forest
    products. *Forest Products Abstracts*, **10**, 299–322.
Laitila J, Nuutinen Y (2015) Efficiency of integrated grinding and screening of stump
    wood for fuel at roadside landing with a low-speed double-shaft grinder and a
    star screen. *Croatian Journal of Forest Engineering*, **36**, 19–32.
Malinen J, Nousiainen V, Palojärvi K, Palander T (2014) Prospects and challenges of
    timber trucking in a changing operational environment in Finland. *Croatian Jour-
    nal of Forest Engineering*, **35**, 91–100.
Mola-Yudego B, Picchi G, Röser D, Spinelli R (2015) Assessing chipper productivity
    and operator effects in forest biomass operations. *Silva Fennica*, **49**, 1–14.
Nurmi J (1990) Polttohakkeen varastointi suurissa aumoissa [Longterm storage of
    fuel chips in large piles]. *Folia Forestalia*, **767**, 18. (In Finnish with English sum-
    mary).
Nurmi J (1999) The storage of logging residue for fuel. *Biomass and Bioenergy*, **17**, 41–
    47.
Nurminen T, Heinonen J (2007) Characteristics and time consumption of timber
    trucking in Finland. *Silva Fennica*, **41**, 471–487.
Palander T, Voutilainen J (2013) Modelling fuel terminals for supplying a combined
    heat and power (CHP) plant with forest biomass in Finland. *Biosystems Engineer-
    ing*, **114**, 135–145.
Raitila J, Korpinen O-J (2016) The concept of terminals in the supply chain. In:
    Energy Biomass Supply Chain Concepts Including Terminals (eds Sikanen L, Kor-
    pinen O-J, Tornberg J, Saarentaus T, Leppänen K, Jahkonen M), pp. 6–7. Luke
    reports. Luonnonvarakeskus (Luke), Helsinki.
Raitila J, Sikanen L (2016) Dry matter losses. In: *Energy Biomass Supply Chain Concepts
    Including Terminals* (eds Sikanen L, Korpinen O-J, Tornberg J, Saarentaus T,
    Leppänen K, Jahkonen M), pp. 63–65. Luke reports. Luonnonvarakeskus (Luke),
    Helsinki.
Raitila J, Virkkunen M (2016) Terminals in Sweden and Finland. In: *Energy Biomass
    Supply Chain Concepts Including Terminals* (eds Sikanen L, Korpinen O-J, Tornberg
    J, Saarentaus T, Leppänen K, Jahkonen M), pp. 8–10. Luke reports. Luonnonvara-
    keskus (Luke), Helsinki.
Routa J, Asikainen A, Björheden R, Laitila J, Röser D (2013) Forest energy procure-
    ment - state of the art in Finland and Sweden. *WIREs Energy and Environment*, **2**,
    602–613.
Routa J, Kolström M, Ruotsalainen J, Sikanen L (2015) Precision measurement of for-
    est harvesting residue moisture change and dry matter losses by constant weight
    monitoring. *International Journal of Forest Engineering*, **26**, 71–83.

Spinelli R, Visser R (2009) Analyzing and estimating delays in wood chipping operations. *Biomass Bioenergy*, **33**, 429–433.

Stampfer K, Kanzian C (2006) Current state and development possibilities of wood achip supply chains in Austria. *Croatian Journal of Forest Engineering*, **27** , 135–145.

Strandström M (2016) Metsähakkeen tuotantoketjut Suomessa vuonna 2015. Metsätehon tuloskalvosarja 7/2016. 20 slides. (In Finnish).

Väätäinen K, Anttila P, Laitila J, Nuutinen Y, Asikainen A (2014) Aines- ja energiapuun kaukokuljetuksen tulevaisuuden haasteet ja teknologiat. *Metlan Työraportteja*, **291**, 31. (In Finnish with English summary).

Venäläinen P, Poikela A (2016) Scenarios of energy wood transport fleet and backhaulage. Metsätehon tuloskalvosarja 2b/2016. 66 slides.

Virkkunen M, Kari M, Hankalin V, Nummelin J (2015) Solid biomass fuel terminal concepts and a cost analysis of a satellite terminal concept. VTT.69 p.

Virkkunen M, Raitila J, Korpinen O-J (2016) Cost analysis of a satellite terminal for forest fuel supply in Finland. *Scandinavian Journal of Forest Research*, **31**, 175–182.

Windisch J, Väätäinen K, Anttila P, Nivala M, Laitila J, Asikainen A, Sikanen L (2015) Discrete-event simulation of an information-based raw material allocation process for increasing the efficiency of an energy wood supply chain. *Applied Energy*, **149**, 315–325.

Wolfsmayr UJ, Rauch P (2014) The primary forest fuel supply chain: a literature review. *Biomass and Bioenergy*, **60**, 203–221.

7

Time-dependent climate impact and energy efficiency of combined heat and power production from short-rotation coppice willow using pyrolysis or direct combustion

NICLAS ERICSSON[1] (iD), CECILIA SUNDBERG[1,2], ÅKE NORDBERG[1], SERINA AHLGREN[1] and PER-ANDERS HANSSON[1]

[1]*Department of Energy and Technology, Swedish University of Agricultural Sciences (SLU), SE 750 07, Uppsala, Sweden,* [2]*KTH Royal Institute of Technology, School of Architecture and the Built Environment, Department of Sustainable Development, Environmental Science and Engineering, Unit of Industrial Ecology, Stockholm, Sweden*

Abstract

A life cycle assessment of a Swedish short-rotation coppice willow bioenergy system generating electricity and heat was performed to investigate how the energy efficiency and time-dependent climate impact were affected when the feedstock was converted into bio-oil and char before generating electricity and heat, compared with being combusted directly. The study also investigated how the climate impact was affected when part of the char was applied to soil as biochar to act as a carbon sequestration agent and potential soil improver. The energy efficiencies were calculated separately for electricity and heat as the energy ratios between the amount of energy service delivered by the system compared to the amount of external energy inputs used in each scenario after having allocated the primary energy related to the inputs between the two energy services. The energy in the feedstock was not included in the external energy inputs. Direct combustion had the highest energy efficiency. It had energy ratios of 10 and 36 for electricity and heat, respectively. The least energy-efficient scenario was the pyrolysis scenario where biochar was applied to soils. It had energy ratios of 4 and 12 for electricity and heat, respectively. The results showed that pyrolysis with carbon sequestration might be an option to counteract the current trend in global warming. The pyrolysis system with soil application of the biochar removed the largest amount of CO_2 from the atmosphere. However, compared with the direct combustion scenario, the climate change mitigation potential depended on the energy system to which the bioenergy system delivered its energy services. A system expansion showed that direct combustion had the highest climate change mitigation potential when coal or natural gas were used as external energy sources to compensate for the lower energy efficiency of the pyrolysis scenario.

Keywords: biochar, climate impact metrics, land use change, LCA, pyrolysis, Salix, soil organic carbon, SRC, willow

Introduction

Climate change is a natural phenomenon that can be observed in paleoclimate records stretching back millions of years in time (Hansen and Sato, 2012). However, the rate at which the temperature has changed since the mid-20th century is unprecedented in modern human history. The major part of this change can be attributed to increased concentrations of carbon dioxide (CO_2) and other greenhouse gases (GHG) in the atmosphere. The use of fossil resources has historically contributed approximately 75% of the anthropogenic emissions of CO_2, while land use change is responsible for the remainder (Denman *et al.*, 2007). Reducing the use of fossil resources and reversing the trend of carbon (C) losses from soils are two important steps to prevent further temperature increase.

Biomass can be used to replace fossil fuels and is a major source of renewable energy in Sweden, corresponding to 23% of the total energy supplied to the Swedish energy system in 2013 (Swedish Energy Agency, 2015). It can be used as fuel in combined heat and power (CHP) plants for the simultaneous generation of electricity and heat. Cogeneration of electricity and heat is an energy-efficient use of the feedstock in the Nordic climate due to the relatively high heat demand.

Biomass can be pretreated in a decentralized way using pyrolysis to improve its storage and handling properties. Pyrolysis is a thermochemical process in which the biomass is converted into bio-oil, noncondensable gases and char at temperatures between 300 and 800 °C (Brown, 2011). All three product fractions can be used for energy service generation, both on site or at a centralized CHP. Char from pyrolysis has also received attention due to its agronomic properties (Sohi *et al.*, 2010) and recalcitrance when applied to soils (Liang *et al.*, 2008). In this context, it is commonly

Correspondence: Niclas Ericsson
e-mail: niclas.ericsson@slu.se

referred to as biochar. Biochar can be a potential soil improver and C sequestration agent. Application of biochar to arable soils has been suggested as a way to counteract the effects of GHG emissions (Gaunt and Lehmann, 2008). There is, however, a trade-off between the use of char for energy service generation and for C sequestration, as biochar (Pourhashem et al., 2013). Furthermore, the biomass can also be used as raw material for nonenergy products where it can replace fossil sources (Gallezot, 2012). This creates a situation with multiple trade-offs in which the function of the product cannot always be used to compare the climate impact of land use systems.

The advantages of pretreating biomass with pyrolysis could benefit short-rotation coppice (SRC) willow bioenergy systems, as the raw biomass is wet, bulky and biologically active (Noll and Jirjis, 2012). Willow is a fast-growing, high-yielding, lignocellulosic plant species that can be readily used for cofiring with other biomass in existing CHP plants. It has been established on more than 14 000 ha of land in Sweden since the 1980s. In Sweden, large agricultural areas are not being used for productive purposes (SJV, 2008). A large share of these are unused cropland and excess grassland. Using part of this land for energy crop production could offer both an extra source of income to farmers and increase the share of renewable electricity and heat in the Swedish energy system.

The environmental sustainability of bioenergy systems is often assessed using life cycle assessment (LCA) methodology (ISO 14040, 2006; ISO 14044, 2006). The climate impact is commonly characterized in LCA using the global warming potential (GWP) (IPCC, 1991). Common LCA practice is to multiply the net life cycle emission of different GHGs with their respective characterization factor (CF), after which they are summed to calculate the total climate impact of the system. To derive the CFs, it is necessary to choose a time horizon over which the impacts are being assessed. There are two issues related to time inherent in this way of determining the climate impact: (i) it disregards when the impact actually takes place, treating all emissions as if they were emitted at the beginning of the assessment period (Peters et al., 2011), and (ii) the choice of time horizon is inherently subjective and effects the relative weight between different GHGs (Fuglestvedt et al., 2010). As a direct consequence of the modelling choice, a CF derived this way cannot possibly capture any impact from time-variable GHG fluxes where the net emission is 0 (Cherubini et al., 2011). The timing of emission may affect the evaluation of bioenergy systems from a climate impact perspective (e.g. Kendall et al., 2009; O'Hare et al., 2009; Levasseur et al., 2010).

There are several approaches for including timing of emissions in bioenergy scenarios in LCAs (e.g. Zetterberg et al., 2004; Levasseur et al., 2010; Kendall, 2012; Porsö et al., 2016; Pourhashem et al., 2016). All approaches include a characterization model which converts the emissions to climate impacts. To be able to describe time-dependent impacts, it is necessary to record the timing of emissions in a time-distributed inventory. The impacts can then be described either by introducing a time-dependent weighting function in the elaboration of the CFs (e.g. Courchesne et al., 2010; Levasseur et al., 2010; Kendall, 2012) or through the use of a time-dependent indicator (e.g. Zetterberg et al., 2004; Peters et al., 2011; Pourhashem et al., 2016). The use of an absolute, instantaneous and time-dependent indicator can compliment the use of CFs in LCA as they contribute to information on both timing and rate of change, as well as provide a visual representation of how the climate impact evolves over time.

A common indicator to describe the time-dependent climate impact is the radiative forcing (RF), which describes a perturbation in the earth's energy balance due to the emission or removal of a climate forcer (Fuglestvedt et al., 2003). The global mean surface temperature change (ΔT_S) can also be used for this purpose. Peters et al. (2011) compared the use of RF to ΔT_S as time-dependent indicators in an LCA of different transportation modes, showing differences in the timing and rate of change between the two impact indicators. These differences are due to the inertia of the climate system which delays the response in ΔT_S from a perturbation of the radiative balance when heat is being exchanged between the atmosphere and terrestrial sinks, especially the deep oceans (Berntsen and Fuglestvedt, 2008).

A few studies have used ΔT_S as an indicator for the time-dependent climate impacts of bioenergy systems (e.g. Cherubini et al., 2013; Ericsson et al., 2013, 2014; Hammar et al., 2014, 2015; Giuntoli et al., 2015; Ortiz et al., 2016; Porsö et al., 2016), but to our knowledge, no such study has been presented on pyrolysis-based bioenergy systems. However, several authors have investigated the climate impact and energy efficiency of pyrolysis-based bioenergy systems from a life cycle perspective based on GWP (Gaunt and Lehmann, 2008; Roberts et al., 2010; Hammond et al., 2011; Hanandeh, 2012; Ibarrola et al., 2012; Wang et al., 2013; Peters et al., 2015; Thornley et al., 2015). The biochar has been observed to have a large impact on the results in these studies.

Prior research (Ericsson et al., 2014) examined the time-dependent impacts from C returned to the field in an anaerobic digestion-based bioenergy system. They were observed to be large and act at a different time scale than the impacts of other biogenic C stock changes and GHGs in the system. Similar effects can be expected in pyrolysis-based bioenergy systems where biochar is applied to soils. Furthermore, the biochar-energy trade-off can be expected to vary with time as the biochar will slowly decay once applied to soils.

The aim of this study was to improve the understanding of the time-dependent climate impact and its relation to the energy efficiency in willow-based bioenergy systems generating electricity and heat, with and without the production of biochar for C sequestration in soils. To achieve this, a case study focusing on a Swedish willow plantation used for electricity and heat generation was conducted and the time-dependent climate impact was determined using ΔT_S as an instantaneous and time-variable indicator. Both decentralized pretreatment with pyrolysis before energy service generation and direct combustion of the willow feedstock were considered. The char from the pyrolysis process was used either for energy service generation or as biochar through soil application. The multiple use of biomass in nonenergetic sectors was also considered by including two different functional units in the assessment: one output based that permits a fair assessment of energy systems and one input based that enables the comparison of systems having products that cannot be compared based on functional equivalence.

Materials and methods

Scenarios and methodology

Life cycle assessment (ISO 14040, 2006; ISO 14044, 2006) was used to analyse the climate impact and energy efficiency of a bioenergy system generating electricity and heat from SRC willow feedstock. The electricity and heat were assumed to be fed into the electric grid and a local district heating (DH) system, respectively.

The impact on climate change was assessed with time-dependent climate impact methodology (Ericsson et al., 2013), using the contribution to $\Delta T_S(n)$ from each scenario as an indicator of the climate impact.

A 60 hectare (ha) SRC willow plantation was modelled to determine the annual net GHG emissions from the feedstock production system. All emissions related to the production of the biomass, feedstock conversion, transportation and final energy service generation were recorded in a time-distributed life cycle inventory. In this inventory, the net emissions of GHGs were recorded for each year of the study period. The change in atmospheric CO_2 levels due to carbon stock changes in the live biomass, the SOC and the biochar pools was included in the assessment. Carbon taken up by growing biomass was recalculated to CO_2 and recorded as negative emissions in the inventory.

Two feedstock conversion scenarios, pyrolysis and direct combustion, were compared with a reference scenario where natural gas was used in a large-scale CHP to generate the same amount of electricity and heat as in the most energy-efficient bioenergy scenario (Fig. 1). The pyrolysis scenario consisted of

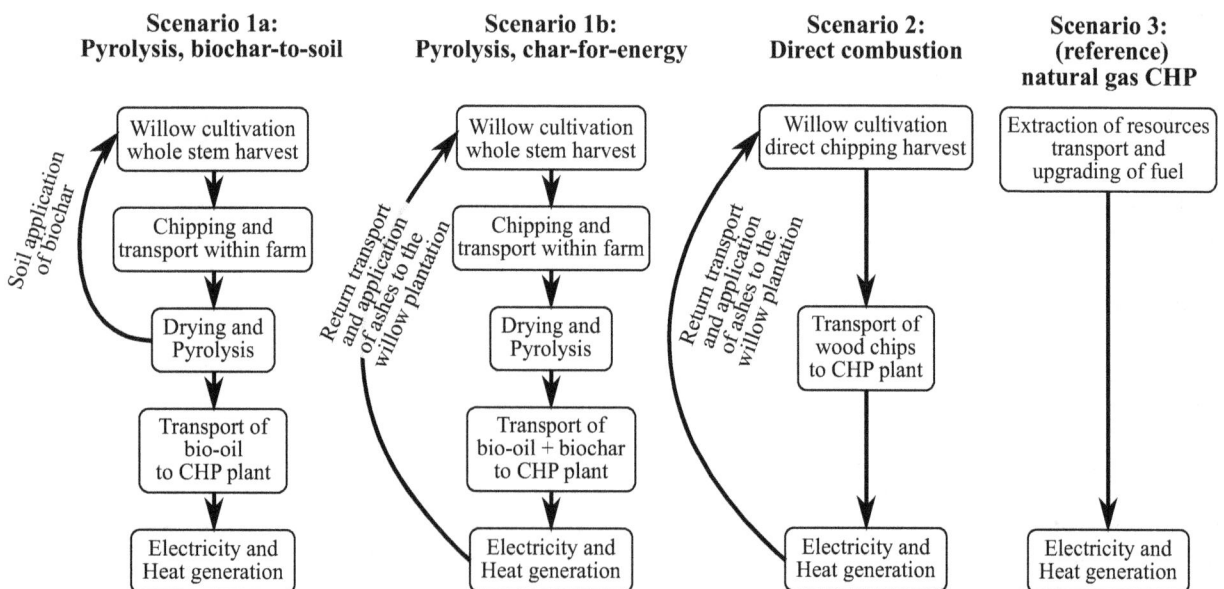

Fig. 1 The different feedstock conversion scenarios. In scenario 1, two ways of using the char were compared: (a) soil application (biochar) and (b) energy service generation. In scenario 2, the raw willow chips were used for energy service generation directly in a large-scale CHP. In scenario 3, heat and electricity were generated in a combined cycle natural gas CHP plant. This scenario was used as a reference to the energy services delivered in scenarios 1 and 2.

two cases, where alternative end use of the produced char was considered: using the char as biochar, i.e. applying it to the soil where it acted as a carbon sequestration agent, or using the char to generate energy services in a large-scale CHP.

- Scenario 1: The willow feedstock was pyrolyzed at the farm to generate bio-oil and char. The bio-oil fraction was then transported to a large-scale CHP, where it was co-combusted with heavy fuel oil to generate electricity and heat. The two end uses of the char were as follows:

  (a) Biochar-to-soil: The biochar was returned to the soil of the willow plantation to act as a carbon sequestration agent and potential soil improver.

  (b) Char-for-energy: The char was transported to a large-scale CHP plant where it was co-combusted with biomass in a biomass boiler to generate electricity and heat.

- Scenario 2: The willow feedstock was transported, without pretreatment to a large-scale CHP plant, where it was co-combusted with biomass of other origin to generate electricity and heat.

- Scenario 3: In the reference scenario, the same amount of electricity and heat as in the most energy-efficient scenario was generated in a combined cycle natural gas (NG)-fed CHP plant.

The CHP plants in scenario 1 and scenario 2 were assumed to use back-pressure steam turbines, having a power-to-heat ratio of 0.45, while the combined cycle CHP plant in scenario 3 was assumed to have a power-to-heat ratio of 0.95 (EU, 2012).

Scope. The assessment included the three major GHGs contributing to global warming: CO_2, nitrous oxide (N_2O) and methane (CH_4).

The primary energy use and emissions from all stages of production were included within the system boundaries. Production stages included were cultivation and harvesting of the willow feedstock, preprocessing and pyrolysis, transportation of the raw biomass and pyrolysis products as well as electricity and heat generation. The return transport of bottom ashes generated at the CHP plant as well as the biochar produced in the pyrolysis plant and their application to the soil were also included within the system boundaries. Handling of the fly ash was not included in the inventory as it cannot be used on productive land due to its possibly high heavy metal content. The handling and deposition of the relatively small amount of fly ash and flue gas cleaning residue generated were assumed not to contribute significantly to the climate impact and energy use in the studied scenarios.

Activities and losses associated with the distribution and use of the generated electricity and heat, after being fed into the electric grid and DH distribution system, were not included within the system boundaries.

The construction and decommissioning of the capital goods were not included in any of the scenarios as it has been shown in environmental product declarations (EPD) that infrastructure is responsible for <2% of the GHG emissions associated

with the energy services generated in similar CHPs (EPD, 2013, 2015), without taking biogenic C stock changes into account. The authors are not aware of any EPDs on pyrolysis systems, but previous studies have shown that contribution from the biogenic C stock changes dominates the climate impact in similar SRC bioenergy systems (Ericsson *et al.*, 2013, 2014). It was therefore assumed that the contribution from the capital goods in this case would not significantly change the relative outcome of the comparison between the scenarios.

Electricity used for preprocessing of the biomass and operation of the pyrolysis plant was assumed to be taken from the grid. Net energy delivered in scenario 1 was calculated by subtracting this amount of electricity, including distribution losses of an additional 7.5% for the used electricity (Gode *et al.*, 2011), from the gross electricity delivered to the grid.

Functional units. The climate impact from different scenarios might depend on the function of the system and the consequent functional unit (FU) used. The goal of the LCA was twofold: on the one hand, it aims to guide decisions on the use of willow biomass in order to mitigate climate change, and on the other hand, it aims to provide information on climate impacts as a consequence of the use of land.

In this study, the SRC willow-based bioenergy system was considered to have two functions. Generation of electricity and heat was the main function of the system. It could also serve the important function of mitigating climate impact by representing an optional land management that may potentially act as a carbon sink, independent of its function as an energy system. This later function is rather related to the system using land, where energy generation is but one of many optional uses.

Using an energy service as the FU describes the relative impact of the electricity or heat compared with that of other sources. It is a relevant FU when assessing the impacts caused by consumption of the delivered energy. The relative land use efficiency of the system can be described by relating the results to the area used to produce the willow feedstock for the electricity and heat generation. Such a FU may be better suited when land is a restricted resource, and the focus is to assess the climate and energy benefits from different types of land use.

When using an energy service as the FU, the amount of energy delivered from the system has to be equal in all scenarios. The energy delivered can be made equal in all scenarios using system expansion. System expansion is, however, not performed when land use is the FU, as the main function considered in this case is not that of generating energy, but the use of land. The amount of land used therefore has to be the same in all scenarios, while the amount of delivered energy might be different. This may lead to differences in the relative climate change mitigation potential between scenarios when different FUs are being used.

As neither of these two functions can be said to be more important *a priori* when assessing the climate impact and energy efficiency of a system, both the area and the energy services were used as FUs in this study. One hectare of willow cultivation was chosen to represent the area, while 1 kWh of electricity delivered to the grid and 1 MJ of heat delivered to the local district heating (DH) system were used to represent the energy services.

External energy service generation. Scenarios 1 and 2 were expected to deliver different amounts of electricity and heat from the same amount of feedstock due to the extra conversion step introduced in the pyrolysis process. The climate impact could therefore not be directly compared without taking the energy efficiency into account.

The reference flows were made equal in all scenarios using true system expansion when assessing the climate impact per kWh of electricity and per MJ of heat. The amounts of electricity and heat per ha of land were made equal in all scenarios by adding electricity and heat to the less energy-efficient scenarios. These scenarios were assigned emissions and primary energy use from external energy service generation, taking place outside the system (Fig. 2), here after referred to as external energy sources.

This method of system expansion differs from that recommended in the ISO standard (ISO 14044, 2006), which subtracts avoided emissions from the more energy-efficient system. The choice of method does not affect the relative comparison between different scenarios in a specific study. The only visible difference is that the absolute values appear smaller for all scenarios when the ISO method is used, as it credits the more energy-efficient system with avoided emissions rather than penalizing the less efficient system with indirect emissions.

The effect of the choice of energy sources on the climate impact and energy efficiency of the electricity and heat was investigated in scenario 1. The external energy sources

compared were cogenerated electricity and heat from NG- and hard coal (HC)-fuelled CHP plants, electricity from wind power (WP) and heat from household waste (HHW).

All CHP plants were assumed to have a total energy efficiency of 90% (lower heating value, LHV). The NG plant was assumed to use the same technology as in scenario 3. The HC plant was assumed to use a back-pressure steam turbine with a power-to-heat ratio of 0.45 (EU, 2012). The HHW boiler was assumed to have a heat efficiency of 0.8. The primary energy factors, energy ratios and GWP_{100} of the external energy sources can be found in the (Appendix S1: Table S1).

Allocation between electricity and heat. Electricity and heat represent two different types of energy services which can be generated independently of each other and fulfil different functions. In this study, the emissions and primary energy (PE) were therefore allocated between them to make the results comparable to other studies of electricity and/or heat generation.

The separate production reference method was used to perform the allocation (Beretta *et al.*, 2012; Swedenergy, 2012) Eqn (1). According to this method, the allocation (α) between electricity and heat is different for different fuels (i). It is affected by the relative amount of electricity and heat generated at the CHP plant (E_x, x:electricity or heat), without including flue gas condensation or internal electricity demand. It is also affected by the efficiency by which electricity and heat can be generated separately, using the same fuel, at stand-alone power-only (η_{el_i}) and heat-only plants (η_{heat_i}), optimized to generate as much electricity and heat as possible, respectively.

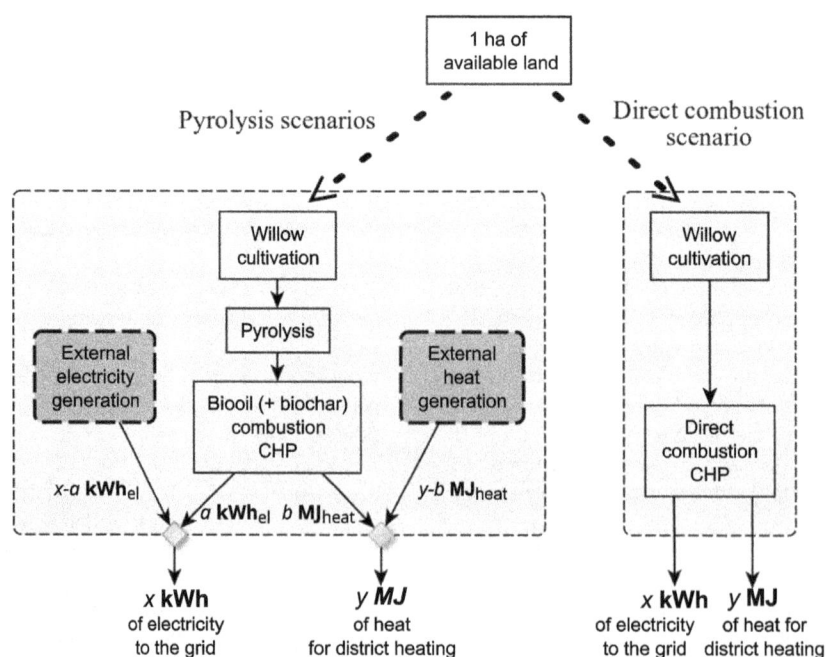

Fig. 2 Energy service generation per ha was made equal in all scenarios using system expansion. Electricity and heat generated by external sources were added to scenario 1. It was assigned the primary energy and GHG emissions associated with this energy. Dashed dark boxes represent the external sources of electricity and heat.

$$\alpha_{x_i} = \frac{\frac{E_x}{\eta_{x_i}}}{\frac{E_{el}}{\eta_{el_i}} + \frac{E_{heat}}{\eta_{heat_i}}} \qquad (1)$$

Harmonized efficiency reference values (η_{el_i} & η_{heat_i}) for the separate production of electricity and heat from representative fuels were taken from EU (2011). Bio-oil and biochar are not present among the representative fuels. In this study, the harmonized efficiency reference value for wood fuels was used to represent the direct combustion of willow chips, the reference value for biofuels was used to represent the bio-oil, and the reference value for lignite was used to represent the biochar. In all cases, values for heat generation using steam/hot water were used. The harmonized efficiency reference values and allocation factors used in this study can be found in the (Appendix S2: Tables S2 and S3).

System description

Cultivation system. The SRC willow plantation was assumed to be established on set-aside agricultural land in central Sweden that had been under fallow for a period of 20 years. The rotation period of the willow plantation was 25 years, including eight subsequent 3-year coppicing cycles and 1 year of annual crops between each rotation to reduce pressure from perennial weeds. The activities associated with this year were not included in the bioenergy scenarios, as nonbioenergy goods were produced. One-third of the total area was established each year during the first 3 years, giving the total study period a length of 53 years. All coppicing cycles, except the first, were expected to yield 30 t dry matter (DM) per ha at harvest. This corresponds to an annual growth rate of 10 t of DM per ha. The yield of the first cutting cycle was reduced by one-third of full growth rate.

Two rotations were modelled, starting from ploughing of the soil to prepare for the willow in the year before planting the seedlings. Weed control was carried out prior to ploughing and several times during the establishment year, both mechanically and through the use of herbicides. Each rotation was terminated by applying herbicides to kill off remaining plants before cutting up the roots using a rotary cultivator. Shallow soil preparation was performed prior to the second rotation using a disc harrow. Seedling production and planting were also included in the cultivation activities.

The amount of nutrients supplied was identical in all three scenarios. Fertilizer was assumed to be applied to achieve recommended levels of nutrients (Aronsson and Rosenqvist, 2011). The amount of fertilizer was adjusted for the level of nutrients returned to the field with the biochar and ash in scenarios 1 and 2. The fertilizer levels, application method and amount of biochar and ash can be found in the (Appendix S3).

Harvest and postharvest operations were included. In scenario 1a and 1b, whole stem harvest was used to facilitate storage of the biomass at the field side without excessive loss of dry matter while supplying the pyrolysis unit with a steady supply of biomass throughout the year. In scenario 2, a direct chipping harvester was used as the biomass was removed from the field immediately and combusted within 3 months after harvest. The bio-oil in scenario 1a and 1b and the char used for

energy generation in scenario 1b, as well as the fresh willow chips in scenario 2, were transported to a large-scale CHP to generate electricity and heat. A more detailed description of the processes in the harvest and postharvest system can be found in the (Appendix S3).

Pyrolysis step. In scenario 1a and 1b, the feedstock was pyrolyzed at the farm to generate bio-oil and char. The pyrolysis system and the mass and energy balance modelling are only briefly described here. A more detailed description is given in the (Appendices S4 and S5, Eqn (S1)).

An auger reactor located at the farm was assumed to be used to pyrolyze the willow biomass, which was chipped at the field side using a mobile wood chipper before transporting the willow from the field to the pyrolysis unit.

The biomass was dried to a moisture content of 10% (dry basis) before entering the auger reactor using recovered heat from flue gases and process streams.

The bio-oil was recovered in a condenser following the reactor, and the char was collected and quenched at the end of the auger rector. Ten per cent of the tar fraction in the bio-oil (non-water fraction) was assumed to pass through the condenser in aerosol form and be combusted together with the noncondensable (NC) gases to provide the heat needed for pyrolysis and drying of the feedstock prior to entering the reactor. In this study, an empirical model based on a large number of slow and fast pyrolysis studies was used to calculate the product distribution (Neves *et al.*, 2011). The model took the temperature of pyrolysis and the elemental composition of the biomass as input. The auger was modelled to process 220 kg of DM per h at a temperature of 525 °C with a solid retention time in the hot zone of 3 min. More details on the composition of the input used in the mass balance model, as well as the composition and energy content of the pyrolysis products, can be found in the (Tables S4–S8).

Energy service generation. In all bioenergy scenarios, electricity and heat were generated in a central CHP plant located 30 km from the farm. All scenarios, including the reference case, had an overall conversion efficiency from fuel entering the CHP to electricity and heat of 90% (LHV).

As flue gases from biomass, bio-oil and biochar contain a considerable amount of steam, the CHPs in scenarios 1 and 2 were assumed to use flue gas condensation technology, recovering 90% of the potential heat in the flue gases. No flue gas condensation was assumed in the reference case as the excess air used in a NG combined cycle lowers the dew point of the exhaust gas to below the return temperature of the district heating network, making flue gas condensation for this purpose implausible.

Data on primary energy use and emissions associated with the extraction, distribution and use of NG in the reference case were taken from Gode *et al.* (2011). Only the primary energy use and emissions related to the use of willow in co-combustion were included in the bioenergy scenarios as the aim of this study was to investigate the contribution to the climate impact from the willow biomass, not the co-combusted fuels, in the bioenergy system.

Energy efficiency

The external energy ratio (ER) (Murphy *et al.*, 2011) was used to assess the energy efficiency in this study. To avoid confusion with the external energy sources used in the system expansion, the ER is simply referred to as the energy ratio for the remainder of this study. It was defined as the ratio between the energy delivered and all external energy inputs used to generate this energy Eqn (2). Note that the energy in the fuel is not included in this definition of energy inputs, thereby excluding losses taking place in the energy conversion process at the CHP. These losses do, however, indirectly influence the ER by decreasing the amount of useful electricity and heat delivered to the grid.

$$ER = \frac{\text{Delivered energy}}{\text{Energy input}} \qquad (2)$$

The ER indicates the amount of useful electricity and heat delivered in relation to the external energy inputs used in the production and delivery of the fuels to the CHP. It was calculated separately for electricity and heat after allocating the energy inputs for each scenario between the two energy services.

Greenhouse gas and carbon fluxes

Emissions from the technical system and non-C emissions. The GHG emissions from the technosphere were assigned to the year in which activities took place. Upstream emissions were assigned to the same year as the main activity.

Primary energy use and emissions from return transport and application of the bottom ash, as well as the handling of the biochar, were assigned to the year following harvest.

Direct and indirect N_2O emissions were calculated using emission factors from the IPCC guidelines for national greenhouse gas inventories (IPCC, 2006). According to these, 1% of the N in the applied fertilizer and in the decomposing biomass was assumed to be converted to N_2O. Indirect emissions were calculated assuming that 30% of the N fertilizer applied was leached and that 0.75% of this fraction was converted to N_2O. Direct and indirect N_2O emissions were assigned to the year in which the fertilizer was applied and the biomass litter was generated.

Biogenic carbon fluxes. The C fluxes between the atmosphere and the biosphere were calculated for three different compartments: live biomass, SOC and biochar. The net annual flux for each year of the study period was based on the net annual C stock change and was calculated using different approaches for each compartment. Only C stock changes occurring during the study period were included in the assessment. This was carried out as these emissions depend on the decisions made by the farmer choosing to cultivate the willow. All emissions taking place prior to and after the end of the study period were consequently ascribed to the preceding and subsequent cultivation systems.

Live biomass. Net annual C stock change in the live biomass was calculated based on the expected yield, interannual growth rate and C allocation pattern of the willow (Rytter, 2001). The

C stored in the stems was considered to be combusted within a short time frame and returned to the atmosphere in the year of harvest.

SOC. The annual SOC stock changes were modelled using the ICBMr model (Andrén *et al.*, 2004), adjusted for a SRC willow system (Ericsson *et al.*, 2013; Fig. 3).

ICBMr is conceptually divided into two pools, a young (*Y*) and an old (*O*) pool. Fresh input (*i*) enters a *Y* pool. A fraction of the young pool is broken down every year. Part of this fraction enters the *O* pool, and the rest is returned to the atmosphere as CO_2. The humification factor (*h*) determines the fraction of the aboveground (subscript *a*) and belowground (subscript *b*) C leaving the *Y* pool every time step that will enter the *O* pool. The original ICBMr model was modified to fit the willow system by keeping above- and belowground input in separate *Y* pools and multiplying the *h* values of the Y_b pool by a factor of 2.3 (Kätterer *et al.*, 2011). The equations and parameter values used in this study can be found in Eqns (S2) and (S3), and Table (S9).

The initial SOC stock was calculated by performing a spin-up simulation of 1000 years using only annual crop input followed by 20 years of fallow input (Appendix S6: Table S10). Total initial SOC level was 85 t of C per ha.

Biochar. The biochar was assumed to be applied to the soil of the willow plantation, where it was subsequently mineralized and returned to the atmosphere through physical and biochemical processes. As SOC and biochar exhibit very different physical and biochemical properties, and the ICBMr model

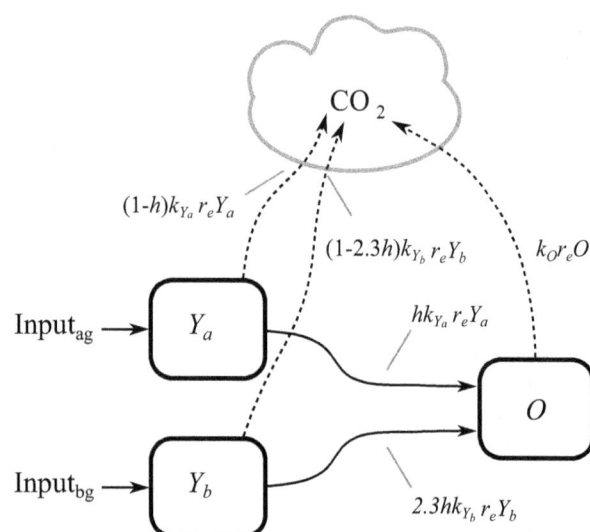

Fig. 3 Schematic illustration of the Introductory Carbon Balance Model (ICBM) used to calculate changes in soil organic carbon (SOC) stocks. The arrow labels indicate how the humification factor (*h*), soil climate parameter (r_e) and the degradation rate constant (*k*) govern the carbon (C) flows from the above-(*a*) and belowground (*b*) young (*Y*) C pools to the old (*O*) C pool, and eventually to the atmosphere.

was not developed to calculate the decay of biochar, the exponential decay model presented in Zimmerman (2010) was used to calculate the net annual CO_2 emissions due to decomposition of the biochar.

The model presented by Zimmerman (2010) does not separate the carbon into different fractions, but exhibits similar characteristics to double exponential decay models over the time frame of this study (53 years, Fig. 4). A description of the biochar decay model used in this study can be found in the (Appendix S5, Eqn (S4)).

Climate impact characterization

In this study, the contribution to the global mean surface temperature change as a function of time ($\Delta T_S(n)$) was used as an indicator for the time-dependent climate impact (Aamaas et al., 2013). The $\Delta T_S(n)$ was chosen as an indicator for its ability to express the impact on the climate at varying points in time, thereby indicating both the timing and rate of change. Furthermore, the temperature might be easier to relate to as a metric than RF to many people. It also describes a physical response in the climate system which lays closer to impacts on ecosystems and human society than RF. These are ultimately what concerns human society and decision-makers.

When calculating the contribution to the global mean surface temperature change ($\Delta T_S(n)$), the individual temperature response ($\Delta T_S(t)$) from each annual (i) net GHG (x) emission was calculated and summed for every year of the study period (n) Eqn (3). The net annual emissions of every GHG therefore had to be specified in a time-distributed life cycle inventory. A detailed explanation of the methodology can be found in Ericsson et al. (2013).

$$\Delta T_S(n) = \sum_{x=1}^{3} \sum_{i=1}^{n} \Delta T_S^{x_i}(t) \qquad [\text{K}] \qquad (3)$$

The equations used to calculate $\Delta T_S^{x_i}$ in this study can be found in Eqns (S5–S7).

The evaluation period of the time-dependent climate impact was set to 100 years, beginning in the first year of the study period. The difference between the evaluation period and the study period was that activities and carbon stock changes giving rise to GHG emissions only took place during the study period, while the effects of these emissions on the climate were evaluated up until the end of the evaluation period. This is important when using $\Delta T_S(n)$ as an indicator together with time-dependent climate impact methodology as the major contributions from emissions occurring close to the end of the study period would otherwise not be recognized (Ericsson et al., 2014). The upper limit of the evaluation period was set to 100 years because a longer time frame increases the uncertainty and decreases the usefulness of the metric results. The later needs to be kept in mind when interpreting the results from response functions that reach far into the future. The IPCC decided not to include GWP values with a longer TH than 100 years in AR5 for this specific reason (Myhre et al., 2013).

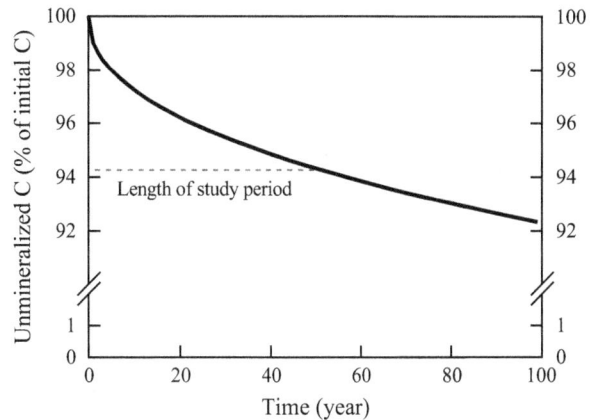

Fig. 4 Carbon (C) fraction remaining in the soil as a function of time after biochar application.

Sensitivity analysis

The sensitivity of the temperature response to changes in some of the assumptions and choices made in scenario 1a was tested using electricity and heat as FUs. The assumptions tested included the following: increasing the transportation distance to 100 km; excluding the use of flue gas condensation in the biomass CHP; assuming a 30% yield increase in the second rotation due to a new clone with a higher nutrient use efficiency and growth rate; a 50% reduction in field N_2O emissions due to biochar application (Cayuela et al., 2014); and a reduction in biochar stability in the soil, leading to 80% of the C in the original biochar remaining in the soil 100 years after application.

As some of the parameters influenced the energy yield, system expansion was applied in all cases making the electricity and heat output from1 ha of land equal to that of scenario 2 with an unaltered yield. The external energy sources used in the system expansion were assumed to originate from a natural gas-fuelled combined cycle CHP.

Results

Energy efficiency

The amount of energy delivered in scenario 1a and 1b was considerably lower than in scenario 2, while the energy input was only marginally smaller (Table 1). The smaller energy input was mainly due to reduced transport requirement between the farm and the CHP. The conversion efficiency from feedstock energy (HHV) into electricity and heat at the CHP was 42% in scenario 1a and 67% in scenario 1b, compared with 88% in scenario 2.

Scenario 2 used its energy inputs (not including the feedstock energy) a little more efficiently than scenario 1b when producing electricity and heat while it was almost twice as efficient as scenario 1a and scenario 3 (Table 1). This was not entirely reflected in the ER of the generated electricity and heat because of the difference in allocation of the energy inputs caused by the higher electricity-to-heat ratio in the combined cycle used in scenario 3, compared with that of the steam turbines used in the bioenergy scenarios. The heat delivered from the natural gas fed combined cycle CHP system in scenario 3 also had a lower ER than that of the willow scenarios due to the higher amount of energy input required to produce and transport the fuel and the use of flue gas condensation technology in the biomass CHPs.

When performing the system expansion in scenario 1a and 1b, the ER of the heat decreased in most cases. This was due to the relatively low ER of the external energy sources used compared to the ER of the heat in the unexpanded systems.

Greenhouse gas and carbon fluxes

Emissions from technical system and non-C emissions. The GHG emissions from the production system, including induced N_2O emissions from decomposing biomass, were very similar in scenario 1a and 1b and scenario 2 (Fig. 5). The small difference in CO_2 and CH_4 emissions could be attributed to the difference in the amount of transport work required between the farm and the CHP.

Scenario 1a and 1b and scenario 2 emitted much less CH_4 than scenario 3. The high CH_4 emissions in scenario 3 originated from distribution losses and unburned gas when flaring. In total, 11 g CH_4/kg NG fuel was emitted (Gode *et al.*, 2011).

The relatively high N_2O emissions in scenario 1a and 1b, as well as in scenario 2, originated from decomposing biomass (51%), fertilizer-induced emissions (31%) and the production and application of fertilizers (18%). Other sources in the system, such as machine operations, transportation, waste handling and related upstream emissions accounted for only 0.02% of the total N_2O emissions in these scenarios.

Of the external energy sources, HC contributed much of the CO_2 and N_2O emissions in scenario 1a and 1b (Fig. 6). Natural gas contributed much more CH_4 than all the other sources due to the distribution losses and flaring emissions mentioned earlier.

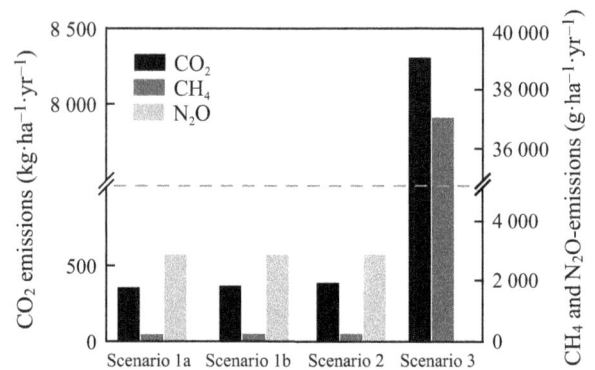

Fig. 5 Part of the inventory results showing the emissions from the technical system, including upstream emissions and non-C emissions, i.e. induced N_2O emissions.

Table 1 Electricity and heat delivered, total energy input and energy ratios (ER) of the electricity and heat in the scenarios studied. Scenario 1a and 1b is given both excluding (unexpanded) and including (NG,HC,WP+HHW) the external energy service generation of the expanded system

| | Energy delivered | | | | |
	El $\frac{GJ}{ha \cdot yr}$	Heat $\frac{GJ}{ha \cdot yr}$	Energy input $\frac{GJ}{ha \cdot yr}$	ER el	ER heat
Scenario 1a (unexpanded)	7	58	5.7	3	19
Scenario 1b (unexpanded)	18	86	5.8	7	28
Scenario 2	34	102	6.1	10	36
Scenario 3	34	102	12.0	8	13
Scenario 1a NG*				6	16
Scenario 1a HC†				4	12
Scenario 1a WP+HHW‡				8	20
Scenario 1b NG*				7	24
Scenario 1b HC†				5	20
Scenario 1b WP+HHW‡				10	27

*Natural gas electricity and heat. Data are based on the Rya combined cycle CHP (Gode *et al.*, 2011).
†Hard coal electricity and heat. Data are based on five Danish hard coal powered CHPs (Gode *et al.*, 2011).
‡Wind power and household waste heat. Data for wind power are based on the wind power production of Vattenfall, and the data for household waste heat are based on production, distribution and use of HHW with low organic content in a Swedish waste-fuelled CHP (Gode *et al.*, 2011).

Biogenic carbon fluxes. The live biomass C pool increased temporarily due to changing the land use to willow. On average, 14 700 kg of C per ha was stored in the live biomass aboveground pool in the willow plantation. The carbon stock increased and decreased periodically with the 3-year coppicing cycle. At harvest, all the C stored in the biomass was either converted into fuel for electricity and heat generation and returned to the atmosphere as CO_2 or applied as biochar to the soil. Further, CO_2 was withdrawn from the atmosphere and stored temporarily as C in the coarse roots of the willow plants, holding on average 1400 kg of C per ha. The C of the coarse roots was removed mechanically and incorporated into the soil at the end of each rotation. The coarse roots were then treated as input to the SOC model. The net sequestration of biomass as well as coarse roots was zero over an entire rotation.

Annual carbon input from leaves and fine roots and occasional input from coarse roots at the end of each rotation contributed to the increase in SOC stocks. On average, 390 kg of C per (ha per yr) was accumulated in each of the scenarios due to SOC stock changes. The SOC growth rate decreased with time as the system approached a new state of SOC equilibrium. The rate of growth was 430 and 360 kg of C per (ha per yr) during the first and second rotation, respectively.

In scenario 1a, the biochar C pool in the soil increased at a rate of 1300 kg of C per (ha per yr). Due to the recalcitrance of biochar, the rate of growth did not vary by more than 2% between the first and second rotations. The amount of biochar C remaining in the soil at the end of the study period was calculated to be 94% of the biochar C applied over the course of the study period.

Climate impact

Per hectare of land (no system expansion). When assessing scenario 1a and 1b and scenario 2 based on their function of mitigating climate impacts by representing optional land use systems, scenario 1a contributed much more to a decrease in ΔT_S than scenario 1b (Fig. 7) and scenario 2, whose effects on $\Delta T_S(n)$ resembled each other. This was explained almost entirely by the temperature response from the CO_2 removed by the growing willow and kept out of the atmosphere for a long time as a consequence of diverting and applying C to the soil in the form of relatively stable biochar (Fig. 8).

The contribution from the SOC to ΔT_S was of a similar magnitude, but of opposite sign to the combined contribution from the production system and the N_2O emissions (Fig. 8). They cancelled each other out almost perfectly and were similar in scenario 1a and 1b and in scenario 2. The total temperature response in scenario 1b and in scenario 2 therefore resembled that caused by

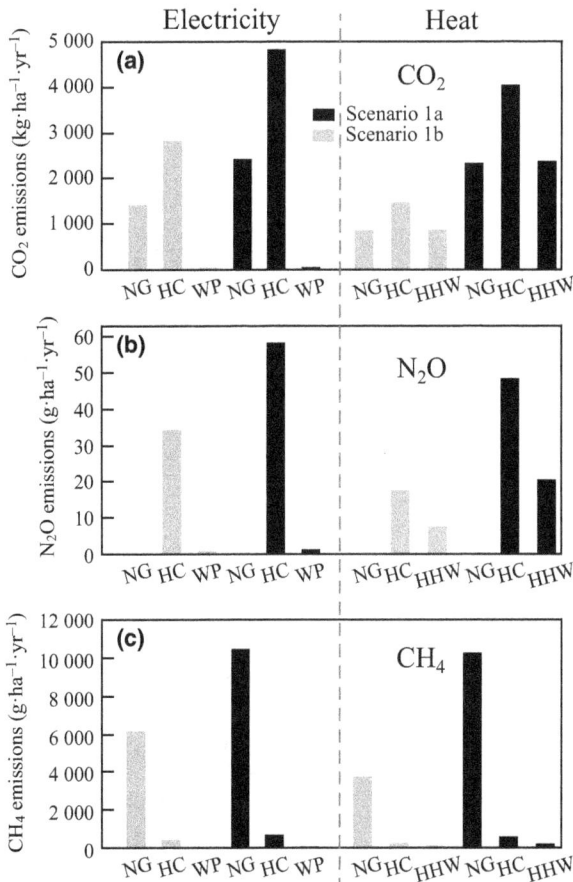

Fig. 6 Part of the inventory results showing the contribution of CO_2 (a), N_2O (b) and CH_4 (c) from the external energy sources in scenario 1a (biochar-to-soil) and 1b (char-to-energy). External energy sources included were natural gas (NG), hard coal (HC), wind power (WP) and household waste (HHW).

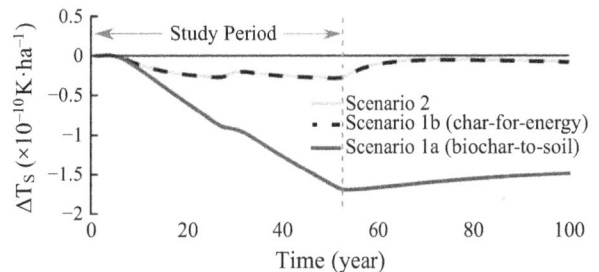

Fig. 7 Contribution to the global mean surface temperature change (ΔT_S) per hectare (ha) of willow plantation in scenario 1a and 1b and in scenario 2. Note: Indirect climate impacts due to different amounts of heat and power produced in the scenarios are not included.

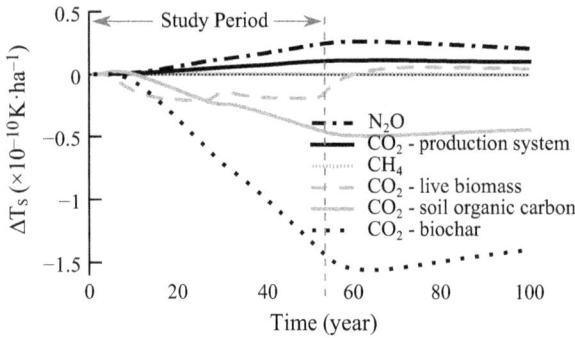

Fig. 8 Contribution to the global mean surface temperature change (ΔT_S) per hectare (ha) of willow plantation from each of the system components of the SRC willow bioenergy system in scenario 1a.

the CO_2 fluxes in the live biomass, while it was dominated by the biochar in scenario 1a (Fig. 7).

Per unit of energy service (with system expansion). When assessing the climate impact of all scenarios based on their function of delivering energy services, the impact from the external energy sources was included in scenario 1a and 1b. The biomass-based scenarios (1a, 1b

and 2) always performed better than the NG-based scenario (3) (Fig. 9a–d).

In this study, the direct combustion used in scenario 2 was always a more favourable option from a climate change mitigation perspective than the char-for-energy alternative in scenario 1b. In this case, the impact from the external electricity and heat required to compensate for the lower energy efficiency in scenario 1b was higher than the difference between the pyrolysis system (scenario 1b) and the direct combustion system (scenario 2). The contribution to ΔT_S from scenario 1b was higher than that of scenario 1a for all external energy sources used in the system expansion, except for the heat produced using HC. With a less GHG-intensive external power source, such as WP or HHW, applying the biochar to the soil, as in scenario 1a, became a better option from a climate change mitigation perspective than the direct combustion of scenario 2 (Fig. 9a and c). The contribution to ΔT_S in scenario 1a was much lower than in scenario 2 when WP was used due to the impact from the CO_2 kept out of the atmosphere when applying biochar to the soil (Fig. 9a). However, scenario 1a was very sensitive to the external energy sources used in the system expansion due to its low energy efficiency.

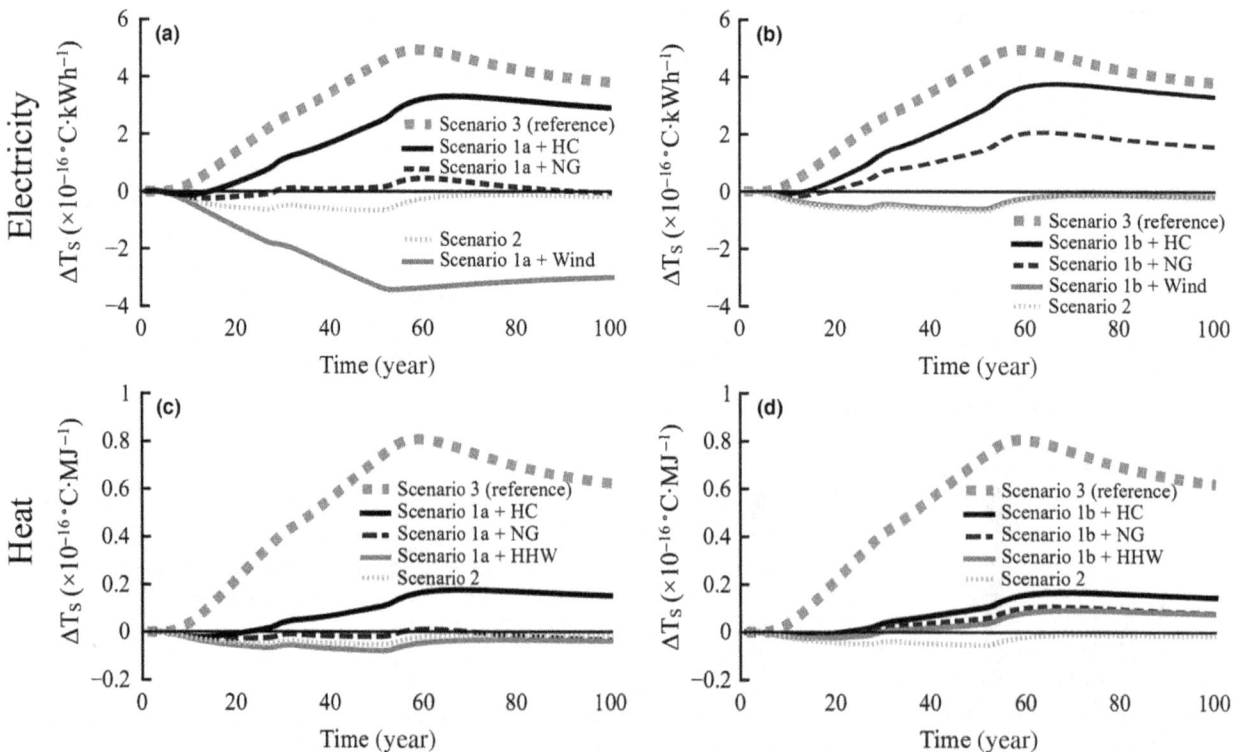

Fig. 9 Temperature response per kWh electricity (a and b) and MJ of heat (c and d) for all scenarios, including external electricity and heat from hard coal (HC), natural gas (NG), wind power (WP) and household waste (HHW) in scenario 1a (a and c) and scenario 1b (b and d). The temperature response for scenarios 2 and 3 (reference) is the same in (a) and (b), as well as in (c) and (d), respectively.

When HC was used as the external energy source, the emissions allocated to the heat in scenario 1a contributed more to an increase in ΔT_S than in scenario 1b (Fig. 9c,d).

The amount of external electricity and heat relative to the amount of delivered electricity and heat also influenced the apparent timing of the temperature response in scenario 1a and 1b. When examining the HC and NG response curves of these two scenarios (Fig. 9a,c and b, d), the temperature response of the heat was slightly delayed compared to that of the electricity. This could be explained by the external heat making up a smaller share of the total amount of heat delivered compared to the external electricity share of the total electricity delivered. As a consequence, the HC and NG contributed less to the total temperature response of the heat than to that of the electricity system. As the contribution from the external energy sources to the increase in ΔT_S of the heat was less pronounced, the temperature increase in the response curve also appeared delayed when compared to the electricity.

Sensitivity analysis

The two assumptions with the largest effect on the results in scenario 1a were the yield level and the use of flue gas condensation in the biomass CHP. Both of these influenced the energy efficiency. Increased yield lowered the contribution to ΔT_S from both the electricity and the heat by decreasing the need for external energy and also increasing the amount of C in the biochar pool (Figs 10 and 11). By excluding the use of flue gas condensation, the need for external heat increased, and as a consequence, so did the contribution to ΔT_S from heat (Fig. 11).

The contribution to ΔT_S from a reduction in N_2O emissions and stability of biochar was of almost equal

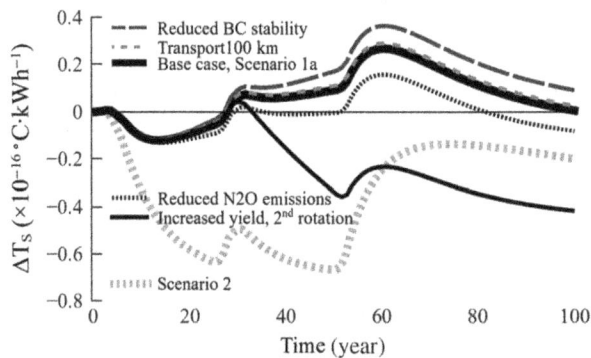

magnitude, but of opposite sign, for both electricity and heat. Their contribution was significantly smaller than that of the assumed yield increase and exclusion of flue gas condensation.

However, if biogenic C stock changes had not been included in this study, the impact from reduced N_2O emissions would not have been insignificant. The changes contributed approximately 70% of the temperature response from the production system, which excludes the response from SOC, live biomass and biochar C stock changes.

Given the magnitude of the contribution from the biochar to the temperature response of the system, it was expected that the stability of the biochar would be an important parameter. However, in this scenario analysis, the degradation rate was increased by 2.6-fold without a large impact on the contribution to ΔT_S from the system. This indicates that biochar with much lower stability than that used in this study might still offer significant climate change mitigation benefits in pyrolysis systems with soil application of biochar.

Discussion

The results of the present study show that using land to grow SRC willow for bioenergy in central Sweden can have a cooling influence on the global mean surface temperature. The time-dependent model and climate impact indicator used in this study showed that this influence will prevail for at least 50 years after the end of the cultivation of the willow (Fig. 7). Converting the feedstock into bio-oil and char using pyrolysis before generating electricity and heat, as in scenario 1b, did not lead to a significant change in the contribution to ΔT_S from the bioenergy system compared with combusting the raw willow chips directly in a large-scale CHP, as in scenario 2. However, from a land use

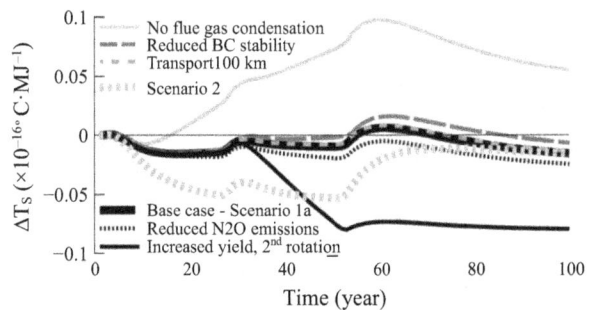

Fig. 10 Influence of some of the assumptions made in scenario 1a on the temperature response from the electricity. Natural gas generation of electricity was used as the external energy source to make the electricity delivered equal to scenario 2, without increased yields, in all cases.

Fig. 11 Influence of some of the assumptions made in scenario 1a on the temperature response from the heat. Natural gas generation of heat was used as the external energy source to make the heat delivered equal to scenario 2, without increased yields, in all cases.

perspective, the cooling effect can increase substantially if the char is applied to soil in the form of biochar, as in scenario 1a. This is due to the large C sink created by the biochar, which is very stable and keeps part of the C sequestered by the willow out of the atmosphere for a very long time. When the climate impact of bioenergy systems is compared, it is however advisable to include the indirect effects that different amounts of energy services deliver.

Converting the feedstock to bio-oil and char using pyrolysis inevitably reduces the energy efficiency of the system. When comparing the climate impact of the electricity and heat in scenario 1a and 1b to scenarios 2 and 3, the outcome was very dependent on the external electricity and heat sources considered in the system expansion. The system expansion performed to compensate for the lower energy efficiency of the pyrolysis system showed that the direct combustion system in scenario 2 contributed more to a decrease in ΔT_S than when pyrolysis was applied and all products were used for energy service generation (scenario 1b), regardless of the external energy source used to generate the electricity and heat in the expanded system. When using GHG-intensive sources (e.g. hard coal and natural gas), the expanded system even contributed to an increase in ΔT_S, but in all cases, the pyrolysis bioenergy system in scenario 1a and 1b was a better option from a climate change mitigation perspective than using the electricity and heat produced in the natural gas-fuelled combined cycle CHP in scenario 3.

In a situation where a choice can be made between the use of a pyrolysis bioenergy system in which biochar is applied to soils, as in scenario 1a, or where the biomass is used for energy generation, as in scenarios 1b and 2, substantial climate change mitigation benefits can potentially be achieved using a biochar system if the external energy sources that are needed to compensate for the lower energy efficiency of the pyrolysis-biochar system, compared to direct combustion, have lower GHG intensities than fossil fuels. This is in accordance with the climate mitigation potential study of biochar systems globally by Woolf et al. (2010).

The sensitivity analysis in this study showed the importance of the biomass yield for the climate change mitigation potential of a SRC willow-based bioenergy system producing and applying biochar to soils. This emphasizes the importance of good management practices to achieve high yields for the climate impact of the system, and also the importance of choosing high-yielding clones when establishing a new willow plantation. The temperature response curve for an increased biomass yield also showed that this might be a good strategy when trying to achieve continuous long-term climate change mitigation from the generated electricity

and heat (Figs 10 and 11). The climate change mitigation effort effectively gets a boost every time yield improvements are achieved due to its influence on both C stocks and energy efficiency of the system.

The use of flue gas condensation was also an important parameter for the climate impact of the heat delivered by the system as it increases the energy efficiency, potentially avoiding the use of fossil fuels. However, most Swedish CHP plants use flue gas condensation, as it is a profitable investment (Axby et al., 2000). The stability of the biochar had a remarkably low influence on the climate impact of the system in this study, given its share of the total contribution to ΔT_S. This may be due to the short time period used in this study. Climate impacts from biochar act on much longer time scales than those from SOC and live biomass due to the high stability of the biochar. Biochar stability may still be important for the long-term climate impact, but this was not studied here.

Much more information can be conveyed by the graphical representation of the time-dependent climate impact used in this study than what can be achieved by the use of a simple index, such as the GWP. This is why a time-dependent impact assessment methodology may serve as a useful complement to traditional impact assessment methods in LCA. It is, however, a more time-consuming endeavour to set up the time-distributed inventory model and gather data with the necessary temporal resolution to assess the temporal dimension of impacts in an LCA.

Several authors have proposed ways of integrating temporal information in LCI databases (Collinge et al., 2013; Beloin-Saint-Pierre et al., 2014; Tiruta-Barna et al., 2016). These might be useful to add temporal information to the background system and make the use of time-dependent impact indicators easier and less resource-consuming. However, the LCA community and database developers need to agree on how to store temporal information in a database format without bloating the storage space, and still provide useful data. In the present study, a more pragmatic approach was used towards the emissions from the background system. These were simply assigned to the year in which the associated main event took place in the foreground system. There is still plenty of temporal information in the foreground system to motivate the use of a time-dependent LCA methodology in many systems that have life cycles that extend over many years, which is typically the case in short-rotation forestry.

Under all circumstances, SRC willow offers an opportunity for individual farmers to become part of the energy supply chain. The possibility of using the feedstock for either direct combustion or bio-oil and biochar production represents an opportunity which offers flexibility that could make the producers of the biomass less sensitive to fluctuations in market price. The commercial applications for bio-oil and biochar are currently limited, but substantial research has been carried out over the last 30 years, and the potential applications are numerous (Qian *et al.*, 2015). Pyrolysis systems might not always be beneficial from an energy efficiency perspective, but other benefits such as improved handling, storage and fuel properties, as well as increased energy density and product flexibility, might be more important when choosing this energy conversion technology. In all cases, the SRC willow-based bioenergy systems included in this study offer a way of generating electricity and heat while at the same time contributing to a decrease in the global mean surface temperature change. This study was, however, limited to the energy efficiency and climate impact of willow-based pyrolysis bioenergy systems. These systems will also have other types of environmental impacts which should be scrutinized using a full life cycle approach before passing verdict on their environmental sustainability.

Acknowledgements

We are grateful to the STandUP for energy programme for financial support.

References

Aamaas B, Peters GP, Fuglestvedt JS (2013) Simple emission metrics for climate impacts. *Earth System Dynamics*, 4, 145–170.

Andrén O, Kötterer T, Karlsson T (2004) ICBM regional model for estimations of dynamics of agricultural soil carbon pools. *Nutrient Cycling in Agroecosystems*, 70, 231–239.

Aronsson P, Rosenqvist H (2011) Gödslingsrekommendationer för salix 2011. Report, Swedish University of Agricultural Sciences, Uppsala.

Axby F, Gustafsson JO, Nyström J, Johansson K (2000) Study of flue gas condensing for biofuel fired heat and power plants. Technical Report 719, Värmeforsk, Stockholm.

Beloin-Saint-Pierre D, Heijungs R, Blanc I (2014) The ESPA (Enhanced Structural Path Analysis) method: a solution to an implementation challenge for dynamic life cycle assess-ment studies. *The International Journal of Life Cycle Assessment*, 19, 861–871.

Beretta GP, Iora P, Ghoniem AF (2012) Novel approach for fair allocation of primary energy consumption among cogenerated energy-intensive products based on the actual local area production scenario. *Energy*, 44, 1107–1120.

Berntsen T, Fuglestvedt J (2008) Global temperature responses to current emissions from the transport sectors. *Proceedings of the National Academy of Sciences of the United States of America*, 105, 19154–19159.

Brown RC (ed.) (2011) *Thermochemical Processing of Biomass. Conversion into Fuels, Chemicals and Power*. Wiley Series in Renewable Resources, John Wiley & Sons, Ltd, Chippenham, UK.

Cayuela M, van Zwieten L, Singh B, Jeffery S, Roig A, Sánchez-Monedero M (2014) Biochars role in mitigating soil nitrous oxide emissions: a review and meta-analysis. *Agriculture, Ecosystems & Environment*, 191, 5–16.

Cherubini F, Peters GP, Berntsen T, Strømman AH, Hertwich E (2011) CO_2 emissions from biomass combustion for bioenergy: atmospheric decay and contribution to global warming. *GCB Bioenergy*, 3, 413–426.

Cherubini F, Bright RM, Strømman AH (2013) Global climate impacts of forest bioenergy: what, when and how to measure? *Environmental Research Letters*, 8, 014049.

Collinge WO, Landis AE, Jones AK, Schaefer LA, Bilec MM (2013) Dynamic life cycle assessment: framework and application to an institutional building. *The International Journal of Life Cycle Assessment*, 18, 538–552.

Courchesne A, Bécaert V, Rosenbaum R, Deschênes L, Samson R (2010) Using the lashof accounting methodology to assess carbon mitigation projects with life cycle assessment. *Journal of Industrial Ecology*, 14, 309–321.

Denman K, Brasseur G, Chidthaisong A et al. (2007) Couplings between changes in the climate system and biogeochemistry. In: *Climate Change 2007: The Physical Science Basis* (eds Solomon S, Qin D, Manning M et al.), chapter 7, pp. 499–588. Cambridge University Press, Cambridge, UK.

Environmental Product Declaration (EPD) (2013) Environmental Product Declaration of Rizziconi combined-cycle gas turbine plant, Italy.

Environmental Product Declaration (EPD) (2015) Environmental Product Declaration of Domat/Ems wood-fired power plant.

Ericsson N, Porsö C, Ahlgren S, Nordberg Å, Sundberg C and Hansson PA (2013) Timedependent climate impact of a bioenergy system - methodology development and application to Swedish conditions. *GCB Bioenergy*, 5, 580–590.

Ericsson N, Nordberg Å, Sundberg C, Ahlgren S and Hansson PA (2014) Climate impact and energy efficiency from electricity generation through anaerobic digestion or direct combustion of short rotation coppice willow. *Applied Energy*, 132, 86–98.

EU (2011) Commission implementing decision of 19 december 2011 establishing harmonised efficiency reference values for separate production of electricity and heat in application of directive 2004/8/ec of the european parliament and of the council and repealing commission decision 2007/74/ec. commision implementing decision L343/91, European commision.

EU (2012) Directive 2012/27/EU of the european parliament and of the council of 25 october 2012 on energy efficiency, amending directives 2009/125/EC and 2010/30/EU and repealing directives 2004/8/EC and 2006/32/EC. Directive L315/1, European comission.

Fuglestvedt J, Berntsen T, Godal O, Sausen R, Shine K and Skodvin T (2003) Metrics of climate change: assessing radiative forcing and emission indices. *Climatic Change*, 58, 267–331.

Fuglestvedt J, Shine K, Berntsen T et al. (2010) Transport impacts on atmosphere and climate: metrics. *Atmospheric Environment*, 44, 4648–4677.

Gallezot P (2012) Conversion of biomass to selected chemical products. *Chemical Society Reviews*, 41, 1538–1558.

Gaunt JL, Lehmann J (2008) Energy balance and emissions associated with biochar sequestration and pyrolysis bioenergy production. *Environmental Science & Technology*, 42, 4152–4158.

Giuntoli J, Caserini S, Marelli L, Baxter D, Agostini A (2015) Domestic heating from forest logging residues: environmental risks and benefits. *Journal of Cleaner Production*, 99, 206–216.

Gode J, Martinsson F, Hagberg L, Öman A, Höglund J, Palm D (2011) Miljöfaktaboken 2011. Estimated emission factors for fuels, electricity, heat and transport in Sweden. Technical Report 1183, Värmeforsk, Stockholm.

Hammar T, Ericsson N, Sundberg C, Hansson PA (2014) Climate impact of willow grown for bioenergy in Sweden. *BioEnergy Research*, 7, 1529–1540.

Hammar T, Ortiz C, Stendahl J, Ahlgren S, Hansson PA (2015) Time-dynamic effects on the global temperature when harvesting logging residues for bioenergy. *BioEnergy Research*, 8, 1912–1924.

Hammond J, Shackley S, Sohi S, Brownsort P (2011) Prospective life cycle carbon abatement for pyrolysis biochar systems in the UK. *Energy Policy*, 39, 2646–2655.

Hanandeh AE (2012) Carbon abatement via treating the solid waste from the australian olive industry in mobile pyrolysis units: LCA with uncertainty analysis. *Waste Management & Research*, 31, 341–352.

Hansen J, Sato M (2012) Paleoclimate implications for human-made climate change. In: *Climate Change* (eds Berger A, Mesinger F, Sijacki D), pp. 21–47. Springer, Vienna.

Ibarrola R, Shackley S, Hammond J (2012) Pyrolysis biochar systems for recovering biodegradable materials: a life cycle carbon assessment. *Waste Management*, 32, 859–868.

IPCC (1991) *Climate Change - The IPCC Scientific Assessment*. Cambridge University Press, Cambridge, Great Britain, New York, NY, USA and Melbourne, Australia, repr. edition.

IPCC (2006) 2006 IPCC guidelines for national greenhouse gas inventories, prepared by the national greenhouse gas inventories programme.

ISO 14040 (2006) Environmental management - life cycle assessment - principles and framework (ISO 14040:2006). ISO standard EN ISO 14040:2006 ISO/TC 207, CMC, European Comittee for Standardization, Brussels.

ISO 14044 (2006) Environmental management - life cycle assessment - requirements and guidelines (ISO 14044:2006). ISO standard EN ISO 14044:2006 ISO/TC 207, CMC, European Comittee for Standardization, Brussels.

Kätterer T, Bolinder MA, Andrén O, Kirchmann H, Menichetti L (2011) Roots contribute more to refractory soil organic matter than above-ground crop residues, as revealed by a long-term field experiment. *Agriculture, Ecosystems and Environment*, 141, 184–192.

Kendall A (2012) Time-adjusted global warming potentials for lca and carbon foot-prints. *The International Journal of Life Cycle Assessment*, **17**, 1042–1049.

Kendall A, Chang B, Sharpe B (2009) Accounting for time-dependent effects in bio-fuel life cycle greenhouse gas emissions calculations. *Environmental Science & Technology*, **43**, 7142–7147.

Levasseur A, Lesage P, Margni M, Deschênes L, Samson R (2010) Considering time in lca: dynamic lca and its application to global warming impact assessments. *Environmental Science and Technology*, **44**, 3169–3174.

Liang B, Lehmann J, Solomon D et al. (2008) Stability of biomass-derived black car-bon in soils. *Geochimica et Cosmochimica Acta*, **72**, 6069–6078.

Murphy DJ, Hall CAS, Dale M, Cleveland C (2011) Order from chaos: a preliminary protocol for determining the eroi of fuels. *Sustainability*, **3**, 1888–1907.

Myhre G, Shindell D, Breón FM et al. (2013) *Anthropogenic and Natural Radiative Forc-ing*, book section 8. Cambridge University Press, Cambridge, United Kingdom and New York, NY, USA, 659–740.

Neves D, Thunman H, Matos A, Tarelho L, Gómez-Barea A (2011) Characterization and prediction of biomass pyrolysis products. *Progress in Energy and Combustion Science*, **37**, 611–630.

Noll M, Jirjis R (2012) Microbial communities in large-scale wood piles and their effects on wood quality and the environment. *Applied Microbiology and Biotechnol-ogy*, **95**, 551–563.

O'Hare M, Plevin RJ, Martin JI, Jones AD, Kendall A, Hopson E (2009) Proper accounting for time increases crop-based biofuels' greenhouse gas deficit versus petroleum. *Environmental Research Letters*, **4**, 024001.

Ortiz CA, Hammar T, Ahlgren S, Hansson PA, Stendahl J (2016) Time-dependent global warming impact of tree stump bioenergy in Sweden. *Forest Ecology and Management*, **371**, 5–14.

Peters GP, Aamaas B, T Lund M, Solli C, Fuglestvedt JS (2011)$dummy$Alternative "global warming" metrics in life cycle assessment: a case study with existing transportation data. *Environmental Science and Technology*, **45**, 8633–8641.

Peters JF, Iribarren D, Dufour J (2015) Biomass pyrolysis for biochar or energy appli-cations? A life cycle assessment. *Environmental Science & Technology*, **49**, 5195–5202.

Porsö C, Mate R, Vinterbäck J, Hansson PA (2016) Time-dependent climate effects of eucalyptus pellets produced in Mozambique used locally or for export. *BioEnergy Research*, **9**, 942–954.

Pourhashem G, Spatari S, Boateng AA, McAloon AJ, Mullen CA (2013) Life cycle environmental and economic tradeoffs of using fast pyrolysis products for power generation. *Energy & Fuels*, **27**, 2578–2587.

Pourhashem G, Adler PR, Spatari S (2016) Time effects of climate change mitigation strategies for second generation biofuels and co-products with temporary carbon storage. *Journal of Cleaner Production*, **112**, 2642–2653.

Qian K, Kumar A, Zhang H, Bellmer D, Huhnke R (2015) Recent advances in utiliza-tion of biochar. *Renewable and Sustainable Energy Reviews*, **42**, 1055–1064.

Roberts KG, Gloy BA, Joseph S, Scott NR, Lehmann J (2010) Life cycle assessment of biochar systems: estimating the energetic, economic, and climate change poten-tial. *Environmental Science & Technology*, **44**, 827–833.

Rytter RM (2001) Biomass production and allocation, including fine-root turnover, and annual N uptake in lysimeter-grown basket willows. *Forest Ecology and Man-agement*, **140**, 177–192.

SJV (2008) Kartläggning av mark som tagits ur produktion. Report 08:7, Swedish Board of Agriculture.

Sohi SP, Krull E, Lopez-Capel E, Bol R (2010) A review of biochar and its use and function in soil. *Advances in Agronomy*, **105**, 47–82.

Swedenergy (2012) Miljövärdering 2012. Guide för allokering i kraftvärmeverk och fjärrvärmens elanvändning.

Swedish Energy Agency (2015) Energy in sweden 2015. Report ET015:19, Swedish Energy Agency.

Thornley P, Gilbert P, Shackley S, Hammond J (2015) Maximizing the greenhouse gas reductions from biomass: the role of life cycle assessment. *Biomass and Bioen-ergy*, **81**, 35–43.

Tiruta-Barna L, Pigné Y, Gutiérrez TN, Benetto E (2016) Framework and computa-tional tool for the consideration of time dependency in life cycle inventory: proof of concept. *Journal of Cleaner Production*, **116**, 198–206.

Wang Z, Dunn JB, Han J, Wang MQ (2013) Effects of co-produced biochar on life cycle greenhouse gas emissions of pyrolysis-derived renewable fuels. *Biofuels, Bio-products and Biorefining*, **8**, 189–204.

Woolf D, Amonette JE, Street-Perrott FA, Lehmann J, Joseph S (2010) Sustainable biochar to mitigate global climate change. *Nature Communications*, **1**, 56.

Zetterberg L, Uppenberg S, Åhman M (2004)$dummy$Climate impact from peat utilisation in Sweden. *Mitigation and Adaptation Strategies for Global Change*, **9**, 37–76.

Zimmerman AR (2010) Abiotic and microbial oxidation of laboratory-produced black carbon (biochar). *Environmental Science and Technology*, **44**, 1295–1301.

8

Climate impact assessment of willow energy from a landscape perspective: a Swedish case study

TORUN HAMMAR[1], PER-ANDERS HANSSON[1] and CECILIA SUNDBERG[1,2]

[1]Department of Energy and Technology, Swedish University of Agricultural Sciences (SLU), SE-750 07, Uppsala, Sweden,
[2]Division of Industrial Ecology, Department of Sustainable Development, Environmental Science and Engineering, KTH Royal Institute of Technology, SE-100 44, Stockholm, Sweden

Abstract

Locally produced bioenergy can decrease the dependency on imported fossil fuels in a region, while also being valuable for climate change mitigation. Short-rotation coppice willow is a potentially high-yielding energy crop that can be grown to supply a local energy facility. This study assessed the energy performance and climate impacts when establishing willow on current fallow land in a Swedish region with the purpose of supplying a bio-based combined heat and power plant. Time-dependent life cycle assessment (LCA) was combined with geographic information system (GIS) mapping to include spatial variation in terms of transport distance, initial soil organic carbon content, soil texture and yield. Two climate metrics were used [global warming potential (GWP) and absolute global temperature change potential (AGTP)], and the energy performance was determined by calculating the energy ratio (energy produced per unit of energy used). The results showed that when current fallow land in a Swedish region was used for willow energy, an average energy ratio of 30 MJ MJ^{-1} (including heat, power and flue gas condensation) was obtained and on average 84.3 Mg carbon per ha was sequestered in the soil during a 100-year time frame (compared with the reference land use). The processes contributing most to the energy use during one willow rotation were the production and application of fertilizers (~40%), followed by harvest (~35%) and transport (~20%). The temperature response after 100 years of willow cultivation was $-6\cdot10^{-16}$ K MJ^{-1} heat, which is much lower compared with fossil coal and natural gas ($70\cdot10^{-16}$ K MJ^{-1} heat and $35\cdot10^{-16}$ K MJ^{-1} heat, respectively). The combined GIS and time-dependent LCA approach developed here can be a useful tool in systematic analysis of bioenergy production systems and related land use effects.

Keywords: bioenergy, geographic information system, global warming, land use, life cycle assessment, Salix, soil organic carbon, spatial variation

Introduction

The high consumption of fossil fuels during the past century has generated large emissions of greenhouse gases (GHGs), which have contributed to global warming. Several climate change mitigation targets have been adopted worldwide, most recently in the Paris Agreement signed by the member countries of the United Nations Framework Convention on Climate Change (UNFCCC, 2015). One strategy to reduce GHG emissions and mitigate climate change is to move towards a more bio-based economy, by replacing fossil energy with bioenergy. In addition to climate change mitigation, bioenergy can play an important role in securing the energy supply in a region when locally produced biomass is utilized. However, there are concerns about shifting problems from one area to another, especially regarding potential negative land use effects (both direct and indirect) when increasing utilization of biomass for energy purposes, which may alter carbon stocks or displace land use for food production (Fargione et al., 2008; Searchinger et al., 2008).

One energy crop that has shown potential to generate bioenergy while increasing soil organic carbon (SOC) is short-rotation coppice willow (Rytter, 2012; Ericsson et al., 2013; Zetterberg & Chen, 2015). Willow is a potentially high-yielding crop that can be harvested after only a few years due to its high growth rate (Karp & Shield, 2008; Djomo et al., 2011). The productivity has high importance for both the energy return and the SOC content, because a higher carbon input from leaf litter and root turnover can build up the carbon stock. Growing willow on available agricultural land can be one strategy to provide a local community with a continuous supply of bioenergy. Climate impact assessments of willow are usually performed on stand level (e.g. Ericsson et al. (2014); Hammar et al. (2014); Porsö & Hansson

Correspondence: Torun Hammar
e-mail: torun.hammar@slu.se

(2014)), but assessments of the climate impact of this strategy need to consider the variation within a landscape, as soil texture and water availability are important for willow productivity (Krzyżaniak *et al.*, 2015). Field size and transport distance also vary within regions, affecting the energy return.

In this study, a life cycle assessment (LCA) of willow establishment on current fallow land in a Swedish region was carried out. Only fallow land according to Swedish statistics was selected (which is around 5% of total crop land in Sweden) to avoid possible indirect land use effect of displaced land (Statistics Sweden, 2015). LCA is a standardized method for assessing the environmental impacts of a product or service during its whole lifespan (ISO 14040, 2006; ISO 14044, 2006). The climate metric most commonly used for assessing climate impact in LCA is global warming potential (GWP) (Cherubini & Strømman, 2011; Hauschild *et al.*, 2012), which converts GHG emissions into CO_2 equivalents (IPCC, 2007). When applying this metric to bioenergy systems, the biogenic carbon fluxes are usually set to zero; that is, bioenergy is considered carbon neutral, because the CO_2 released from bioenergy utilization has previously been captured from the atmosphere and/or will be recaptured again during regrowth.

While GWP has some benefits (e.g. enabling comparisons with previous studies), the metric also has limitations; for example, it does not consider the timing of the GHG fluxes, including temporal SOC changes, which have been shown to be of major importance for the overall climate impact of bioenergy (Brandão *et al.*, 2011; Zetterberg & Chen, 2015). The climate metric absolute global temperature change potential (AGTP), also referred to as ΔT, considers the yearly emissions of GHGs and their specific effect on the radiative balance, which affects the global mean surface temperature (Ericsson *et al.*, 2013; Myhre *et al.*, 2013a). The AGTP metric was applied in this study, because it captures the dynamics of biogenic carbon (i.e. fluxes of carbon between the atmosphere, biomass and soil).

Geographic information system (GIS) was used to identify available land and soil properties in the study region. The GIS methodology has been used previously to assess different aspects of bioenergy, for example to determine optimal placement of bioenergy facilities (Ekman *et al.*, 2013; Thomas *et al.*, 2013), assess land availability for short-rotation woody crops (Aust *et al.*, 2014; Abolina *et al.*, 2015) and calculate biomass potential at different spatial scales (Castellano *et al.*, 2009; Fiorese & Guariso, 2010; Wightman *et al.*, 2015). GIS modelling has also been incorporated into LCA to assess the GWP of bioenergy systems (Gasol *et al.*, 2011), with some studies including changes in soil carbon stocks (van der Hilst *et al.*, 2012; Humpenöder *et al.*, 2013;

Monteleone *et al.*, 2015), commonly using IPCC emissions factors for direct land use change (Goglio *et al.*, 2015). However, to our knowledge, the time-dependent LCA method has not previously been combined with the landscape dynamic approach for energy forestry.

The overall aim of this study was to assess the climate effects of increased production of willow energy in a specific region, considering existing land use, soil conditions and geographical location of the region. Specific objectives were to determine:

1. the climate impact per unit of produced energy that can be expected from increased production of biomass in the form of willow grown on existing fallow, given the conditions in a larger area of land
2. the energy balance achieved in different willow systems
3. the effects on climate impact and energy balance of choosing particular fields for willow (due to spatial variations in terms of initial carbon content, transport distance, yield).

The county of Uppsala (located in east-central Sweden; Fig. 1) was chosen as the study region, as in a Swedish perspective, it has a relatively high share of energy forestry [about 1800 ha (Statistics Sweden, 2015)]. There is also potential to increase this amount, as around 10% of the arable land in the region is currently under fallow (Statistics Sweden, 2015), of which about 70% is perennial (i.e. minimum 3 years) (SCB, 2015). In addition, a new bio-based combined heat and power (CHP) plant is planned for the region, making it suitable as a case study area.

Materials and methods

An LCA was performed to determine the climate impact and energy performance of the willow system. The climate impact was assessed in terms of temperature response over time, to capture the temporal dynamics of GHGs. The soil carbon balance was modelled by the ICBM model, a carbon balance model adapted for agricultural soils (Andrén & Kätterer, 1997). ArcGIS was used to identify available land (which was defined as fallow land in this study) and soil properties in the study region and to map transport routes. All fluxes of the three major GHGs (CO_2, N_2O and CH_4) and use of primary energy for the willow procurement chains were included in the LCA, which was performed using the software MATLAB (version R2012b, The MathWorks, Inc., Natick, MA, USA). The energy performance of the willow systems was assessed by calculating the energy ratio (ER), which measures the energy output per unit energy input (Djomo *et al.*, 2011).

System boundaries

Only fields in the study region of Uppsala currently under fallow on mineral soils were included in the study. The time

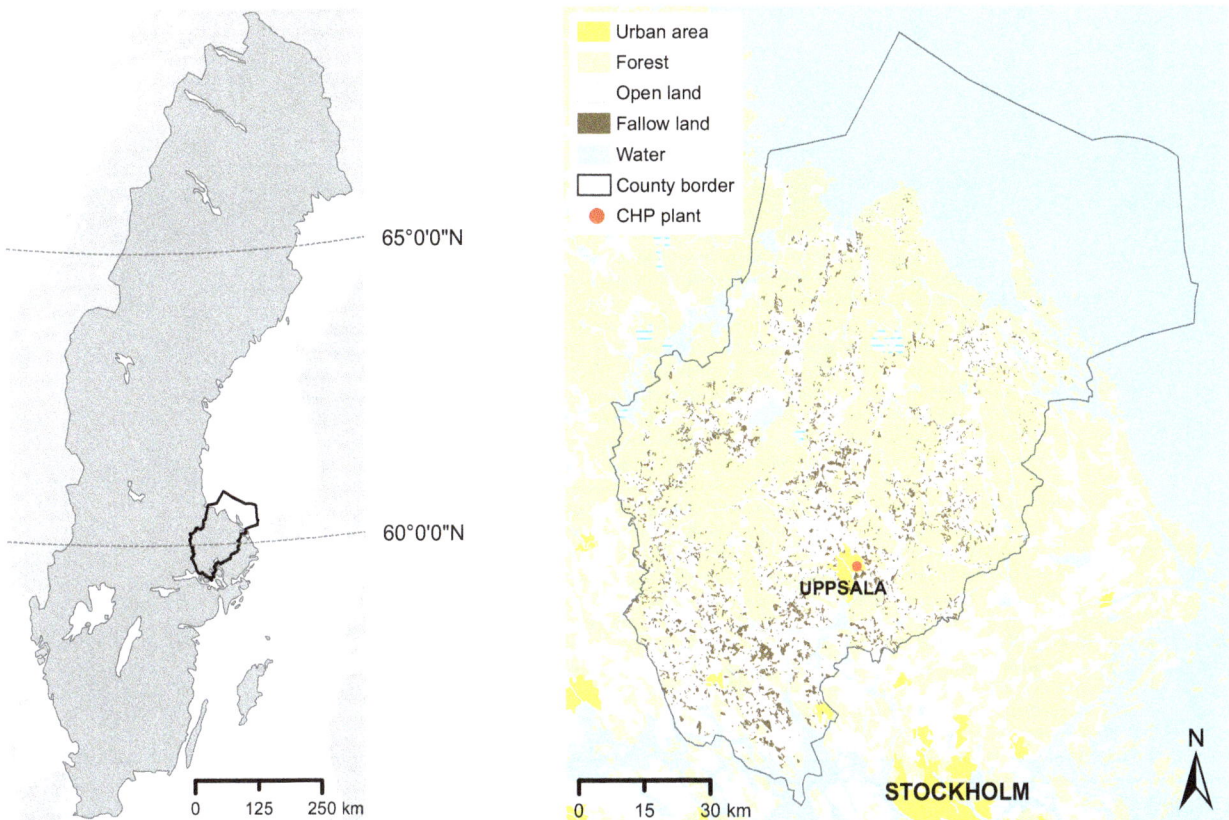

Fig. 1 Left figure: map of Sweden indicating the study region (Uppsala County). Background map © Lantmäteriet. Right figure: distribution of fallow land (brown dots) in Uppsala County. Crop and field information © Swedish Board of Agriculture; background map: overview map 1:1 000 000 © Lantmäteriet.

frame for the study was 100 years, which corresponds to four willow coppice cycles.

The system boundaries included processes related to the willow procurement chain, land use and energy conversion at a CHP plant (Fig. 2). The impact of a one unit increase in energy produced from willow was assessed and only direct land use effects were included as the land was assumed to be initially unused (i.e. fallow). Direct land use effects were defined as the impact of land transformation from the existing land use green fallow to willow cultivation and the continuous effect of altered land use. The climate impact was allocated between heat and power (see Impact allocation). To provide a continuous biomass supply to the CHP plant, all fields were randomly divided into three groups, which were harvested at one-year time steps (within a three-year cutting cycle).

Two functional units were used: (1) 1 hectare (ha) of land, and (2) average heat (MJ) generated at the CHP plant. The per hectare unit was used in the inventory analysis to show the land use change effect on carbon stocks. The heat functional unit was used in the climate impact assessment to show the temperature response when continuously generating heat from willow biomass. A sensitivity analysis was performed where effect of transport distance, yield level and initial SOC content was studied (Table 1).

System description

Procurement chain. The willow plantations were assumed to have a cutting cycle of three years; that is, the willow was harvested and chipped directly at the site every three years. The willow was then regrown and harvested every three years for 25 years, after which the stumps were broken up and new seedlings were planted. The willow procurement chain included the processes site preparation, planting, application of herbicides and pesticides, fertilization, harvesting, chipping, transportation and storage, after which the willow chips were combusted at a CHP plant located in Uppsala.

The yield level in the base scenario was set to 20 Mg dry matter (DM) ha^{-1} for the first harvest and 30 Mg DM ha^{-1} for subsequent harvests in the rotation for all fields (Hollsten et al., 2013). The willow chips were stored for 30 days before combustion with a DM loss of 3%. The production of inputs (seedlings, herbicides, pesticides and fertilizers) was included in the analysis. The willow system was as defined by Hammar et al. (2014), which is based on previous studies of willow production in Sweden (Börjesson, 2006; Nilsson & Bernesson, 2008). Updated data were used for the production of mineral fertilizers (Fossum, 2014). For more details on used input data, see Table S2.

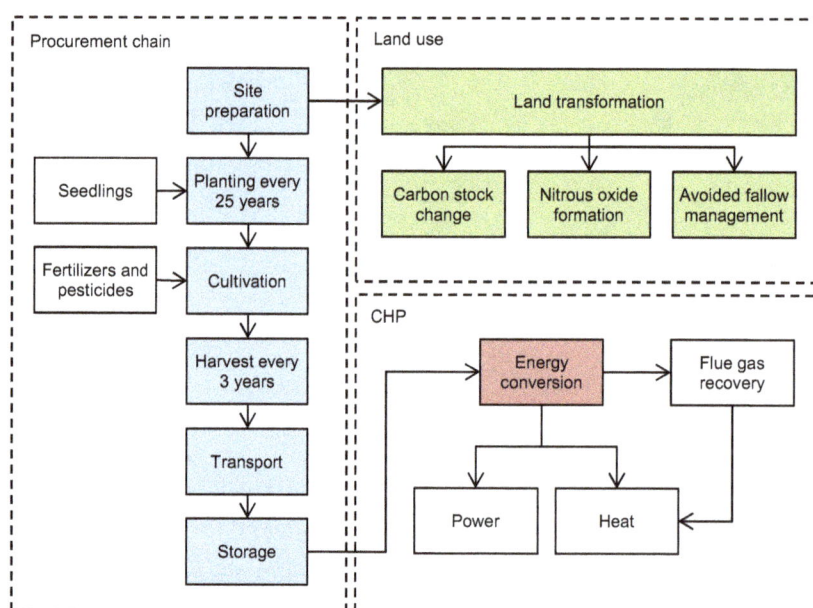

Fig. 2 System boundaries of the study showing processes included in the willow procurement chain, energy conversion at a combined heat and power (CHP) plant and direct land use change effects.

Table 1 Description of sensitivity analysis

Scenario	Description
Base scenario	See System description
Transport ≤60 km	Only including fields located 0–60 km from the CHP plant
Transport ≤30 km	Only including fields located 0–30 km from the CHP plant
Yield −25%	Yield decreased by 25% for all fields and years
Yield +25%	Yield increased by 25% for all fields and years
Yield rand.	Random yield for all fields, ±20% with the same average as the base scenario (i.e. 20 Mg DM first harvest, 30 Mg DM subsequent harvest)
Yield rand. top10%	The top 10% of fields (with random yield) giving the lowest climate impact
Low SOC	Fields with initial SOC content <150 Mg carbon per ha
High SOC	Fields with initial SOC content ≥150 Mg carbon per ha

CHP, combined heat and power; SOC, soil organic carbon; DM, dry matter.

Land use. The direct land use effect of willow cultivation was expressed in two ways: (1) by only considering the willow carbon stock development over time, or (2) by counting the net effect compared with the reference land use, that is the yearly difference between willow and green fallow. The reason for using two forms of expression was to clarify the impact of willow cultivation alone and the impact of the chosen reference

land use. Avoided use of fossil fuel from cutting the fallow once every year was included in the net land use effect (Table S3). The fallow biomass was assumed to be left in the field after cutting and emissions of CO_2 and N_2O due to the change in land use were included. To assess soil carbon flux for the willow plantations, yearly net primary production (NPP) of the willow stands was calculated based on Rytter (2001) to determine yearly carbon uptake in living biomass and yearly carbon input to the soil via leaf litter and root turnover (Hammar *et al.*, 2014). Direct and indirect N_2O soil emissions from application of fertilizers and biomass input were calculated using IPCC emissions factors (for both willow and fallow) (IPCC, 2006; Ahlgren *et al.*, 2009).

Energy conversion. The bio-based CHP plant was assumed to be located in the same area as the current energy facility in Uppsala and to generate the same amount of heat and electricity as the existing facility, that is 1600 GWh (5760 TJ) heat and 225 GWh (810 TJ) electricity per year (Vattenfall, 2016). A higher heating value (HHV) of 19.9 GJ Mg^{-1} DM (dry and ash free) was used for calculating the lower heating value (LHV) for the willow biomass (Strömberg & Herstad Svärd, 2012), which was adjusted for the specific moisture content (MC) by:

$$LHV_{MC} = (HHV - 2.45 \cdot 0.09 \cdot H_2) \cdot \left(1 - \frac{A}{100}\right) - 2.45$$
$$\cdot \frac{MC}{100 - MC} \quad (MJ\,kg^{-1}\,DM) \quad (1)$$

where LHV_{MC} is the theoretical heat gained from wood chips excluding water condensation heat, and 2.45 is the latent heat of water vaporization at 20 °C (MJ kg^{-1}), A is the ash content, 0.09 represents one part hydrogen and eight parts oxygen in water, and H_2 is the hydrogen content (6% assumed) (Lehtikangas, 1999). An ash content of 1.5% and a moisture content of

50% gave an LHV$_{MC}$ of 15.8 GJ Mg^{-1} DM. Emissions of CH$_4$ and N$_2$O from combusting willow chips were set to 11 and 6 g GJ^{-1} fuel, respectively (Paulrud et al., 2010).

Heat produced from hard coal or natural gas was used as reference. Emissions factors from Gode et al. (2011), which include production, distribution and combustion of the fuels, were used. Conversion efficiency for willow and natural gas was adjusted to account for flue gas condensation (Table 2). Flue gas condensation increases the conversion efficiency (for heat) by 15–35% for woody biomass and 10–15% for natural gas (Swedish EPA, 2005), which can give conversion efficiency values of over 100% when using the lower heating value.

Impact allocation. The climate impact was allocated between heat and power using an efficiency allocation method (also called benefit-sharing method; Martinsson et al., 2012; Olsson et al., 2015). The method allocates emissions between power and heat based on the corresponding amount of power and heat that would have been produced in separate production facilities. The allocation factor for heat (α_h) is calculated as:

$$\alpha_h = \frac{\frac{Q_h}{\eta_h}}{\frac{Q_h}{\eta_h} + \frac{Q_p}{p_h}} \qquad (2)$$

where Q is the energy produced from heat (h) or power (p) and η is the conversion efficiency for separate production (excluding flue gas recovery). The allocation factor for power is calculated in the same way. The conversion efficiency for separate heat and power production was set according to EU (2011) (Table 3).

GIS model

The ARCGIS product (ARCMAP version 10.3, ESRI, Redlands, CA, USA) was used for mapping land use in the study region and to link soil data with specific fields. Information regarding land

use, soil texture and soil organic matter (SOM) was obtained from the Swedish Board of Agriculture. Initial SOM data were available for 880 measurement points in the Uppsala region. The SOM for each field was defined as the SOM value at the closest measurement point. The specific soil properties at each site were used as the base for the carbon balance modelling. Fields smaller than 2 ha in area were excluded from the study according to Swedish management recommendations (Hollsten et al., 2013). Distances between fields and the CHP plant were also calculated using ARCGIS, based on road network data from the Swedish Transport Administration (Trafikverket, 2016).

Soil carbon model

Soil carbon balances were calculated using the ICBM (Introductory Carbon Balance Model; Andrén et al., 2004). The model assumes two soil pools, one young (Y) and one old (O), where the carbon input (i) from litter and roots first enters the young pool. A fraction then returns to the atmosphere by oxidation to CO$_2$, while the rest is transferred to the old pool. This fraction, described by the humification coefficient (h), varies with aboveground (a) and belowground (b) biomass. The carbon input for willow was calculated based on the net primary production from Rytter (2001), and the input from fallow was calculated based on Andrén et al. (2004) and an assumed productivity of 4.8 Mg ha^{-1} (including all biomass) (Aronsson et al., 2009). Carbon in coarse roots and stumps entered the soil pool at the end of each rotation (Table 4).

The relationship between the young and old pool is described by:

$$O(t) = \left(O_{t-1} - \left(\frac{h_a \cdot k_Y}{(k_O - k_Y)} \cdot (Y_{a_{t-1}} + i_{a_{t-1}}) + \frac{h_b \cdot k_Y}{(k_O - k_Y)} \cdot (Y_{b_{t-1}} + i_{b_{t-1}}) \right) \right)$$
$$\cdot \exp^{-k_O \cdot r_e} + \left(\frac{h_a \cdot k_Y}{(k_O - k_Y)} \cdot (Y_{a_{t-1}} + i_{a_{t-1}}) + \frac{h_b \cdot k_Y}{(k_O - k_Y)} \cdot (Y_{b_{t-1}} + i_{b_{t-1}}) \right)$$
$$\cdot \exp^{-k_y \cdot r_e} \qquad (3)$$

where the young pool is described by:

$$Y_{[a,b]}(t) = \left(Y_{[a,b]_{t-1}} + i_{[a,b]_{t-1}} \right) \cdot \exp^{-k_Y \cdot r_e} \qquad (4)$$

and where k_Y and k_O are constants representing the decay rate of the two pools (Andrén & Kätterer, 1997; Andrén et al., 2004). The r_e parameter describes external factors such as soil temperature and water-holding capacity (Karlsson, 2012). The r_e parameter was altered to adjust the model for differences in soil texture (Table S1). The total SOC content each year is the sum of the two pools. The SOM content was converted to SOC

Table 2 Conversion efficiency (%) for willow, hard coal and natural gas when combusted in a combined heat and power plant (Börjesson et al., 2010), including increase due to assumed flue gas recovery

	Willow	Hard coal	Natural gas
Heat	55	55	45
Power	30	30	40
Flue gas recovery	20	0	10
Total efficiency	105	85	95

Table 3 Conversion efficiencies for separate heat and power production (excluding flue gas recovery; EU, 2011), and allocation factors of emissions and climate impact between heat (α_h) and power (α_p) production for willow and the two reference fuels hard coal and natural gas

	Conversion efficiencies (%)			Allocation factors (%)		
	Willow	Hard coal	Natural gas	Willow	Hard coal	Natural gas
Heat	78	80	82	44	50	42
Power	33	44	53	56	50	58

Table 4 Annual carbon input (i) to the soil from willow (in base scenario) and green fallow (Mg C per year and ha; Hammar *et al.*, 2014)

	Aboveground (i_a)	Belowground (i_b)
Green fallow	0.7	1.4
Willow		
1st cycle (year 1–3)	0.6, 1.2, 0.9	1.5, 2.8, 2.5
2nd–8th cycle (year 4–24)	1.2, 1.9, 1.6	2.6, 4.1, 3.6
Year 25		2.1

by division by a factor of 1.7 (60% of SOM is SOC). The SOC content was converted from fraction to mass by:

$$SOC(Mg\,C\,ha^{-1}) = \frac{SOC(\%)}{100} \cdot \rho \cdot V \quad (5)$$

where ρ is the specific bulk density for each soil texture, and V is the volume of topsoil (25 cm depth) in 1 ha (10 000 m^2). The bulk density values for the different soil textures were set according to Kätterer *et al.* (2006).

Climate model

The climate impact was assessed using AGTP, as defined by the IPCC (Myhre *et al.*, 2013a). This climate metric considers the temperature change resulting from a radiative imbalance of the globe, that is radiative forcing (RF), due to a pulse emission of a GHG. Each GHG has a specific radiative efficiency, meaning that the gases have different abilities to change the balance between the incoming solar radiation and the outgoing terrestrial radiation. The GHGs also remain in the atmosphere for varying lengths of time before they decay. N$_2$O and CH$_4$ have an average perturbation lifetime of 121 and 12.4 years, respectively. CO$_2$ remains in the atmosphere until it is taken up by oceans or the biosphere, while about one-third remains airborne. The perturbation lifetime of CO$_2$ was modelled using the Bern carbon cycle model (Joos *et al.*, 2001, 2013). The indirect effect of CH$_4$ oxidation was included in the climate model. The AGTP (referred to as 'temperature response' in the Results section) is described by:

$$AGTP_x(H) = \int_0^H RF_x(t) R_T(H - t) dt \quad (6)$$

which is a convolution between the radiative forcing (RF) and the climate response function (R_T) due to a unit change in the RF from a pulse emission of gas x. The temperature metric considers the timing of the GHG emissions and their perturbation lifetime and is therefore a very useful metric for displaying time-dependent climate change, unlike the more common GWP metric, which describes the cumulative RF of one gas relative to the cumulative RF of CO$_2$ during a set time frame (Joos *et al.*, 2013). However, GWP in a 100-year perspective (GWP$_{100}$) was also applied in this study to enable comparisons with previous studies. According to Myhre *et al.* (2013b), the GWP$_{100}$ of CH$_4$ and N$_2$O is 28-fold (fossil methane) and 265-fold larger, respectively, than that of CO$_2$ in a 100-year time frame.

Results

Inventory analysis

Field properties. In Uppsala County, about 7200 fields of varying size were reported as being under fallow in 2014, giving a total fallow area of around 14 000 ha (Fig. 1). Of these, about 2100 fields exceeded the cut-off size of 2 ha applied in this study, giving a total area of about 9800 ha. The transport distance between the fields and the CHP plant varied from 3 to 96 km, with an average distance of 43 km (Table 5).

The most common soil texture in the selected fields was clay (26%), followed by clay loam (21%), loam (14%) and loamy sand (12%) (Fig. 3). The area of each soil texture was decreased by around 30% for most soils on only selecting fields ≥2 ha.

Soil carbon balance. The initial SOC content for all fields varied between 19.5 and 447 Mg C ha^{-1}, with an average of 114 Mg C ha^{-1} (Fig. 4). The initial SOC pool was not in steady-state, and the content had a strong influence on changes in carbon stocks (Fig. 5). Soils with an initially high SOC content released carbon both when willow was established (Fig. 5a) and when the land remained as green fallow (Fig. 5b). Fields with initially low SOC content sequestered carbon over time, particularly when willow was established rather than leaving green fallow in place. Thus, the net land use effect of establishing willow on fallow land was net uptake of carbon (Fig. 5c). This effect showed almost no variation between fields due to the assumption of constant willow and fallow productivity. The final net effect on SOC after 100 years varied between 83.1 and 85.2 Mg C ha^{-1}, with an average of 84.3 Mg C ha^{-1}.

Energy balance. The energy supplied by willow biomass each year (from all fields ≥2 ha) was on average 1040 TJ heat and 420 TJ power, which corresponds to ~20% of the heat and ~50% of the power produced at the existing energy facility. This corresponds to around

Table 5 Properties of fields ≥2 ha in Uppsala County ($N = 2083$)

	Area (ha)	Distance* (km)	Initial SOC (%)	Initial SOC (Mg C ha^{-1})
Max	28.3	95.8	12.1	447.0
Min	2.0	3.1	0.6	19.5
Median	3.6	44.0	2.6	88.9
Average	4.7	43.4	3.4	114.1

SOC, soil organic carbon.

*Distances are for one-way transport.

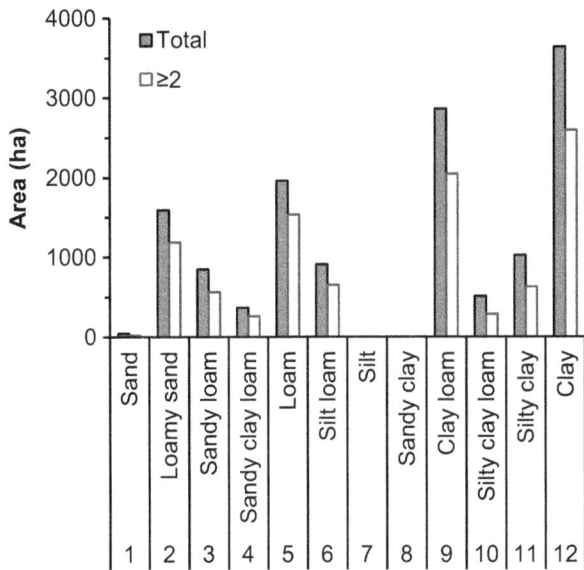

Fig. 3 Distribution of soil texture in all fields in Uppsala County (Total) and in fields ≥2 ha.

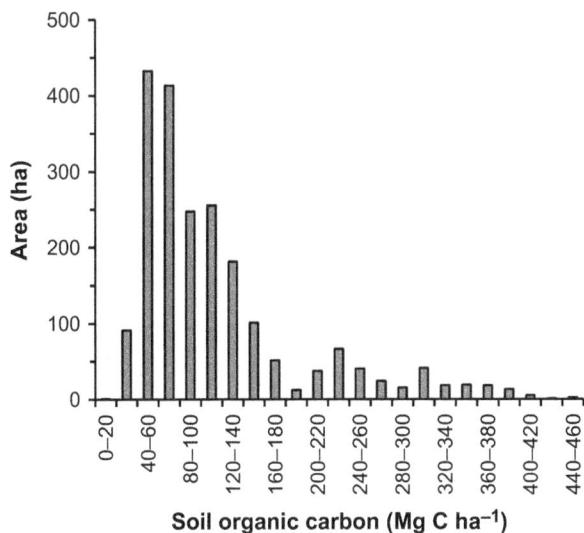

Fig. 4 Frequency distribution of initial soil organic carbon content.

150 GJ ha^{-1} and yr (including heat, power and flue gas condensation). The average primary energy use per hectare and year ranged between 4 to 6.1 GJ, with an average of about 4.9 GJ (including all processes). This resulted in an average energy ratio (ER) ranging from 24 to 36, with an average of 30 MJ MJ^{-1} (including heat, power and flue gas condensation). The production and application of fertilizers gave the highest primary energy use during one rotation period, followed by the processes harvest and chipping, forwarding and transport (Fig. 6).

Climate impact assessment

Global warming potential (GWP). The average GWP$_{100}$ was 7.2 g CO_2-eq per MJ heat for the willow procurement chains in the different fields (excluding biogenic carbon) (Table 6). The GWP$_{100}$ varied mainly due to differences in transportation distance between field and energy facility. Production and use of fertilizers (including soil N_2O emissions) gave the highest GWP$_{100}$, followed by emissions of N_2O and CH_4 from incomplete combustion (Fig. 6). Including biogenic carbon for the willow cultivation gave large variations in GWP$_{100}$ (as shown in Fig. 5a), with an average of −2.3 g CO_2-eq per MJ heat. When the net land use effect was included (Fig. 5c), the variation was small, with an average GWP$_{100}$ of −8.2 g CO_2-eq per MJ heat. The GWP was thus smaller when accounting for avoided emissions from the reference land use of green fallow. In comparison, the GWP$_{100}$ for fossil coal and natural gas was 116 and 59.7 g CO_2-eq per MJ heat, respectively (no land use emissions included).

Temperature response. The climate impact of willow energy, in terms of temperature response over time, varied greatly when only including SOC changes for the willow cultivation (and no comparison with green fallow) (Fig. 7a). Fields with high initial SOC content released carbon from the soil when willow was established (Fig. 5a), which gave a positive temperature response (i.e. warming effect). However, the reference land use green fallow would release even more carbon from those fields, which means that the net effect of growing willow was a negative temperature response for all fields (i.e. a cooling effect) (Fig. 7b). On harvesting all fields in the landscape to continuously supply the local CHP plant, the temperature response was negative (cooling effect) (Fig. 7c). The final temperature response after 100 years was −6·10^{-16} K MJ^{-1} heat (including the net land use effect).

The climate impact and energy return of the individual fields varied due to varying field properties. Prioritizing the best performing fields (in terms of lowest climate impact) improved the climate change mitigation potential per MJ heat (Fig. 8). However, this meant that the total heat production was also lower. There was no trade-off between maximizing total heat production and lowering the climate impact, as all fields (including net land use effect) showed negative climate impacts (i.e. cooling effect) (Fig. 8b). When fields giving the highest climate impact were utilized (bottom 10% or 50%), the temperature response was positive (i.e. warming effect) when considering the willow SOC only (Fig. 8a). This means that the choice of field plays a greater role when

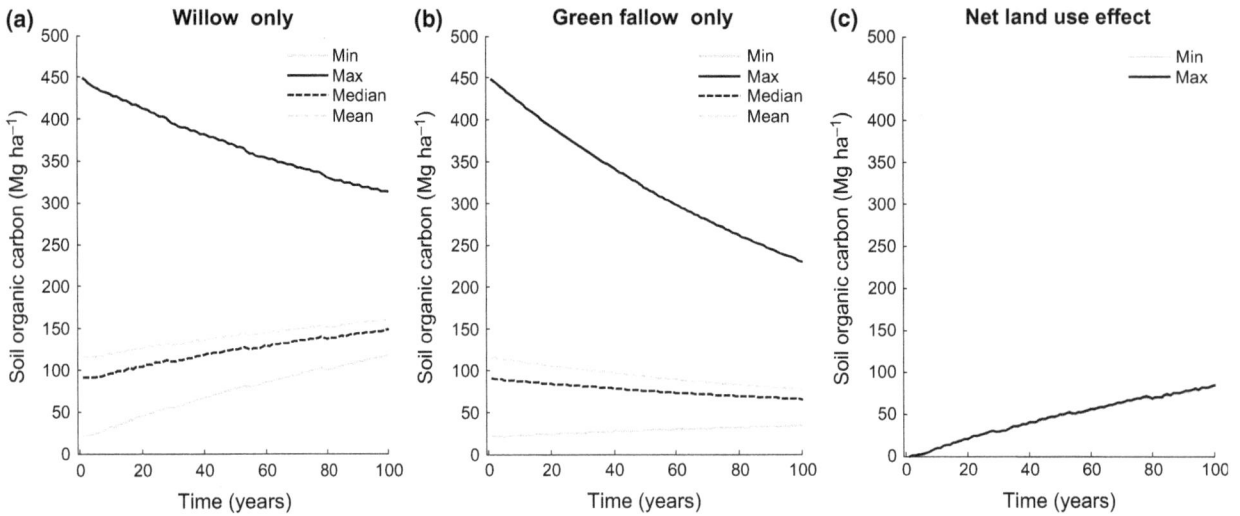

Fig. 5 Changes in soil organic carbon (SOC) content for all fields ($N = 2083$) when including (a) willow cultivation only and (b) reference land use only (continued green fallow); and (c) the net effect of transforming green fallow into willow cultivation and the continuing effect over time.

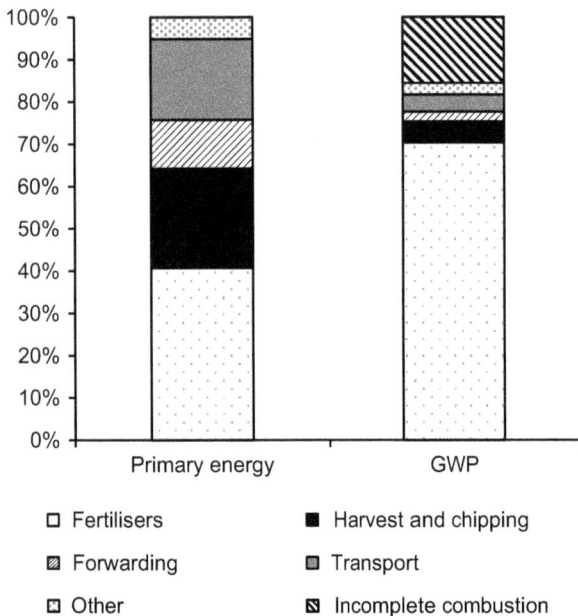

Fig. 6 Distribution of primary energy use during one willow rotation (MJ ha^{-1}) and global warming potential (GWP$_{100}$) from procurement chain (g MJ^{-1} heat), based on average transport distance (43 km). Other includes field preparation, production and planting of seedlings, production and application of pesticides and stump removal. Incomplete combustion includes emissions of nitrous oxide and methane. No soil organic carbon changes are included.

excluding the impact of avoided carbon emissions from the reference land use green fallow and might have an even larger impact if another alternative land use were considered.

Table 6 Global warming potential (GWP$_{100}$, g CO$_2$-eq per MJ heat) for willow systems during 100 years of cultivation on different fields ($N = 2083$). The procurement chain included fossil greenhouse gas emissions and nitrous oxide emissions from the soil

	Procurement chain (excluding SOC)	Procurement chain including SOC (willow only)	Procurement chain including SOC (net land use effect)
Min	6.9	−10.4	−8.6
Max	7.6	25.2	−7.8
Average	7.2	−2.3	−8.2

SOC, soil organic carbon

Continuously growing willow for energy over a landscape gave a much smaller climate impact than using fossil coal or natural gas ($\sim$70$\cdot$10^{-16} K MJ^{-1} heat and $\sim$35$\cdot$10^{-16} K MJ^{-1} heat after 100 years, respectively) (Fig. 9).

Sensitivity analysis

The sensitivity analysis showed that willow yield level had the largest influence on the temperature response over the whole landscape, while initial SOC content and transport distance made a minimal difference when considering the average effect over the whole landscape (Table 7). However, yearly heat production was highly affected by initial SOC content and transport distance, as fewer fields were assumed to be cultivated in these scenarios. When fields were assumed to be located within 60 or 30 km from the energy facility (compared

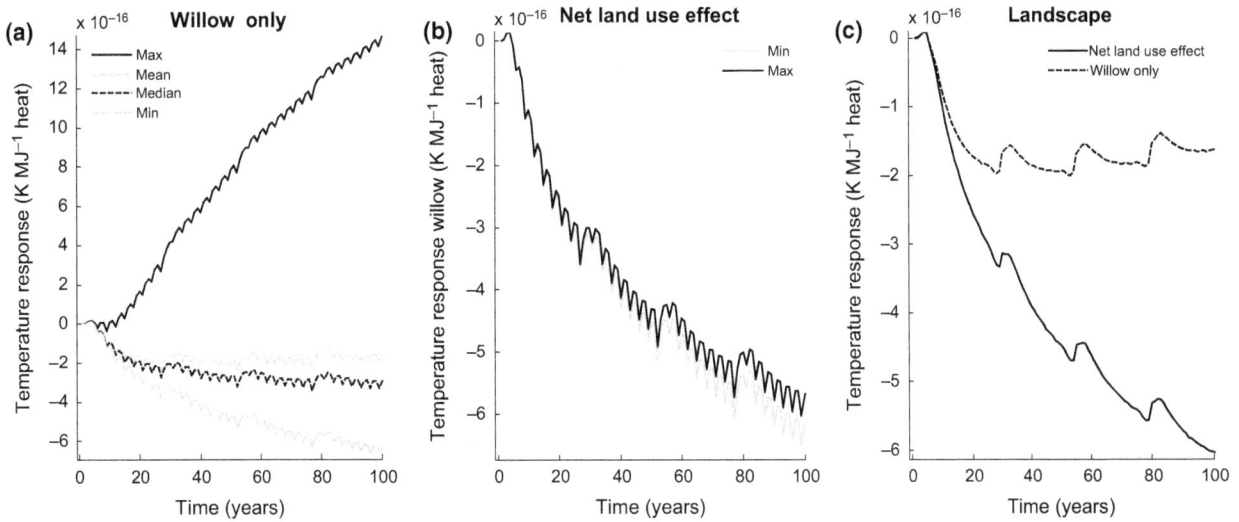

Fig. 7 Temperature response of all fields ($N = 2083$) including (a) willow procurement chains and soil organic carbon (SOC) changes for willow cultivation only; (b) willow procurement chains and net land use effect (i.e. difference between willow and green fallow); and (c) the combined effect of willow cultivation over a landscape (considering willow SOC only or the net land use effect), where the willow stands were harvested in groups of three with one-year time steps.

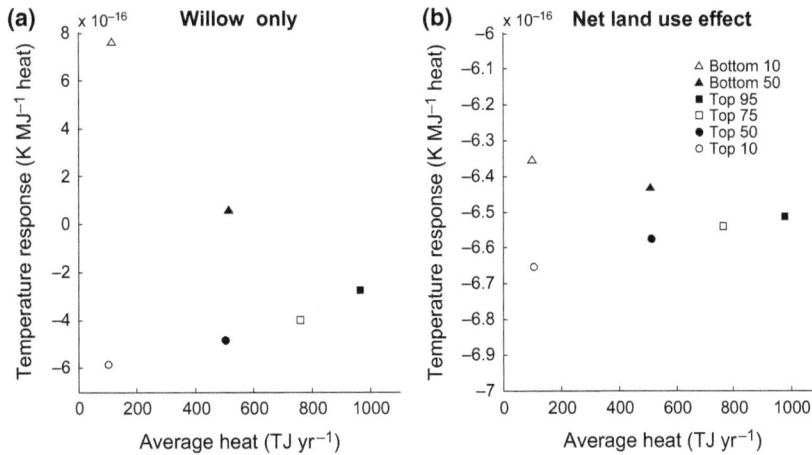

Fig. 8 Final temperature response after 100 years of utilizing willow energy harvested over a landscape and average heat produced per year when considering: (a) willow procurement chains and willow soil organic carbon (SOC) only; or (b) willow procurement chains and the net SOC effect, that is the difference between willow and the reference land use green fallow. The diagrams show the effect when selecting the best performing fields (top 10–95%) and worst performing fields (bottom 10–50%) in terms of climate impact (Note scale differences).

with 96 km in the base scenario), the average GWP_{100} (only including fossil GHGs) was lowered by 0.5–2.1% per MJ of heat compared with the base scenario, while the amount of heat produced was lowered by 17–71%. Selecting fields based on initial carbon content gave either a slightly lower climate impact (low initial SOC) or higher climate impact (high initial SOC), while the energy production was lowered by 20–80%.

Higher yield gave the largest climate benefit and energy output (Fig. 10). When the yield was set randomly, the temperature response over the landscape

was slightly lower than for the base scenario and the average net SOC effect was somewhat higher. On choosing the top 10% (with random yields) of fields (with the smallest climate impact), the temperature response over the landscape was lowered by 5%, but heat production decreased by 88%.

Discussion

This study examined the climate effects of supplying a local CHP plant with willow biomass during a 100-year

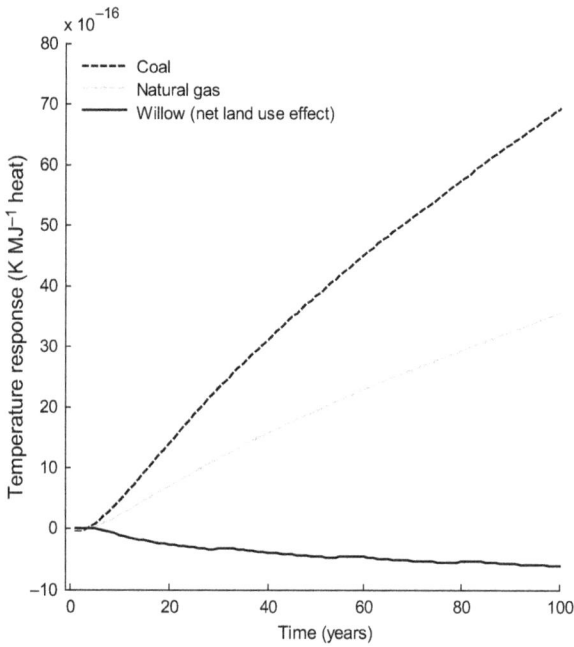

Fig. 9 Temperature response of willow energy (landscape perspective including net land use effect, i.e. difference between willow and green fallow), fossil coal and natural gas per MJ heat and year. The net effect of land use change in the fossil reference systems was zero; that is, no emissions from land use were included for coal and natural gas.

time frame, with the aim of analysing the potential effects of regionally produced energy considering spatial variations. If all fallow land (≥2 ha) in the study region were utilized for producing willow energy, ~20% and ~50% of the yearly heat and power production, respectively, at the energy facility could be produced from willow chips. The average climate impact, in terms of temperature response over the 100-year period, would be negative (i.e. a cooling effect) due to carbon sequestration in living biomass and soil. In other words, under the assumptions in this study, willow cultivation could both generate energy and mitigate climate change even when considering spatial variations in a landscape.

The major contributor to the primary energy use in the willow procurement chains was the production and use of fertilizers (~40%) followed by the willow harvest (~35%) (including chipping and forwarding), which is in line with previous studies on willow energy (Djomo et al., 2011). When excluding SOC changes, the production and application of fertilizers gave the largest contribution to the GWP_{100} due to soil N_2O emissions (in total ~70%) (Fig. 6). Including SOC changes gave negative average GWP_{100} of −2.3 or −8.2 g CO_2-eq MJ^{-1} heat (depending on if only willow cultivation or net land use

Table 7 Sensitivity analysis based on net land use effect, that is the difference between willow cultivation and the reference land use green fallow

Parameter	Base scenario	Transport ≤60 km	Transport ≤30 km	Yield −25%	Yield +25%	Yield rand	Yield top 10%	SOC <150	SOC ≥150
No. of fields	2083	1705	598	2083	2083	2083	209	1668	415
Final average net SOC (Mg C ha^{-1})	84.3	+0.0%	+0.1%	−32%	+32%	+0.9%	+24%	+0.1%	−0.4%
Final temperature response landscape (K MJ^{-1} heat)	−6·10^{-16}	−0.4%	−1.7%	+7.7%	−4.6%	−0.1%	−5.0%	−0.2%	+0.8%
Heat (TJ yr^{-1})	1040	−17%	−71%	−25%	+25%	+0.5%	−88%	−20%	−80%
Average GWP$_{100}$ excluding SOC (g CO$_2$-eq MJ^{-1} heat)	7.2	−0.5%	−2.1%	+1.1%	−1.1%	+0.0%	−1.7%	−0.1%	+0.4%

SOC, soil organic carbon; GWP, global warming potential.

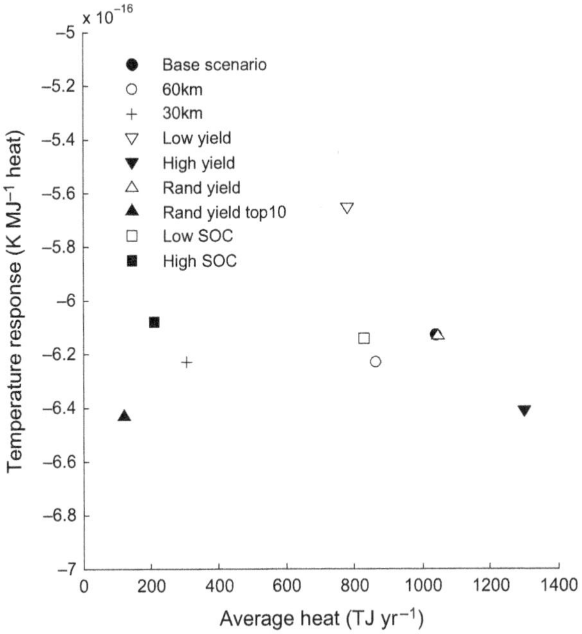

Fig. 10 Final temperature response after 100 years of willow cultivation and yearly average heat produced over a landscape in the different scenarios, including willow procurement chains and the net land use effect, that is the difference between willow cultivation and the reference land use green fallow. Rand yield, randomly allocated yield level; SOC, soil organic carbon.

effect was included). According to a literature review by Djomo et al. (2011), the GWP of willow energy ranged from −2.7 to −4.7 g CO_2-eq MJ^{-1} fuel (including SOC). In Zetterberg & Chen (2015), the GWP of willow was reported as −2.0 g CO_2-eq MJ^{-1} fuel and the temperature change after 100 years was estimated to −5·10^{-16} K MJ^{-1} fuel. The same value in this study was −6·10^{-16} K MJ^{-1} heat; however, the values are problematic to compare because functional units and system boundaries differ.

From a landscape perspective, growing willow on all available fields for energy purposes resulted in a negative temperature response (i.e. cooling effect) (Fig. 7c). From a single stand perspective, there was large variation when only considering willow SOC changes, which could potentially result in a positive temperature response (i.e. warming effect) if the worst fields were chosen (Fig. 8a). When including the net land use effect (i.e. avoided SOC fluxes from green fallow), a negative temperature response (i.e. cooling effect) was shown for all fields (Fig. 8b). In conclusion, the reference land use had a large influence on the results, especially when considering single stands.

A sensitivity analysis was also performed (considering the net land use effect) to study the effect of transport distance, initial SOC content and yield level. Transport distance had the largest influence on fossil GHG

emissions, which was reflected in the GWP$_{100}$ (Table 7), but the influence of transport on the overall temperature response was low. By selecting, for example fields located close to the CHP plant, the climate mitigating potential could be improved, but this would also generate less energy (Fig. 10). Therefore, the best option from a climate mitigation perspective would be to utilize all fields to generate as much energy as possible.

The yield level is an important factor for the energy return and the climate impact of willow. When commercial willow was introduced in Sweden, high yields were expected but unfortunately were not obtained in practice (Mola-Yudego et al., 2015). The reasons for the unexpectedly low yields may have been the combination of poor management practices and use of low productivity soils (Dimitriou et al., 2011; Mola-Yudego, 2011). Today, new improved clones have been developed, in terms of resistance to diseases, insects and frost damage as well better stem characteristics and coppice responses (Karp & Shield, 2008). This in combination with better stand management guidelines for willow farmers increases the prospect of achieving better willow yields. Mola-Yudego (2011) studied the production trend of commercial willow plantations in Sweden (year 1986–2000) and found an annual increment of 2.06 Mg DM per yr and ha per decade. They concluded that higher yields could be expected in a near future due to the development of new willow varieties. The willow yield likely varies with soil texture and water availability in the study region (Nord-Larsen et al., 2015), but since found yield models for the specific region were based on old statistic with very low yield, it was not considered to be applicable for this study where the aim was to assess the available best willow practice. Instead, the yield was kept constant for all fields in a base scenario, while yield variations were assessed in a sensitivity analysis.

The net primary production of willow also has a large influence on the soil carbon stock, as the carbon balance is affected by the initial carbon content and the carbon input from above- and belowground biomass. Moreover, the productivity of the reference land use also affects the results when the net land use effect is considered. In this study, the productivity of green fallow was kept constant for all fields and scenarios, but it would also be important to determine the influence of soil texture on the reference land use. Long-term field measurements of both willow and fallow would be useful both for validating the carbon balance model and for decreasing uncertainty in willow energy production estimates.

A landscape perspective was applied to continuously supply the bio-based CHP plant with feedstock. However, willow is harvested during the winter when the soil carrying capacity is high due to frost, so to obtain a continuous energy supply during the whole year other feedstocks are required, as it is not appropriate to store willow chips for a longer period (Strömberg & Herstad Svärd, 2012; Dimitriou & Rutz, 2015). Willow is also generally cofired with other fuels, to decrease related problems such as sintering and coating (Strömberg & Herstad Svärd, 2012). To meet the total demand of the energy facility (and the study region), additional feedstocks (and land) are therefore required.

No indirect land use changes (iLUC) were included in this study under the assumption that only fallow land was utilized. An increased demand for energy or food crops in the future may increase the pressure on land. There is, however, no agreed international accounting standard for including iLUC in LCA, but so-called iLUC factors have been developed for biofuels (Ahlgren & Di Lucia, 2014). These factors are calculated with theoretical models based on market predictions and have resulted in a wide range linked with a high degree of uncertainty. Including iLUC in LCA may according to Finkbeiner (2014) damage the reliability of natural science based LCA results.

In conclusion, this study showed that supplying an energy facility yearly with willow biomass grown on fallow land had a negative temperature response (i.e. cooling effect) when considering spatial variations in a landscape. The climate change mitigation potential was improved by selecting the best performing fields (e.g. in terms of highest SOC increase), but all fields needed to be utilized to generate as much energy as possible. Another conclusion was that the choice of reference land use played a major role for the results. Combined GIS and time-dependent LCA method proved to be a useful way to assess land use change in bioenergy systems. Moreover, the approach can be further improved with better data availability in the future and expanded to include and compare different types of biomass.

Acknowledgements

This research was funded by COMPLEX (project 308601) and the Swedish Energy Agency (project 41976-1). The authors gratefully acknowledge the contribution of Thomas Kätterer and Martin Bolinder (Dept. of Ecology, SLU) for their expertise in soil carbon dynamics and Anders Larsolle (Dept. of Energy and Technology, SLU) for his help with GIS modelling.

References

Abolina E, Volk TA, Lazdina D (2015) GIS based agricultural land availability assessment for the establishment of short rotation woody crops in Latvia. *Biomass and Bioenergy*, **72**, 263–272.

Ahlgren S, Di Lucia L (2014) Indirect land use changes of biofuel production – a review of modelling efforts and policy developments in the European Union. *Biotechnology for Biofuels*, **7**, 1–10.

Ahlgren S, Hansson P-A, Kimming M, Aronsson P, Lundkvist H (2009) *Greenhouse gas emissions from cultivation of agricultural crops for biofuels and production of biogas from manure - Implementation of the Directive of the European Parliament and of the Council on the promotion of the use of energy from renewable sources. Revised according to instructions for interpretation of the Directive from the European Commission 30 July 2009.* Dnr SLU ua 12-4067/08. Uppsala.

Andrén O, Kätterer T (1997) ICBM: The introductory carbon balance model for exploration of soil carbon balances. *Ecological Applications*, **7**, 1226–1236.

Andrén O, Kätterer T, Karlsson T (2004) ICBM regional model for estimations of dynamics of agricultural soil carbon pools. *Nutrient Cycling in Agroecosystems*, **70**, 231–239.

Aronsson H, Stenberg M, Rydberg T (2009) *Kväve- och Fosforutlakning Från två Växtföljder på Lerjord med Grön- och Stubbträda.* Sveriges lantbruksuniversitet (SLU), Uppsala.

Aust C, Schweier J, Brodbeck F, Sauter UH, Becker G, Schnitzler J-P (2014) Land availability and potential biomass production with poplar and willow short rotation coppices in Germany. *GCB Bioenergy*, **6**, 521–533.

Börjesson P (2006) *Livscykelanalys av Salixproduktion (Life Cycle Assessment of Willow Production, english abstract).* Rapport nr 60. Lund: Institutionen för teknik och samhälle Avdelningen för miljö- och energisystem.

Börjesson P, Tufvesson L, Lantz M (2010) *Livscykelanalys av svenska biodrivmedel (Life Cycle Assessment of Biofuels in Sweden).* In Swedish with English summary. 70. Lund, Sweden.: LUNDS TEKNISKA HÖGSKOLA.

Brandão M, Milà i Canals L, Clift R (2011) Soil organic carbon changes in the cultivation of energy crops: Implications for GHG balances and soil quality for use in LCA. *Biomass and Bioenergy*, **35**, 2323–2336.

Castellano PJ, Volk TA, Herrington LP (2009) Estimates of technically available woody biomass feedstock from natural forests and willow biomass crops for two locations in New York State. *Biomass and Bioenergy*, **33**, 393–406.

Cherubini F, Strømman AH (2011) Life cycle assessment of bioenergy systems: State of the art and future challenges. *Bioresource Technology*, **102**, 437–451.

Dimitriou I, Rutz D (2015) *Sustainable Short Rotation Coppice – A Handbook.* WIP Renewable Energies, Munich, Germany.

Dimitriou I, Rosenqvist H, Berndes G (2011) Slow expansion and low yields of willow short rotation coppice in Sweden; implications for future strategies. *Biomass and Bioenergy*, **35**, 4613–4618.

Djomo SN, Kasmioui OE, Ceulemans R (2011) Energy and greenhouse gas balance of bioenergy production from poplar and willow: a review. *GCB Bioenergy*, **3**, 181–197.

Ekman A, Wallberg O, Joelsson E, Börjesson P (2013) Possibilities for sustainable biorefineries based on agricultural residues - A case study of potential straw-based ethanol production in Sweden. *Applied Energy*, **102**, 299–308.

Ericsson N, Porsö C, Ahlgren S, Nordberg Å, Sundberg C, Hansson P-A (2013) Time-dependent climate impact of a bioenergy system – methodology development and application to Swedish conditions. *GCB Bioenergy*, **5**, 580–590.

Ericsson N, Nordberg Å, Sundberg C, Ahlgren S, Hansson P-A (2014) Climate impact and energy efficiency from electricity generation through anaerobic digestion or direct combustion of short rotation coppice willow. *Applied Energy*, **132**, 86–98.

EU (2011) *COMMISSION IMPLEMENTING DECISION of 19 December 2011 establishing harmonised efficiency reference values for separate production of electricity and heat in application of Directive 2004/8/EC of the European Parliament and of the Council and repealing Commission Decision 2007/74/EC.*

Fargione J, Hill J, Tilman D, Polasky S, Hawthorne P (2008) Land clearing and the biofuel carbon debt. *Science*, **319**, 1235–1238.

Finkbeiner M (2014) Indirect land use change – Help beyond the hype? *Biomass and Bioenergy*, **62**, 218–221.

Fiorese G, Guariso G (2010) A GIS-based approach to evaluate biomass potential from energy crops at regional scale. *Environmental Modelling & Software*, **25**, 702–711.

Fossum J-P (2014) Calculation of Carbon Footprint of Fertilizer Production. Available at: http://yara.com/doc/122597_2013_Carbon_footprint-of_AN_Method_of_calculation.pdf (accessed 3 December 2015).

Gasol CM, Gabarrell X, Rigola M, González-García S, Rieradevall J (2011) Environmental assessment: (LCA) and spatial modelling (GIS) of energy crop implementation on local scale. *Biomass and Bioenergy*, **35**, 2975–2985.

Gode J, Martinsson F, Hagberg L, Öman A, Höglund J, Palm D (2011) *Miljöfaktaboken 2011. Uppskattade emissionsfaktorer för bränslen, el, värme och transporter (Environmental fact book 2011. Estimated emission factors for fuels, electricity, heat and transport (in Sweden with english abstract). ANLÄGGNINGS- OCH FÖRBRÄNNINGSTEKNIK 1183.).* Stockholm, Sweden.

Goglio P, Smith WN, Grant BB, Desjardins RL, McConkey BG, Campbell CA, Nemecek T (2015) Accounting for soil carbon changes in agricultural life cycle assessment (LCA): A review. *Journal of Cleaner Production*, **104**, 23–39.

Hammar T, Ericsson N, Sundberg C, Hansson P-A (2014) Climate impact of willow grown for bioenergy in Sweden. *BioEnergy Research*, **7**, 1529–1540.

Hauschild MZ, Goedkoop M, Guinée J et al. (2012) Identifying best existing practice for characterization modeling in life cycle impact assessment. *The International Journal of Life Cycle Assessment*, **18**, 683–697.

van der Hilst F, Lesschen JP, van Dam JMC, Riksen M, Verweij PA, Sanders JPM, Faaij APC (2012) Spatial variation of environmental impacts of regional biomass chains. *Renewable and Sustainable Energy Reviews*, **16**, 2053–2069.

Hollsten R, Arkelöv O, Ingelman G (2013) *Handbok för salixodlare (Handbook for willow growers)*. OVR250. Jordbruksverket (Swedish Board of Agriculture). Jönköping, Sweden.

Humpenöder F, Schaldach R, Cikovani Y, Schebek L (2013) Effects of land-use change on the carbon balance of 1st generation biofuels: An analysis for the European Union combining spatial modeling and LCA. *Biomass and Bioenergy*, **56**, 166–178.

IPCC (2006) *2006 IPCC Guidelines for National Greenhouse Gas Inventories, Prepared by the National Greenhouse Gas Inventories Programme*. IGES, Japan.

IPCC (2007) *Contribution of Working Group I to the Fourth Assessment Report of the Intergovernmental Panel on Climate Change, 2007 Cambridge*. Cambridge University Press, UK and New York, NY, USA.

ISO 14040 (2006) ISO 14040:2006. Environmental management. Life cycle assessment – Principle and Framework. Geneva.

ISO 14044 (2006) ISO 14044:2006. Environmental management - Life cycle assessment - Requirements and guidelines. Geneva.

Joos F, Prentice IC, Sitch S et al. (2001) Global warming feedbacks on terrestrial carbon uptake under the Intergovernmental Panel on Climate Change (IPCC) Emission Scenarios. *Global Biogeochemical Cycles*, **15**, 891–907.

Joos F, Roth R, Fuglestvedt JS et al. (2013) Carbon dioxide and climate impulse response functions for the computation of greenhouse gas metrics: a multi-model analysis. *Atmospheric Chemistry and Physics*, **13**, 2793–2825.

Karlsson T (2012) *Carbon and Nitrogen Dynamics in Agricultural Soils. Model Applications at Different Scales in Time and Space*. Diss. Uppsala: Swedish University of Agricultural Sciences.

Karp A, Shield I (2008) Bioenergy from plants and the sustainable yield challenge. *New Phytologist*, **179**, 15–32.

Kätterer T, Andrén O, Jansson PE (2006) Pedotransfer functions for estimating plant available water and bulk density in Swedish agricultural soils. *Acta Agriculturae Scandinavica, Section B — Soil & Plant Science*, **56**, 263–276.

Krzyżaniak M, Stolarski MJ, Szczukowski S, Tworkowski J, Bieniek A, Mleczek M (2015) Willow biomass obtained from different soils as a feedstock for energy. *Industrial Crops and Products*, **75**, 114–121.

Lehtikangas P (1999) *Lagringshandbok för Trädbränslen, 2:A Upplaga (Storage Handbook for Wood Fuels)* (2nd edn). Swedish University of Agricultural Sciences (SLU), Uppsala, Sweden.

Martinsson F, Gode J, Ekvall T (2012) *kraftvärmeallokeringar*.

Mola-Yudego B (2011) Trends and productivity improvements from commercial willow plantations in Sweden during the period 1986–2000. *Biomass and Bioenergy*, **35**, 446–453.

Mola-Yudego B, Díaz-Yáñez O, Dimitriou I (2015) How much yield should we expect from fast-growing plantations for energy? Divergences between experiments and commercial willow plantations. *BioEnergy Research*, **8**, 1769–1777.

Monteleone M, Cammerino ARB, Garofalo P, Delivand MK (2015) Straw-to-soil or straw-to-energy? An optimal trade off in a long term sustainability perspective. *Applied Energy*, **154**, 891–899.

Myhre G, Shindell D, Bréon F-M et al. (2013a) Anthropogenic and natural radiative forcing supplementary material. In: *Climate Change 2013: The Physical Science Basis. Contribution of Working Group I to the Fifth Assessment Report of the Intergovernmental Panel on Climate Change* (eds Stocker TF, Qin D, Plattner G-K, Tignor M, Allen SK, Boschung J, Nauels A, Xia Y, Bex V, Midgley PM), pp. 1–44. Cambridge University Press, Cambridge, UK and New York, NY, USA.

Myhre G, Shindell D, Bréon F-M et al. (2013b) Anthropogenic and natural radiative forcing. In: *Climate Change 2013: The Physical Science Basis. Contribution of Working Group I to the Fifth Assessment Report of the Intergovernmental Panel on Climate*

Change (eds Stocker TF, Qin D, Plattner G-K, Tignor M, Allen SK, Boschung J, Nauels A, Xia Y, Bex V, Midgley PM), pp. 659–740. Cambridge University Press, Cambridge, UK and New York, NY, USA.

Nilsson D, Bernesson S (2008) *Processing biofuels from farm raw materials – A systems study*. Report 001. Department of energy and technology, SLU, Uppsala, Sweden.

Nord-Larsen T, Sevel L, Raulund-Rasmussen K (2015) Commercially grown short rotation coppice willow in Denmark: biomass production and factors affecting production. *BioEnergy Research*, **8**, 325–339.

Olsson L, Wetterlund E, Söderström M (2015) Assessing the climate impact of district heating systems with combined heat and power production and industrial excess heat. *Resources, Conservation and Recycling*, **96**, 31–39.

Paulrud S, Fridell E, Stripple H, Gustafsson T (2010) *Uppdatering av klimatrelaterade emissionsfaktorer (Updated climate related emission factors)*. SMED 92 2010. Swedish Meteorological and Hydrological Institute (SMHI), Norrköping, Sweden.

Porsö C, Hansson P-A (2014) Time-dependent climate impact of heat production from Swedish willow and poplar pellets – In a life cycle perspective. *Biomass and Bioenergy*, **70**, 287–301.

Rytter R-M (2001) Biomass production and allocation, including fine-root turnover, and annual N uptake in lysimeter-grown basket willows. *Forest Ecology and Management*, **140**, 177–192.

Rytter R-M (2012) The potential of willow and poplar plantations as carbon sinks in Sweden. *Biomass and Bioenergy*, **36**, 86–95.

SCB (2015) *Trädesareal 2014 fördelad på kort- och långliggande träda (Fallow land in 2014, divided into annual and perennial fallow)*. Available at: www.scb.se/sv_/Hitta-statistik/Statistik-efter-amne/Miljo/Godselmedel-och-kalk/Godselmedel-och-odlingsatgarder-i-jordbruket/21236/21243/207168/ (accessed 8 December 2012).

Searchinger T, Heimlich R, Houghton RA et al. (2008) Use of U.S. croplands for biofuels increases greenhouse gases through emissions from land-use change. *Science*, **319**, 1238–1240.

Statistics Sweden (2015) *Agricultural Statistics 2015*. Statistics Sweden, Agriculture Statistics Unit, Örebro, Sweden.

Strömberg B, Herstad Svärd S (2012) *Bränslehandboken 2012 (The fuel handbook 2012)*. Anläggnings- och förbränningsteknik VÄRMEFORSK (Thermal Engineering Research Institute). Stockholm, Sweden.

Swedish EPA (2005) *Förbränningsanläggningar för energiproduktion inklusive rökgaskondensering (Incineration plants for energy production including flue gas condensation)*.

Thomas A, Bond A, Hiscock K (2013) A GIS based assessment of bioenergy potential in England within existing energy systems. *Biomass and Bioenergy*, **55**, 107–121.

Trafikverket (2016) www.trafikverket.se/lastkajen.

UNFCCC (2015) *Adoption of the Paris Agreement. Proposal by the President. (FCCC/CP/2015/L.9/Rev.1.)*. United Nations Office at Geneva, Geneva (Switzerland).

Vattenfall (2016) Frågor och svar om nytt kraftvärmeverk i Uppsala (Questions and answers about the new CHP plant in Uppsala). Available at: www.vattenfall.se/sv/fragor-och-svar-om-nytt-kraftvarmeverk-i-uppsala.htm (accessed 13 January 2016).

Wightman J, Ahmed Z, Volk T et al. (2015) Assessing sustainable bioenergy feedstock production potential by integrated geospatial analysis of land use and land quality. *BioEnergy Research*, **8**, 1671–1680.

Zetterberg L, Chen D (2015) The time aspect of bioenergy – climate impacts of solid biofuels due to carbon dynamics. *GCB Bioenergy*, **7**, 785–796.

9

Could biofuel development stress China's water resources?

MENGMENG HAO[1,2], DONG JIANG[1,2] (iD), JIANHUA WANG[3], JINGYING FU[1,2] and YAOHUAN HUANG[1,2]

[1]Institute of Geographical Sciences and Natural Resources Research, Chinese Academy of Sciences, 11A Datun Road, Beijing 100101, China, [2]College of Resources and Environment, University of Chinese Academy of Sciences, Beijing 100049, China, [3]State Key Laboratory of Simulation and Regulation of Water Cycle in River Basin, Department of Water Resources, China Institute of Hydropower & Water Resources Research, Beijing 100038, China

Abstract

Concerns over energy shortages and global climate change have stimulated developments toward renewable energy. Biofuels have been developed to replace fossil fuels to reduce the emissions of greenhouse gases and other environmental impacts. However, food security and water scarcity are other growing concerns, and the increased production of biofuels may increase these problems. This study focuses on whether biofuel development would stress China's water resources. Cassava-based fuel ethanol and sweet sorghum-based fuel ethanol are the focus of this study because they are the most typical nongrain biofuels in China. The spatial distribution of the total water requirement of fuel ethanol over its life cycle process was simulated using a biophysical biogeochemical model and marginal land as one of the types of input data for the model to avoid impacts on food security. The total water requirement of fuel ethanol was then compared with the spatial distribution of water resources, and the influence of the development of fuel ethanol on water resources at the pixel and river basin region scales was analyzed. The result showed that the total water requirement of fuel ethanol ranges from 37.81 to 862.29 mm. However, considering water resource restrictions, not all of the marginal land is suitable for the development of fuel ethanol. Approximately 0.664 million km^2 of marginal land is suitable for the development of fuel ethanol, most of which is located in the south of China, where water resources are plentiful. For these areas, the value of fuel ethanol's water footprint ranges from 0.05 to 11.90 m^3 MJ^{-1}. From the water point of view, Liaoning province, Guizhou province, Anhui province and Hunan province can be given priority for the development of fuel ethanol.

Keywords: biophysical biogeochemical model, cassava-based fuel ethanol, marginal land, sweet sorghum-based fuel ethanol, water requirement, water stress

Introduction

Energy shortages and global climate change are common challenges facing the world today. In December 2015, the Parties to the United Nations Framework Convention on Climate Change (UNFCCC) met at the 21st Conference of the Parties (COP21) and established an agreement to address the challenges of climate change. The Paris agreement determined the global greenhouse gas (GHG) emissions reduction targets, limiting the increase in global average temperature to below 2 °C (Shepherd & Knox, 2016). GHG emissions should lie between approximately 30 and 50 GtCO$_2$-eq yr^{-1} in 2030 in cost-effective scenarios that are likely to limit

Correspondence: Dong Jiang and Jianhua Wang
e-mails: jiangd@igsnrr.ac.cn and wjh@iwhr.com

warming to less than 2 °C this century (Change, 2014). To achieve the purpose of controlling temperature through the reduction of GHG emissions, in addition to decreased energy demand and improved energy efficiency (Schlamadinger et al., 1997; Zhou et al., 2014), development of renewable energy is considered to be one of the most effective ways (Taseska et al., 2011; Akashi & Hanaoka, 2012; Uusitalo et al., 2014; Ozcan, 2016). The renewable energy mainly includes wind energy, solar power and biofuels. Based on REN21's 2016 report, renewables contributed 19.2% of the global energy consumption by humans. This energy consumption is divided into 8.9% coming from traditional biomass, 0.49% from biofuels, 0.24% from solar energy and 0.39% from wind energy (REN21, 2016). Therefore, the development of biofuels is very necessary as one of the most important forms of renewable energy. However, in the development of renewable energy, we must consider its impact on water resources and food production. Related research has shown that the quantity and quality of the water required for different energy production varies significantly according to the process and technology of energy production, from rather negligible quantities of water used for wind and solar electricity generation to vast agricultural-scale water use for the cultivation of biofuel feedstock crops (Dominguez-Faus *et al.*, 2009; Spang *et al.*, 2014; Mengistu *et al.*, 2016). Unlike many other countries, China has a large population but limited arable land resources. It is essential that the development of biofuels does not compete for land with food production or it will effect on food security (Tang *et al.*, 2010; Shuai *et al.*, 2016). Therefore, the effects of biofuels on water and food security should be considered mainly because crops planted for energy need to consume large amounts of water and require large areas of cultivated land (Fraiture *et al.*, 2008).

The water–food–energy nexus has received global attention in recent years (Spang *et al.*, 2014; Ozturk *et al.*, 2015). However, policy objectives related to fresh water and energy are often poorly integrated (Holland *et al.*, 2015). The GHG emissions in China accounted for approximately 29% of the global emissions in 2013; in response to the Paris agreement, the Chinese government proposed to reduce carbon dioxide emissions per unit of GDP by 60–65% in 2030 compared with 2005. To this end, China is vigorously developing renewable energy, especially biofuels, which mainly include fuel ethanol and biodiesel. In China, the biofuel production was 2.7 billion liters including 2.26 billion liters fuel ethanol and 0.45 billion liters biodiesel in 2013 (Zhang *et al.*, 2014). However, these outputs are far behind the Chinese mandated target of 10 million ton (equivalent to 12.7 billion liters) of non-grain-based fuel ethanol and 2 million ton (equivalent to 2.3 billion liters) of biodiesel by 2020 (National Development and Reform Commission, 2007). To achieve this goal, China should make great efforts to develop biofuels, especially fuel ethanol, which accounts for a relatively large proportion of biofuels.

In China, many energy crops are used to produce fuel ethanol. In this study, cassava and sweet sorghum are regarded as the main feedstocks for the production of fuel ethanol. Cassava and sweet sorghum are the main nongrain energy crops and they can grow on the marginal land. Besides, cassava is starchy energy crop and it is abundant in the southern provinces of China (Zhang *et al.*, 2003), while sweet sorghum is carbohydrate energy crop and it is grown in the north of China (Gnansounou *et al.*, 2005). They are the typical nongrain crops-based fuel ethanol in China. With limited cultivated land resources in China (Deng *et al.*, 2006), to avoid impacts on food security, Chinese government

put forward basic principles regarding the development of energy biomass on marginal land. Marginal land has various meanings in different disciplines and therefore the spatial coverage of marginal land differs. According to the definition of marginal land by Ministry of Agriculture (MoA) of China, marginal land is winter-fallowed paddy land and wasteland that may be used to cultivate energy crops. The wasteland is considered in this study which includes shrub land, sparse forest land, grassland (dense grassland, moderate dense grassland and sparse grassland), shoal/bottomland, alkaline land and bare land that may be used to grow energy crops (Cai *et al.*, 2010; Qin *et al.*, 2011; Zhuang *et al.*, 2011; Jiang *et al.*, 2014). There are rich marginal land resources in China to ensure food security, and the total amount of marginal land was approximately 114 million ha in China in 2010 (Jiang *et al.*, 2014). In the future, therefore, the impact of developing fuel ethanol on water resources should be analyzed. China's freshwater reserves ranked fifth in the world, but the per capita freshwater resources are only one-fourth of the world's average (Zhan & Wu, 2014), and water resources vary greatly in space. However, the traditional method often used the statistical data multiplied by the coefficient and the region has only one value. Traditional method does not take into account the spatial differences and cannot explain the problem (Hong *et al.*, 2009; Gheewala *et al.*, 2013). Therefore, the main purpose of this study was to (i) present a distributed process model that can be used to accurately simulate the spatial total water requirements of fuel ethanol in the life cycle process, (ii) analyze water stress through the comparison with water resources at the pixel and river basin region scales, followed by the determination of suitable and unsuitable regions for the development of fuel ethanol and (iii) calculate the spatial distribution of water footprint of fuel ethanol.

Materials and methods

Definition of system boundary

Life cycle assessment (LCA) is a systems approach used to quantify material and energy flows and associated environmental burdens, arising over the production, consumption and disposal or recycling of a specified quantity (functional unit) of product or service(ISO, 2006a,b). The LCA was used to determine the total water consumption of the fuel ethanol in this study. The system boundary includes feedstocks' planting stage, feedstocks' transportation stage, fuel ethanol production stage, fuel ethanol transportation stage and fuel ethanol utilization stage (Cherubini *et al.*, 2009; Murphy & Alissa, 2014). Figure 1 shows that water use for life cycle of fuel ethanol includes water use for energy crops growth and the indirect water use for the production of other materials, electricity

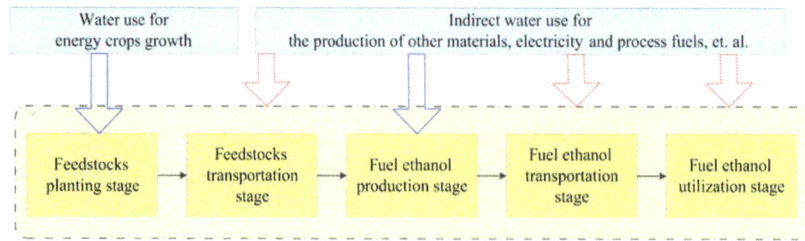

Fig. 1 The system boundary of fuel ethanol and the water use for fuel ethanol. The blue arrow means this part of the water use in the life cycle process of fuel ethanol was considered in this study, while the red arrow means the water use in the stage of fuel ethanol was not considered.

and so on. The water used for feedstocks' planting and fuel ethanol production are considered in this study. The water used in other stages accounts for less than 1% of the total water consumption; therefore, it was disregarded in calculation of total water requirement of fuel ethanol (Gerbens-Leenes *et al.*, 2009a; Su *et al.*, 2014).

Determination of suitable regions for development of fuel ethanol

To determine the suitable regions for development of fuel ethanol based on the water resources, three works were needed which include simulating the total water requirement of fuel ethanol using a biophysical biogeochemical model, optimizing the spatial distribution of total water requirement of fuel ethanol and extracting the suitable regions through the comparison with water resources at the pixel and river basin region scales.

Simulation of the total water requirement of fuel ethanol. From the perspective of the fuel ethanol life cycle, the total

water requirement of fuel ethanol includes the water required for the growth of energy crops, including cassava and sweet sorghum, and the water consumed in the process of fuel ethanol production. The water required for the growth of energy crops is the actual evapotranspiration of energy crops, which is obtained by the GEPIC model. The water requirement for the process of fuel ethanol production is calculated using the distribution of fuel ethanol production and the water consumption per unit mass of fuel ethanol, which is obtained from the fuel ethanol production company. Figure 2 is the technical process of simulation of the total water requirement of fuel ethanol from different energy crops.

Step 1: Preparation of the data and model. In this study, the GEPIC model is used to simulate the spatial distributions of the yield and actual evapotranspiration of energy crops. The GEPIC model is a GIS-based EPIC model designed to simulate the spatial and temporal dynamics of the major processes of the soil–crop–atmosphere management system (Liu *et al.*, 2007a,b). Compared with other models, the GEPIC model has many advantages such as high precision of crop yield

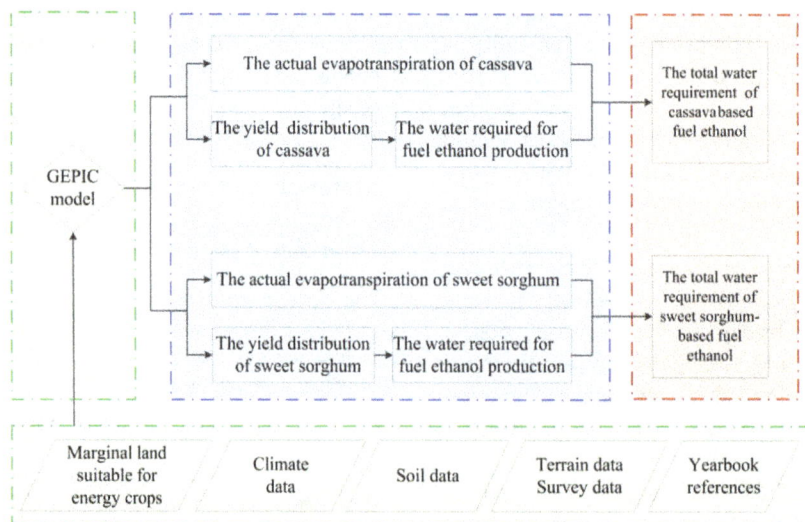

Fig. 2 The technical process of simulation of the total water requirement of fuel ethanol from different energy crops. Three steps were needed to simulate the total water requirement of fuel ethanol which include preparation of the data and model expressed in green, simulation of water demand for fuel ethanol at different stages expressed in blue and calculation of total water requirement for fuel ethanol expressed in red.

simulation, relatively minimal input data and widely used (Dumesnil, 1993; Bernardos *et al.*, 2001; Gassman *et al.*, 2005; Liu *et al.*, 2007b). The GEPIC model takes into account factors relating to weather, hydrology, nutrient cycling, tillage, plant environmental control and agronomics. Therefore, before the model simulation, the data should be prepared, including land use data, climate data, soil data, terrain data and field management data (Jiang *et al.*, 2015). Taking food security into account, the marginal land suitable for energy crops is regarded as the land use data. To ensure the model runs successfully and accurately, the model needs to be localized by first processing the detailed information for the localizing process. The marginal land suitable for cassava and sweet sorghum, the localized parameters for the GEPIC model and model accuracy verification are introduced in our previous paper (Fu, 2015; Jiang *et al.*, 2015).

Step 2: Simulation of water demand for fuel ethanol at different stages. In this research, the total water requirement of fuel ethanol based on each energy crop includes the water requirement in feedstocks' planting stage and fuel ethanol production stage.

The water demand for feedstocks planting. The water requirement in the feedstocks' planting stage refers to the actual evapotranspiration (ET_a) during the energy crops growth which is the sum of actual soil evaporation (E_a) and crop transpiration (T_a). In the GEPIC model, the actual evapotranspiration of each energy crop was calculated using the following formulas (Williams *et al.*, 1989; Ritchie, 1972; Hargreaves & Samani, 1985; Liu, 2009):

$$ET_a = E_a + T_a \tag{1}$$

$$T_a = \min\{ET_0 - I, T_p\} \tag{2}$$

$$E_a = \min\{E_p, E_p(ET_0 - I)/(T_a + E_p)\} \tag{3}$$

When $ET_0 < I$, the actual plant transpiration (T_a) and soil evaporation (E_a) are set to zero.

$$T_p = \begin{cases} ET_0 \text{LAI}/3 & 0 < \text{LAI} < 3 \\ ET_0 & \text{LAI} \geq 3 \end{cases} \tag{4}$$

$$E_p = \max\{(ET_0 - I)\lambda_s, 0\} \tag{5}$$

$$\lambda ET_0 = 0.023 H_0 (T_{mx} - T_{mn})^{0.5}(T_{av} + 17.8) \tag{6}$$

In the above formulas, T_p is the potential transpiration in mm day^{-1}, E_p is the potential soil evaporation in mm day^{-1}, I is the rainfall interception in mm day^{-1}, λ_s is a soil cover index, λ is the latent heat of vaporization in MJ kg^{-1}, LAI is the leaf area index, ET_0 is the reference evapotranspiration in mm day^{-1}, H_0 is the extraterrestrial radiation in MJ m^{-2} day^{-1}, T_{mx}, T_{mn} and T_{av} are the maximum, minimum and mean air temperature for a given day in °C.

The actual evapotranspiration of cassava and sweet sorghum can be simulated using the GEPIC model.

The water demand for fuel ethanol production. In the GEPIC, the yield of energy crops is estimated by multiplying the above-ground biomass at maturity with a water stress-adjusted harvest index for the particular crop (Williams *et al.*, 1989; Jiang *et al.*, 2015). Based on the spatial distribution of energy crops' yield and the conversion coefficient for the energy crop to fuel ethanol, the water demand of fuel ethanol production was calculated (Gerbens-Leenes *et al.*, 2009a; Su *et al.*, 2014). The formula is as following:

$$\text{WCE}_i = \frac{Y_i \times C_i \times W_{pi}}{10 \times \rho} \tag{7}$$

where WCE_i is the water requirement of the fuel ethanol production from ith energy crop per grid in mm, Y_i is the yield of the ith energy crop per grid in tha^{-1}, C_i is the conversion coefficient for the ith energy crop into fuel ethanol, ρ is the water density and the value is 1.0 g cm^{-3} and W_{pi} is the water consumption of the fuel ethanol production process per unit mass in t water t^{-1} ith fuel ethanol type. Different ethanol production technologies are used for different feedstocks, and the water consumed during the fuel ethanol production process also differs. According to a survey of ethanol production company, the water consumption of cassava-based fuel ethanol production is 12.6 t water t^{-1} fuel ethanol, and the water consumption of sweet sorghum-based fuel ethanol production is 9.5 t water t^{-1} fuel ethanol (Luo, 2008; Wei, 2014).

Step 3: Calculation of total water requirement for fuel ethanol. The total water requirement of fuel ethanol based on the energy crop can be then calculated using the following formula:

$$\text{TWC}_i = \text{WCE}_i + ET_{ai} \tag{8}$$

where ET_{ai} is the actual evapotranspiration of the ith energy crop, and TWC_i is the total water requirement of fuel ethanol from the ith energy crop.

Optimization of the spatial distribution of total water requirement of fuel ethanol. The total water requirement of cassava-based fuel ethanol and sweet sorghum-based fuel ethanol was calculated according to the above steps, respectively. There are some areas that are suitable for the development of both cassava-based fuel ethanol and sweet sorghum-based fuel ethanol. Therefore, it is necessary to choose a preferred energy crop to be planted in the development of fuel ethanol. Considering the energy, environmental and economic benefits of cassava-based fuel ethanol and sweet sorghum-based fuel ethanol, cassava-based fuel ethanol is better than that produced from sweet sorghum (Fu, 2015). Therefore, in areas those are suitable for the development of both cassava-based fuel ethanol and sweet sorghum-based fuel ethanol, cassava-based fuel ethanol should be given priority to develop. The spatial distribution of the total water requirement of fuel ethanol was obtained using overlay analysis.

Extraction of the suitable regions through comparison with water resources at the pixel and river basin region scales. The regions suitable for the sustainable development of fuel ethanol at the pixel – river basin region scale from the perspective of water resources can be extracted. Firstly, because the development of fuel ethanol on marginal land is dependent on rainfall, the relationship between precipitation and the water consumption of fuel ethanol at the pixel scale should be considered. In this study, the regions where the precipitation is less than the total water requirement of fuel ethanol are not considered. Then, taking into account the development of fuel ethanol cannot bring pressure on local water resources, the gross amount of water resources and total water consumption (including domestic water, industrial water, agricultural water and water for ecological environment) are introduced at the river basin scale. If total water consumption of the basin plus the total water requirement for fuel ethanol are less than the gross amount of water resources of the river basin, the river basin is not considered to development of fuel ethanol. Finally, the suitable regions for development of fuel ethanol will be extracted based on the precipitation and the gross amount of water resources of the river basin.

Calculation of water footprint of fuel ethanol

In the suitable development area, the development levels are also different. The water footprint was introduced to discuss priorities in the development of fuel ethanol. The concept of water footprint has been introduced as a quantitative indicator of freshwater used for producing a good, or a service (Hoekstra, 2003; Chapagain & Hoekstra, 2008; Wu *et al.*, 2014). It is the sum of all water consumed including both direct and indirect water consumed in the various stages of production and supply chain (Hoekstra *et al.*, 2011; Gheewala *et al.*, 2013; Lampert *et al.*, 2016). The water footprint includes green water footprint, blue water footprint and gray water footprint and life cycle water footprint (Mekonnen & Hoekstra, 2011; Zhang *et al.*, 2014). Over the past decades, water footprint has been used to calculate the water use for a wide range of products especially biofuel crops and biofuels (Gerbens-Leenes & Hoekstra, 2009; Gerbens-Leenes *et al.*, 2009b; Chiu & Wu, 2013; Hernandes *et al.*, 2014; Su *et al.*, 2014; Zhang *et al.*, 2014; Pacetti *et al.*, 2015). However, these studies have only one water footprint value in one area and do not consider the spatial variability of water footprint. In this study, the spatial distribution of life cycle water footprint of fuel ethanol was calculated and it refers to the water consumption per unit of energy, which for fuel ethanol is in $m^3 \, MJ^{-1}$. Based on the optimized spatial distribution of the total water requirement of fuel ethanol and the following formula, the spatial distribution of the water footprint can be calculated (Bhardwaj *et al.*, 2010):

$$WF = \frac{TWC}{E \times Y_e \times 100} \qquad (9)$$

where WF is the water footprint in $m^3 \, MJ^{-1}$, E is the energy value of fuel ethanol ($29.66 \, MJ \, kg^{-1}$) (Bonten & Wösten, 2012; Xia *et al.*, 2012; Anastasakis & Ross, 2015), TWC is the total water total water requirement of fuel ethanol and Y_e is the spatial distribution of the production of fuel ethanol, which is obtained from the spatial distribution of the yield of the energy crop and the conversion coefficient of energy crops to fuel ethanol.

Results

The spatial distribution of total water requirement of fuel ethanol

The spatial distributions of the yield and evapotranspiration of energy crops during the growth of energy crops are obtained using the GEPIC model with marginal land suitable for energy crops and other data. Based on these data and the above methods, the spatial distributions of the total water requirement of fuel ethanol from the different feedstocks are shown in Fig. 3.

Figure 3 shows that there are obvious spatial differences in the total water requirement. Regarding cassava-based fuel ethanol, cassava is suitable for planting on the marginal land in the south of China, especially in Guangxi Zhuang Autonomous Region, Yunnan, Guangdong and Fujian provinces. The total water requirement of cassava-based fuel ethanol ranges from 350.34 to 862.29 mm. In the south region of Yunnan province and the central region of Guangxi Zhuang Autonomous Region, the total water requirement is much greater than that in other regions. The main reason is that the per unit yield of cassava is relatively high in these regions. For sweet sorghum-based fuel ethanol, the marginal land area suitable for the cultivation of sweet sorghum is much greater than that available for cassava, and the total water requirement of sweet sorghum-based fuel ethanol ranges from 37.81 to 839.46 mm. The cause of the total water requirement of sweet sorghum-based fuel ethanol span is relatively large is that the yield gap of sweet sorghum is large. The low values are the abnormal value. The regions which have low value mean are not suitable to plant sweet sorghum in these regions from the water point of view. Low values are mainly distributed in Western Inner Mongolia, Gansu province and Xinjiang Uygur Autonomous Region. In the south region of Shaanxi province and the west region of Hubei province, the total water consumption is much greater than in other regions.

Some areas meet the conditions for the cultivation of both cassava and sweet sorghum. According to our previous study (Fu, 2015), from the respective of energy, the net surplus energy of cassava-based fuel ethanol and sweet sorghum-based fuel ethanol is 5.15 and 0.80 $MJ \, kg^{-1}$, respectively. From the perspective of environmental impact, the environmental impact index of cassava-based fuel ethanol and sweet sorghum-based fuel ethanol is 2.12E-03 population equivalent kg^{-1} fuel ethanol and 2.92E-03 population equivalent kg^{-1} fuel ethanol, respectively. Cassava-based fuel ethanol has less impact on the environment than sweet sorghum-based fuel ethanol. From the point of view of economics, the ratio of output to input of cassava-based

(a)   **(b)**

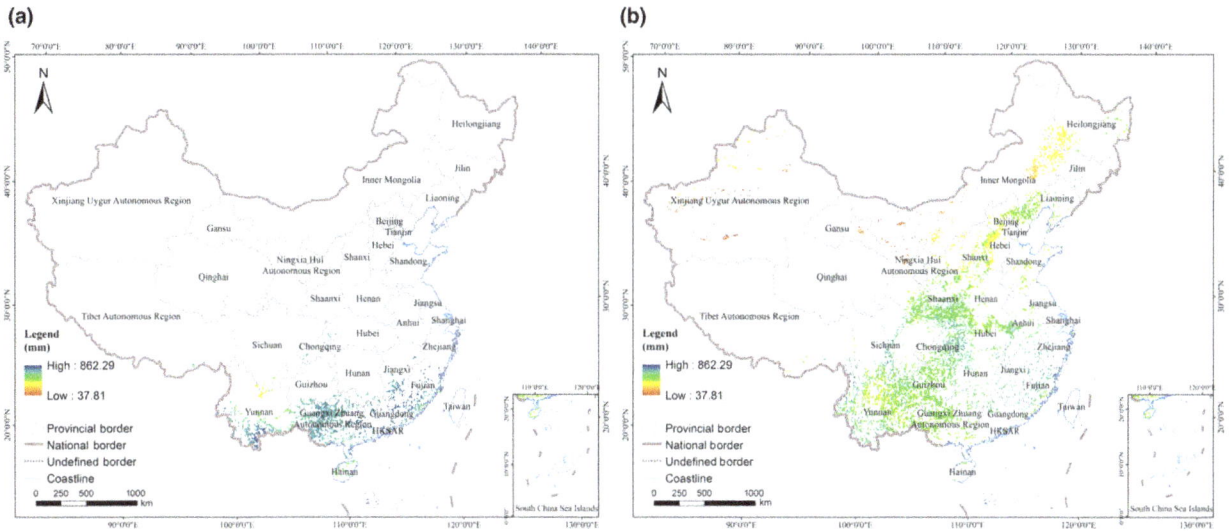

Fig. 3 The spatial distribution of total water requirement of fuel ethanol from different feedstocks' (a) total water requirement of cassava-based fuel ethanol ranges from 350.34 to 862.29 mm, (b) total water requirement of sweet sorghum-based fuel ethanol ranges from 37.81 to 839.46 mm.

fuel ethanol and sweet sorghum-based fuel ethanol is 1.65 and 1.30, respectively. Therefore, cassava-based fuel ethanol should be given priority to develop and the spatial distribution of fuel ethanol water consumption is shown in Fig. 4.

Figure 4 shows that there are great spatial differences of total water requirement for the development of fuel ethanol. The total water requirement of fuel

ethanol ranges from 37.81 to 862.29 mm, and the water demand in the northern part is higher than that in the South. The water requirement of fuel ethanol and the fuel ethanol production in each province are shown in Fig. 5.

Figure 5 shows that there are large differences in the average total water requirement among the provinces. Xinjiang Uygur Autonomous Region is the

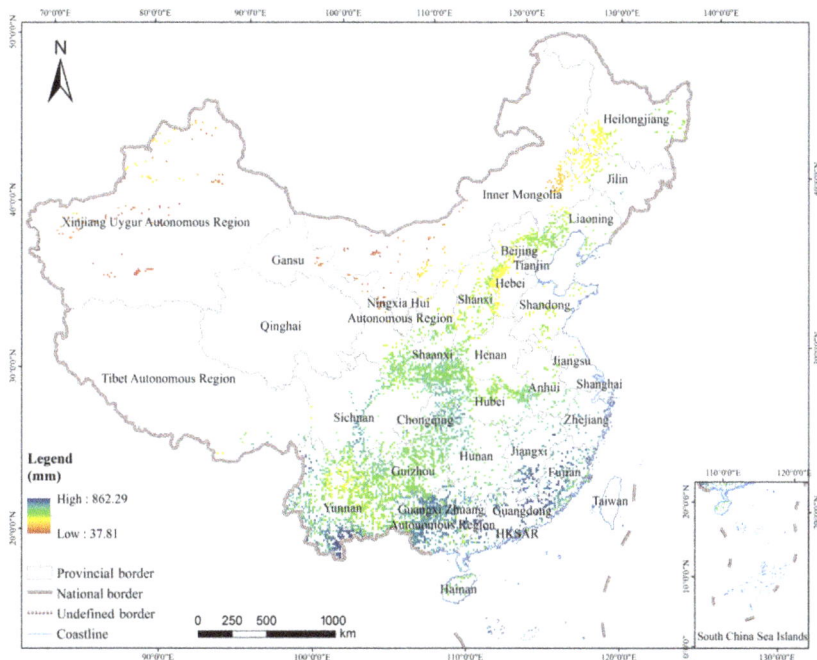

Fig. 4 The spatial distribution of the total water requirement of fuel ethanol. The total water requirement of fuel ethanol has significant spatial differences and it ranges from 37.81 to 862.29 mm. The region which water demand is high expressed in blue.

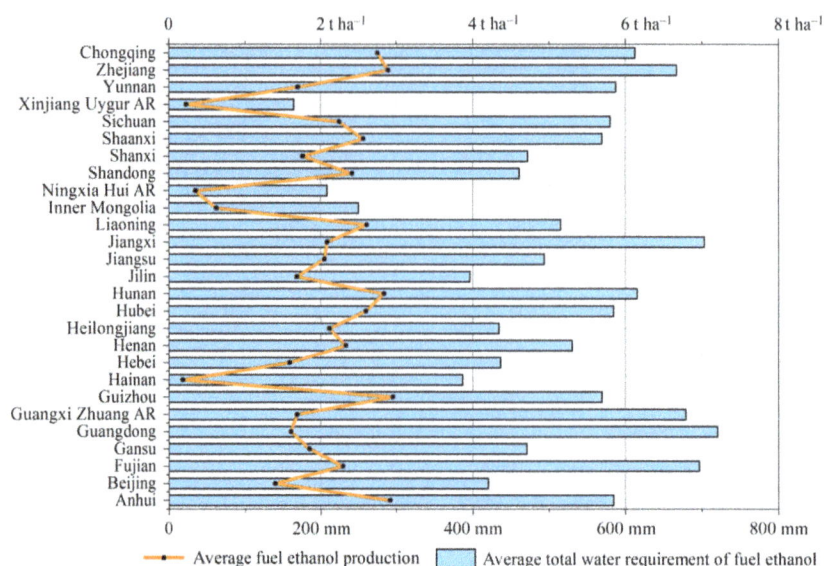

Fig. 5 The average fuel ethanol production and total water requirement of fuel ethanol in each province.

only area in which the average total water consumption is below 200 mm; the reason for this phenomenon is that energy crops do not grow well in this region, causing the crop water consumption to be low. The average total water requirements in Ningxia Hui Autonomous Region and Inner Mongolia are also low, being approximately 208 and 251 mm, respectively. The water requirements in Guangdong province and Jiangxi province are approximately 721 and 703 mm, respectively, and these areas have the highest water requirement. In addition, the average water demands of Chongqing, Zhejiang, Hunan provinces, Guangxi Zhuang Autonomous Region and Fujian province are also high, as they are all above 600 mm. The average water demand in the remaining cities is between 300 and 500 mm. Through the comparison of fuel ethanol production and the total water requirement fuel ethanol in Fig. 5, it is found that there is a positive correlation between fuel ethanol production and the total water requirement fuel ethanol.

Determination of suitable regions for development of fuel ethanol

The regions suitable for the development of fuel ethanol at the pixel scale. In this research, irrigation technology is not used in the development of fuel ethanol, and water requirement is mainly derived from precipitation. To analyze whether precipitation can meet the fuel ethanol demand for water, the relationship between water requirement of fuel ethanol and precipitation was analyzed. Figure 6 is the spatial distribution of precipitation in China.

In China, the change in precipitation from north to south is obvious and gradually increases. There are three obvious isohyets, which are 200, 400 and 800 mm. The 800 mm isohyet is along the Qinling Mountains–Huai River line, west to the southeast edge of the Qinghai Tibet Plateau. This line forms the boundary between the wet area and the semi-wet area. The 400 mm isohyet is the line that extends from the Da Hinggan Mountains, to Zhangjiakou, Lanzhou and Lhasa, to the eastern Himalayas. It is the dividing line between the semi-humid area and the semi-arid area. The 200 mm isohyet is the line extending from the western Inner Mongolia Autonomous Region to the west of the Hexi Corridor and the northern Tibetan Plateau, and it is the boundary between the arid area and the semi-arid area.

To analyze the relationship between precipitation and the total water consumption from the perspective of spatial distribution, the parameter P_n is used to measure this relationship between supply and demand, which is calculated using the following formula:

$$P_n = \frac{PRE_n - TWC_n}{PRE_n} \times 100\% \qquad (10)$$

In this formula, TWC_n is the average total water requirement of the nth grid in mm, and PRE_n is the precipitation in the nth grid in mm. Based on the spatial distributions of precipitation and the total water requirement of fuel ethanol, the spatial distribution of P_n was calculated, and the result is shown in Fig. 7.

Figure 7 shows that there are mainly five regions where the value of P_n is less than zero, which means that the water demand is higher than the precipitation in those regions. The first region includes most areas in

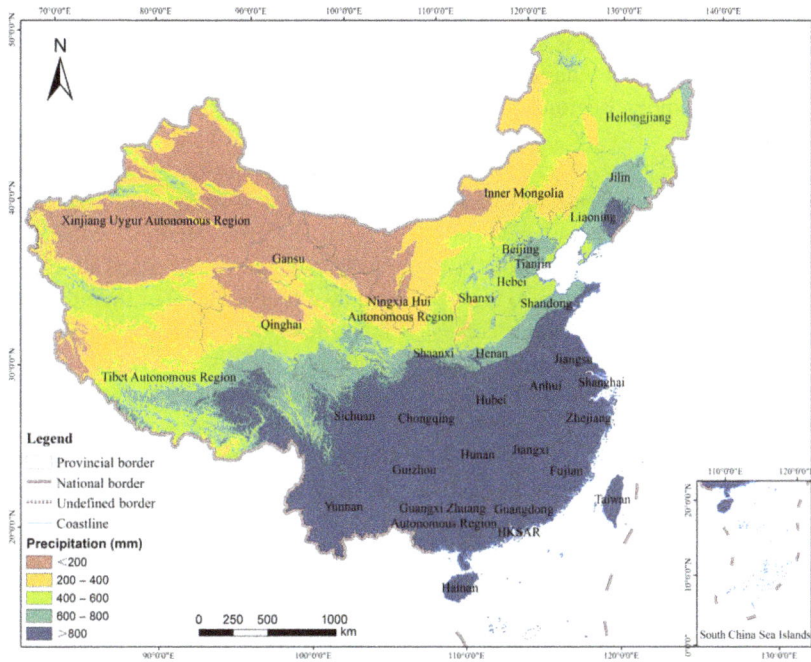

Fig. 6 The spatial distribution of precipitation in China.

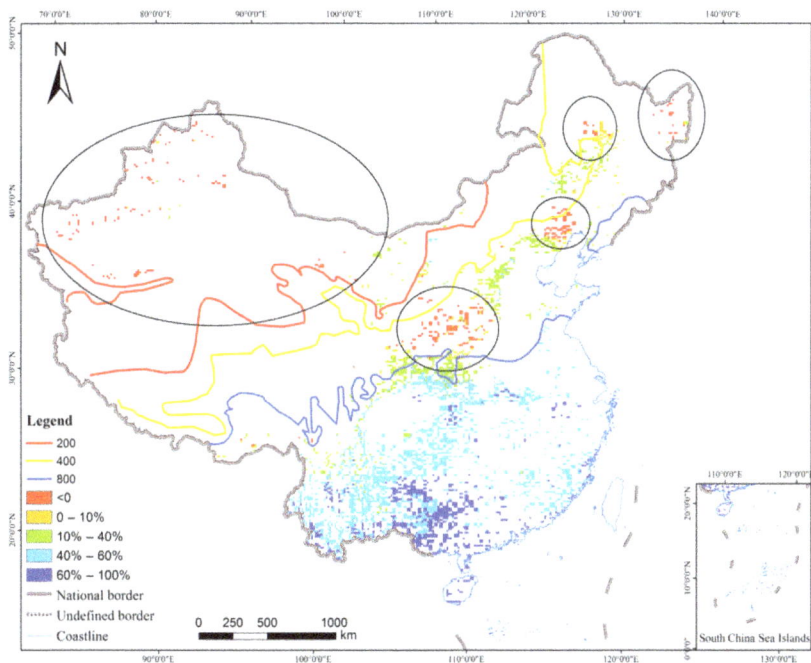

Fig. 7 The spatial distribution of P_n.

Xinjiang Uygur Autonomous Region and the center of Gansu province. The second region occurs at the junction of Shanxi province, Shaanxi province and Gansu province. The third region is in the west of Liaoning province. The last two regions are east and west of Heilongjiang province, respectively. These five regions will not be considered in the development of fuel ethanol. The areas in yellow represent those areas near the five regions mentioned above in which the P_n value is below 10%. In these areas, even though the precipitation can meet the water demands, there are some risks in the development of fuel ethanol. Most of these regions are located north of the 800 mm isohyet. The rainfall in regions located south of the 800 mm isohyet is sufficient compared with the water demand of fuel ethanol, especially in Guangxi Zhuang Autonomous Region and the south of Guizhou province.

The regions suitable for the sustainable development of fuel ethanol at the river basin scale. From the perspective of water security, to determine whether a region is suitable for the development of fuel ethanol, in addition to rainfall, the local water resources and the local water consumption should also be considered. Regarding this issue, the river basin is considered as the research unit, including the Hai River basin, Huai River basin, Liao River basin, northwest China river basins, Pearl River basin, Songhua River basin, southeast China river basins, southwest China river basins, Yangtze River basin and Yellow River basin in China.

Figure 8 shows that the marginal land most suitable for energy crops is distributed in the Yangtze River basin and the Pearl River basin. To analyze the relationship between the supply and demand of water resources, the total water requirement of fuel ethanol (TWC) in each basin was calculated, the gross amount of water resources and total water consumption (including domestic water, industrial water, agricultural water and water for ecological environment) in each basin were obtained from China water resources bulletin 2014 (Ministry of Water Resources of the People's Republic of China, 2015). Then, the relationship between the supply and demand of water resources is shown in Fig. 9.

Figure 9 shows that the gross amount of water is lower than the total water consumption in the Hai River basin, so the Hai River basin is not suitable for the development of fuel ethanol. In the Huai and Liao River basins, the total water consumption is close to the gross amount of water resources; if the water consumption of fuel ethanol is added, the total water consumption will exceed the gross amount of water resources. Therefore, these two basins are also not suitable for the development of fuel ethanol. In the other river basins, the gross amount of water is much greater than the water consumed.

A comprehensive analysis of the above two aspects from the point of view of water resources was conducted to determine whether an area is suitable for the development of fuel ethanol. The result is shown in Fig. 10.

Figure 10 shows that the development of fuel ethanol has obvious regional connectivity from the perspective of water resources, with most of the region in southern China being suitable for the development of fuel ethanol because of the presence of rich water resources. This indicates that the water demands of fuel ethanol will not place pressure on the local water resources. In the northern region of China, the development of fuel ethanol will stress local water resources, as they cannot meet the water demands of the development of fuel ethanol, especially in the strip of land that connects Shanxi province, Hebei province, Beijing, Tianjin and Liaoning province. These regions are therefore not suitable for the development of fuel ethanol. Through

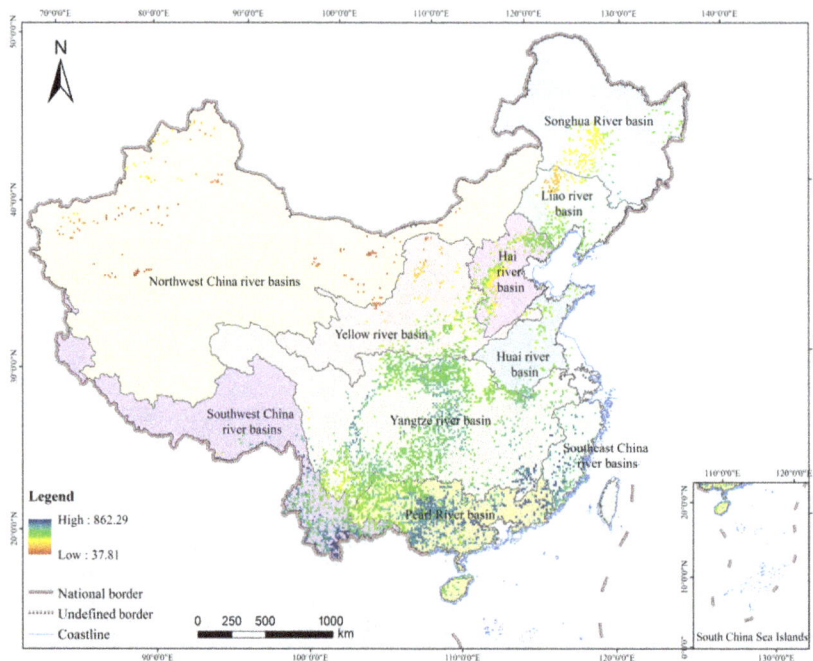

Fig. 8 The overlay chart of total water requirement of fuel ethanol and river basins in China.

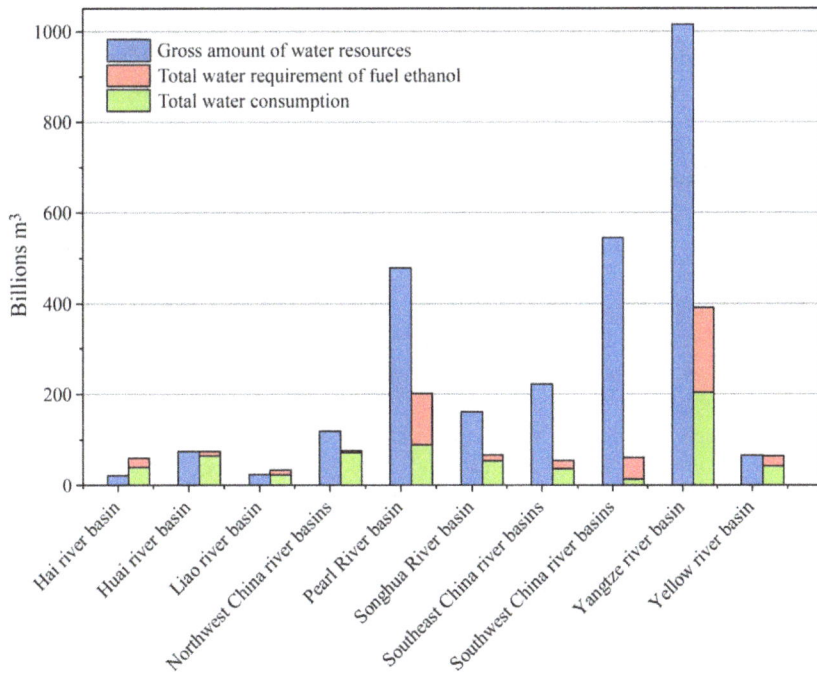

Fig. 9 The relationship between the supply and demand of water resources in each river basin.

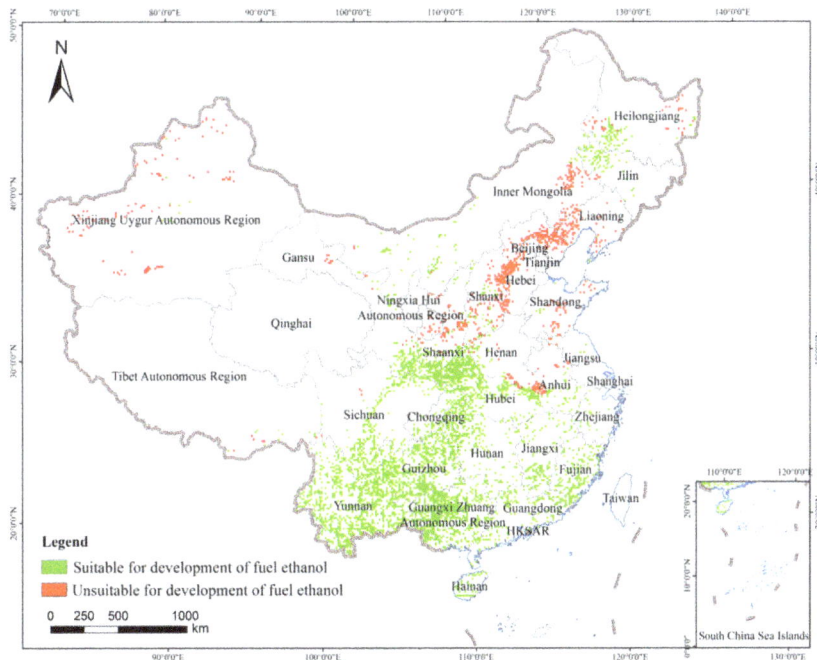

Fig. 10 The suitability distribution of development of fuel ethanol from energy crops.

statistical analysis, the total area of the regions unsuitable for the development of fuel ethanol is approximately 0.138 million km², accounting for 17.24% of the entire region, which has an area of approximately 0.802 million km².

The spatial distribution of water footprint of fuel ethanol

Based on the spatial distribution of the suitable areas, the water footprint of fuel ethanol was calculated, and the results were statistically analyzed by latitude, longitude and province. Figure 11 is the distribution of the water footprint of fuel ethanol.

The spatial distribution of the water footprint of fuel ethanol shows that the value of fuel ethanol's water footprint ranges from 0.05 to 11.90 m³ MJ⁻¹. The span of the water footprint of fuel ethanol is very large. However, Fig. 12 shows that most of the WF values are between 0.05 and 1.0 m³ MJ⁻¹ and only a very small number of outliers. Most of the WF values are reasonable value through compared with the results of other studies. The cause of abnormal value is that the yield of energy crop in these regions is very low. From the economic point of view, these regions are not suitable to plant the energy crop. The high abnormal values are located in the north of China and the south of China, especially in Yunnan province.

From the angle of latitude and longitude, there are three broad peaks in the longitude of the water

footprint at approximately 83°59′E–86°01′E, 100°09′E–102°41′E and 108°47′E–110°13′E. There are two broad peaks in the latitude of the water footprint at approximately 21°28′N–26°35′N and 37°24′N–43°44′N.

From the perspective of regional analysis, the statistical analysis of the water footprint in each province was conducted. The result shows that the average water footprint of fuel ethanol in each province ranges from 0.05 m³ MJ⁻¹ in Liaoning province to 2.85 m³ MJ⁻¹ in Xinjiang Uygur Autonomous Region.

From Fig. 12, we can see that there is an inverse relationship between fuel ethanol production and the water footprint. Therefore, the areas which the fuel ethanol production is high and the water footprint of fuel ethanol is low can be given priority for the development of fuel ethanol. Considering the fuel ethanol production and the water footprint of fuel ethanol (Figs 5 and 11), Liaoning province, Guizhou province, Anhui province

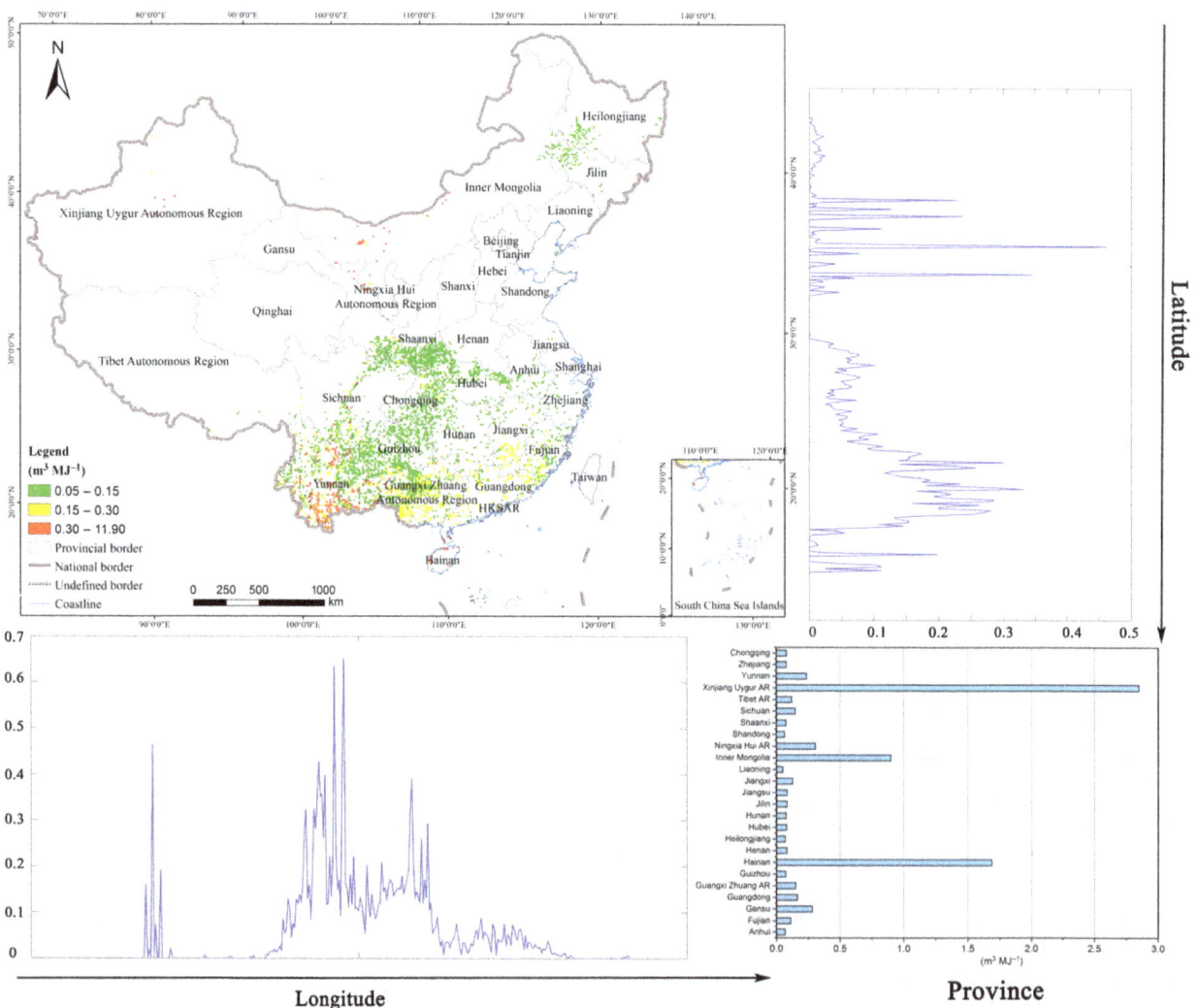

Fig. 11 The distribution of the water footprint of fuel ethanol.

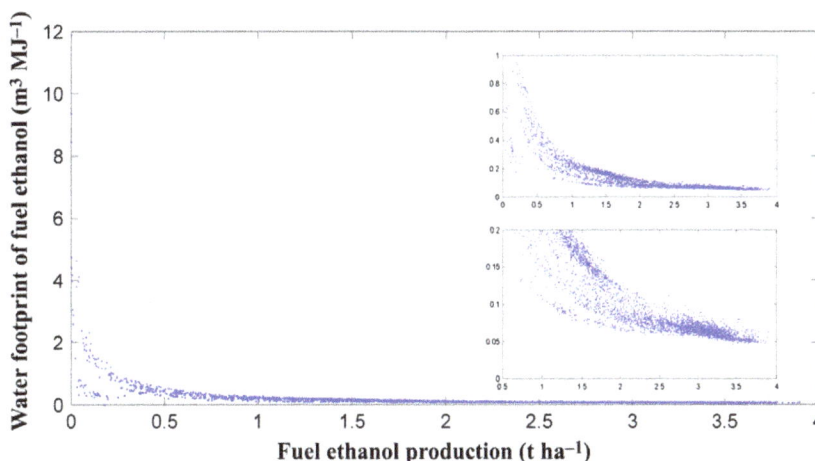

Fig. 12 The relationship between the water footprint of fuel ethanol and fuel ethanol production.

and Hunan province can be given priority for the development of fuel ethanol.

Discussion

This study evaluates from the perspective of water resources whether an area can meet the water requirements of fuel ethanol to determine whether it is suitable for the development of fuel ethanol. For some regions, even though the water resources can meet the water demands for fuel ethanol, the development of fuel ethanol is not recommended in these areas from the perspective of sustainable development. For example, in the Yellow River basin, the total water consumption is 61.43 billion m^3 when the water demand for fuel ethanol is added, accounting for 94% of the gross amount of water resources in the Yellow River, which is 65.37 billion m^3. From a long-term point of view, the development of fuel ethanol will affect the local water security, and the Yellow River basin is not recommended for the development of fuel ethanol.

The water footprint of biofuel varies across both crops and countries based on other studies. The average WF of biomass is 24 m^3 GJ^{-1} in the Netherlands, 58 m^3 GJ^{-1} in USA, 61 m^3 GJ^{-1} in Brazil and 143 m^3 GJ^{-1} in Zimbabwe according to Gerbens-Leenes et al. (2009b). Gerbens-Leenes et al. (2009a) also found that for ethanol, the sugar beet is the most advantageous and the WF is 60 m^3 GJ^{-1}; the sorghum is the most unfavorable and the WF is 400 m^3 GJ^{-1}. Jongschaap et al. (2009) calculated that the WF of bioenergy is 128 m^3 GJ^{-1} in South Africa. Mekonnen & Hoekstra (2011) found that among the crops providing ethanol, sorghum has the largest global average water footprint, with 300 m^3 GJ^{-1}; sugar beet has the smallest global average water footprint,

with 50 m^3 GJ^{-1}. Hong et al. (2009) found that the average WF of cassava-, sugarcane-, sugar beet-, sweet potato-based fuel ethanol in China is 2.64 m^3 L^{-1} ethanol, 1.47 m^3 L^{-1} ethanol, 2.24 m^3 L^{-1} ethanol and 1.83 m^3 L^{-1} ethanol, respectively, which equivalent to 113, 63, 96 and 78 m^3 GJ^{-1}, respectively. This variation is due to differences in crop yields across countries and crops, differences in energy yields across crops and differences in climate and agricultural practices across countries. In this study, the WF of fuel ethanol ranges from 50 to 11 900 m^3 GJ^{-1}. Compared with results from other works, most of the WF values are within a reasonable range, only a very small number of outliers. These outliers are mainly due to the low yield of feedstocks. From the respective of the space, it is not accurate to use statistical data to analyze the influence of fuel ethanol on water resources. The biophysical biogeochemical model was introduced to analyze the influence of the development of fuel ethanol on water resources in space in this study, which can make the results more reasonable.

For the fossil energy carriers, the WF increases in the following order: uranium (0.1 m^3 GJ^{-1}), natural gas (0.1 m^3 GJ^{-1}), coal (0.2 m^3 GJ^{-1}) and finally crude oil (1.1 m^3 GJ^{-1}). For the renewable energy carriers, the WF for wind energy is negligible, for solar thermal energy 0.3 m^3 GJ^{-1}, but for hydropower 22 m^3 GJ^{-1} (Gerbens-Leenes et al., 2009b). The WF of fuel ethanol in this study is much higher than that of fossil energy carriers and other renewable energy carriers, mostly due to the nature of plants to consume water to grow. Therefore, the trend toward larger energy use in combination with an increasing contribution of energy from biomass will enlarge the need for fresh water. China should consider the impact of water resources on the development of fuel ethanol in the future.

Conclusions

In this study, the stress on local water resources caused by the development of fuel ethanol was spatially analyzed through a biophysical biogeochemical model considering climate and crop growth factors. The results showed that there are some regions unsuitable for the development of fuel ethanol, covering an area of approximately 0.138 million km^2. These regions are mostly located in the north of China, especially in the strip of land connecting Shanxi province, Hebei province, Beijing, Tianjin and Liaoning province. In these regions, the development of fuel ethanol will stress the local water resources. In most areas of southern China, the development of fuel ethanol will not stress the local water resources due to the rich water resources in this area. Therefore, the south of China is suitable for the development of fuel ethanol, and the total area is 0.664 million km^2. For these areas, Liaoning province, Guizhou province, Anhui province and Hunan province can be given priority for the development of fuel ethanol.

In the next study, in addition to water resources, the saving of energy and the reduction of emissions should also be considered to determine the appropriate development area for fuel ethanol.

Acknowledgements

This research was supported and funded by the National Natural Science Foundation of China (Grant no. 41571509), and the Ministry of Science and Technology of China (2016Y FC1201300).

References

Akashi O, Hanaoka T (2012) Technological feasibility and costs of achieving a 50% reduction of global GHG emissions by 2050: mid- and long-term perspectives. *Sustainability Science*, **7**, 139–156.

Anastasakis K, Ross AB (2015) Hydrothermal liquefaction of four brown macroalgae commonly found on the UK coasts: an energetic analysis of the process and comparison with bio-chemical conversion methods. *Fuel*, **139**, 546–553.

Bernardos JN, Viglizzo EF, Jouvet V, Lértora FA, Pordomingo AJ, Cid FD (2001) The use of EPIC model to study the agroecological change during 93 years of farming transformation in the argentine pampas. *Agricultural Systems*, **69**, 215–234.

Bhardwaj AK, Zenone T, Jasrotia P, Robertson GP, Chen J, Hamilton SK (2010) Water and energy footprints of bioenergy crop production on marginal lands. *Global Change Biology Bioenergy*, **3**, 208–222.

Bonten LTC, Wösten JHM (2012) Nutrient flows in small-scale bio-energy use in developing countries, Alterra Report 2304, Wageningen, the Netherland.

Cai X, Zhang X, Wang D (2010) Land availability for biofuel production. *Environmental Science & Technology*, **45**, 334–339.

Change IPOC (2014) Climate change 2014 synthesis report. *Environmental Policy Collection*, **27**, 408.

Chapagain AK, Hoekstra AY (2008) The global component of freshwater demand and supply: an assessment of virtual water flows between nations as a result of trade in agricultural and industrial products. *Water International*, **33**, 19–32.

Cherubini F, Bird ND, Cowie A, Jungmeier G, Schlamadinger B, Woessgallasch S (2009) Energy- and greenhouse gas-based LCA of biofuel and bioenergy systems: key issues, ranges and recommendations. *Resources Conservation & Recycling*, **53**, 434–447.

Chiu YW, Wu M (2013) The water footprint of biofuel produced from forest wood residue via a mixed alcohol gasification process. *Environmental Research Letters*, **8**, 35015–35022.

Deng X, Huang J, Rozelle S, Uchida E (2006) Cultivated land conversion and potential agricultural productivity in china. *Land Use Policy*, **23**, 372–384.

Dominguez-Faus R, Powers SE, Burken JG, Alvarez PJ (2009) The water footprint of biofuels: a drink or drive issue? *Environmental Science & Technology*, **43**, 3005–3010.

Dumesnil D (1993).*EPIC User's Guide-Draft*. USDA-ARS, Grassland, Soil and Water Research Laboratory, Temple, TX.

Fraiture CD, Giordano M, Liao YS, Priscoli JD (2008) Biofuels and implications for agricultural water uses: blue impacts of green energy. *Water Policy*, **10**, 67–81.

Fu J (2015) *Assessment of the Non-Grain Based Fuel Ethanol Potential in China*. The Chinese Academy of Sciences, Beijing.

Gassman PW, Williams JR, Benson VW et al. (2005) *Historical Development and Applications of the EPIC and APEX Models*. Working Paper 05-WP-397. Center for Agricultural and Rural Development. Iowa State University. Iowa, USA.

Gerbens-Leenes PW, Hoekstra AY (2009) The water footprint of sweeteners and bioethanol from sugar cane, sugar beet and maize. *Value of Water Research Report*, **40**, 202–211.

Gerbens-Leenes W, Hoekstra AY, Th VDM (2009a) The water footprint of bioenergy. *Proceedings of the National Academy of Sciences of the United States of America*, **106**, 10219–10223.

Gerbens-Leenes PW, Hoekstra AY, Meer TVD (2009b) The water footprint of energy from biomass: a quantitative assessment and consequences of an increasing share of bio-energy in energy supply. *Ecological Economics*, **68**, 1052–1060.

Gheewala SH, Silalertruksa T, Nilsalab P, Mungkung R, Perret SR, Chaiyawannakarn N (2013) Implications of the biofuels policy mandate in Thailand on water: the case of bioethanol. *Bioresource Technology*, **150**, 457–465.

Gnansounou E, Dauriat A, Wyman CE (2005) Refining sweet sorghum to ethanol and sugar: economic trade-offs in the context of north China. *Bioresource Technology*, **96**, 985–1002.

Hargreaves GH, Samani ZA (1985) Reference crop evapotranspiration from temperature. *Transactions of the ASAE*, **1**, 96–99.

Hernandes TAD, Bufon VB, Seabra JEA (2014) Water footprint of biofuels in Brazil: assessing regional differences. *Biofuels, Bioproducts and Biorefining*, **8**, 241–252.

Hoekstra AY (2003) *Virtual Water Trade: Proceedings of the International Expert Meeting on Virtual Water Trade*. IHE, Delft.

Hoekstra AY, Chapagain AK, Aldaya MM, Mekonnen MM (2011) *The Water Footprint Assessment Manual: Setting the Global Standard*. Earthscan, London.

Holland RA, Scott KA, Flörke M, Brown G, Ewers RM, Farmer E et al. (2015) Global impacts of energy demand on the freshwater resources of nations. *Proceedings of the National Academy of Sciences*, **112**, 6707–6716.

Hong Y, Yuan Z, Liu J (2009) Land and water requirements of biofuel and implications for food supply and the environment in China. *Energy Policy*, **37**, 1876–1885.

ISO (2006a) *ISO 14040: Environmental Management – Life Cycle Assessment – Principles and Framework* (2nd edn). ISO, Geneva.

ISO (2006b) *ISO 14044: Environmental Management – Life Cycle Assessment – Requirements and Guidelines*. ISO, Geneva.

Jiang D, Hao M, Fu J, Zhuang D, Huang Y (2014) Spatial-temporal variation of marginal land suitable for energy plants from 1990 to 2010 in china. *Scientific Reports*, **4**, 5816–5816.

Jiang D, Hao MM, Fu JY, Huang YH, Liu K (2015) Evaluating the bioenergy potential of cassava on marginal land using a biogeochemical process model in Guangxi, China. *Journal of Applied Remote Sensing*, **9**, 097699.

Jongschaap REE, Blesgraaf RAR, Bogaard TA, Loo ENV, Savenije HHG (2009) The water footprint of bioenergy from Jatropha curcas L. *Proceedings of the National Academy of Sciences of the United States of America*, **106**, E92–E92.

Lampert DJ, Cai H, Elgowainy A (2016) Wells to wheels: water consumption for transportation fuels in the United States. *Energy & Environmental Science*, **9**, 787–802.

Liu JG (2009) A GIS-based tool for modelling large-scale crop-water relations. *Environmental Modelling & Software*, **24**, 411–422.

Liu J, Williams JR, Zehnder AJB, Yang H (2007a) GEPIC – modelling wheat yield and crop water productivity with high resolution on a global scale. *Agricultural Systems*, **94**, 478–493.

Liu J, Wiberg D, Zehnder AJB, Yang H (2007b) Modelling the role of irrigation in winter wheat yield, crop water productivity, and production in China. *Irrigation Science*, **26**, 21–33.

Luo Q (2008) *Sweet Sorghum*. China's Agricultural Science and Technology Press, Beijing.

Mekonnen MM, Hoekstra AY (2011) The green, blue and grey water footprint of crops and derived crop products. *Hydrology & Earth System Sciences*, **15**, 1577–1600.

Mengistu MG, Steyn JM, Kunz RP *et al.* (2016) A preliminary investigation of the water use efficiency of sweet sorghum for biofuel in South Africa. *Water SA*, **42**, 152–160.

Ministry of Water Resources of the People's Republic of China (2015) *China water resources bulletin 2014*. China Water Power Press, Beijing.

Murphy CW, Alissa K (2014) Life cycle analysis of biochemical cellulosic ethanol under multiple scenarios. *Global Change Biology Bioenergy*, **7**, 1019–1033.

National Development and Reform Commission (2007) Long term development plan for renewable energy. *Renewable Energy Resources*, **25**, 1–5.

Ozcan M (2016) Estimation of Turkey's GHG emissions from electricity generation by fuel types. *Renewable & Sustainable Energy Reviews*, **53**, 832–840.

Ozturk I, Lund H, Kaiser MJ (2015) Sustainability in the food-energy-water nexus: evidence from BRICS (Brazil, the Russian Federation, India, China, and South Africa) countries. *Energy*, **93**, 999–1010.

Pacetti T, Lombardi L, Federici G (2015) Water–energy nexus: a case of biogas production from energy crops evaluated by water footprint and life cycle assessment (LCA) methods. *Journal of Cleaner Production*, **101**, 278–291.

Qin Z, Zhuang Q, Zhu X, Cai X, Zhang X (2011) Carbon consequences and agricultural implications of growing biofuel crops on marginal agricultural lands in China. *Environmental Science & Technology*, **45**, 10765–10772.

REN21 (2016) Global Status Report 2016.

Ritchie JT (1972) Model for predicting evaporation from a row crop with incomplete cover. *Water Resources Research*, **8**, 1204–1213.

Schlamadinger B, Apps M, Bohlin F *et al.* (1997) Towards a standard methodology for greenhouse gas balances of bioenergy systems in comparison with fossil energy systems. *Biomass and Bioenergy*, **13**, 359–375.

Shepherd M, Knox P (2016) The Paris COP21 climate conference: what does it mean for the southeast? *Southeastern Geographer*, **56**, 147–151.

Shuai X, Lewandowski I, Wang XY, Yi Z (2016) Assessment of the production potentials of Miscanthus on marginal land in china. *Renewable & Sustainable Energy Reviews*, **54**, 932–943.

Spang ES, Moomaw WR, Gallagher KS, Kirshen PH, Marks DH (2014) The water consumption of energy production: an international comparison. *Environmental Research Letters*, **9**, 105003.

Su MH, Huang CH, Li WY, Tso CT, Lur HS (2014) Water footprint analysis of bioethanol energy crops in Taiwan. *Journal of Cleaner Production*, **88**, 132–138.

Tang Y, Xie JS, Geng S (2010) Marginal land-based biomass energy production in China. *Journal of Integrative Plant Biology*, **52**, 112–121.

Taseska V, Markovska N, Causevski A, Bosevski T, Pop-Jordanov J (2011) Greenhouse gases (GHG) emissions reduction in a power system predominantly based on lignite. *Energy*, **36**, 2266–2270.

Uusitalo V, Havukainen J, Manninen K *et al.* (2014) Carbon footprint of selected biomass to biogas production chains and GHG reduction potential in transportation use. *Renewable Energy*, **66**, 90–98.

Wei L (2014) *Fuel Ethanol Production Technology*. Chemical Industry Press, Beijing.

Williams JR, Jones CA, Kiniry JR, Spanel DA (1989) The EPIC crop growth model. *Transactions of the ASAE*, **32**, 497–511.

Wu M, Zhang Z (2014) Life-cycle water quantity and water quality implications of biofuels. *Current Sustainable/Renewable Energy Reports*, **1**, 3–10.

Xia XF, Zhang J, Xi BD (2012) *Research on Fuel Ethanol Evaluation and Policy Based on Life Cycle Assessment*. China Environmental Science Press, Beijing.

Zhan Y, Wu L (2014) The United States and China are facing water, energy, food three conflict. *China Economic Report*, **1**, 109–111.

Zhang C, Han W, Jing X, Pu G, Wang C (2003) Life cycle economic analysis of fuel ethanol derived from cassava in southwest China. *Renewable & Sustainable Energy Reviews*, **7**, 353–366.

Zhang T, Xie X, Huang Z (2014) Life cycle water footprints of nonfood biomass fuels in China. *Environmental Science & Technology*, **48**, 4137–4144.

Zhou G, Chung W, Zhang Y (2014) Carbon dioxide emissions and energy efficiency analysis of china's regional thermal electricity generation. *Journal of Cleaner Production*, **83**, 173–184.

Zhuang D, Jiang D, Liu L, Huang Y (2011) Assessment of bioenergy potential on marginal land in China. *Renewable & Sustainable Energy Reviews*, **15**, 1050–1056.

10

Predicting future biomass yield in *Miscanthus* using the carbohydrate metabolic profile as a biomarker

ANNE L. MADDISON[1], ANYELA CAMARGO-RODRIGUEZ[1], IAN M. SCOTT[1], CHARLOTTE M. JONES[1], DAFYDD M. O. ELIAS[2], SARAH HAWKINS[1], ALICE MASSEY[1], JOHN CLIFTON-BROWN[1], NIALL P. MCNAMARA[2], IAIN S. DONNISON[1] and SARAH J. PURDY[1]

[1]*Institute of Biological, Environmental and Rural Sciences, Aberystwyth University, Plas Gogerddan SY23 3EB, UK*, [2]*Centre for Ecology and Hydrology, Lancaster Environment Centre, Library Avenue, Bailrigg, Lancaster LA1 4AP, UK*

Abstract

In perennial energy crop breeding programmes, it can take several years before a mature yield is reached when potential new varieties can be scored. Modern plant breeding technologies have focussed on molecular markers, but for many crop species, this technology is unavailable. Therefore, prematurity predictors of harvestable yield would accelerate the release of new varieties. Metabolic biomarkers are routinely used in medicine, but they have been largely overlooked as predictive tools in plant science. We aimed to identify biomarkers of productivity in the bioenergy crop, *Miscanthus*, that could be used prognostically to predict future yields. This study identified a metabolic profile reflecting productivity in *Miscanthus* by correlating the summer carbohydrate composition of multiple genotypes with final yield 6 months later. Consistent and strong, significant correlations were observed between carbohydrate metrics and biomass traits at two separate field sites over 2 years. Machine-learning feature selection was used to optimize carbohydrate metrics for support vector regression models, which were able to predict interyear biomass traits with a correlation (R) of >0.67 between predicted and actual values. To identify a causal basis for the relationships between the glycome profile and biomass, a ^{13}C-labelling experiment compared carbohydrate partitioning between high- and low-yielding genotypes. A lower yielding and slower growing genotype partitioned a greater percentage of the ^{13}C pulse into starch compared to a faster growing genotype where a greater percentage was located in the structural biomass. These results supported a link between plant performance and carbon flow through two rival pathways (starch vs. sucrose), with higher yielding plants exhibiting greater partitioning into structural biomass, via sucrose metabolism, rather than starch. Our results demonstrate that the plant metabolome can be used prognostically to anticipate future yields and this is a method that could be used to accelerate selection in perennial energy crop breeding programmes.

Keywords: ^{13}C, bioenergy, biomarkers, carbohydrates, cell wall, *Miscanthus*, soluble sugars, starch

Introduction

Miscanthus is a candidate lignocellulosic biofuel crop owing to its high productivity and low chemical input requirements (Visser & Pignatelli, 2001; Somerville *et al.*, 2010). As a C4 grass, it is a close genetic relative of two major biofuel crops, *Zea mays* (maize) and *Saccharum* Sp. (sugarcane; Hodkinson *et al.*, 2002). However, currently, the only commercially grown genotype of *Miscanthus* is a wild accession and not a breeder's line. Therefore, several breeding programmes are now targeting *Miscanthus* for yield and quality improvement. A major hindrance to the improvement of perennial energy crops through

breeding is the long duration for new crosses to reach maturity when they can be assessed for superiority (Purdy *et al.*, 2015). In *Miscanthus*, this is typically in the region of 4 years from when a seed is planted. There is a pressing need to identify new methods to accelerate the selection of elite crosses in *Miscanthus* and other perennial species.

In plant science, numerous studies have demonstrated associations between metabolites and various stress conditions such as increases in proline during chilling (Wanner & Junttila, 1999) or increases in jasmonic acid in response to herbivory (Wang & Wu, 2013). In medicine, metabolic biomarkers are used prognostically, that is to anticipate a future outcome in an asymptomatic individual, an example being the measure of blood cholesterol as a predictor of future heart

Correspondence: Sarah Purdy
e-mail: sap@aber.ac.uk

attack risk. However, in plant science, the metabolome has rarely been used to predict future outcomes in crop species (Steinfath *et al.*, 2010). In *Arabidopsis thaliana*, several studies have successfully correlated biomass with particular metabolites, groupings of metabolites and enzyme activities (when expressed against total protein content; Meyer *et al.*, 2007; Sulpice *et al.*, 2009, 2010, 2013; Scott *et al.*, 2014). By combining a negative correlation with starch and a positive correlation with enzyme activities, approximately a third of the variation in biomass of an *Arabidopsis* inbred family could be accounted for (Sulpice *et al.*, 2010). A notable example of biomarker identification in a crop species is in potato, where the abundance of glucose and fructose was found to positively correlate with discoloration during frying (low chip quality). When either of these hexoses was used as markers to predict chip quality in new crosses, the correlation (R_S) between predicted and measured quality was 0.67 (Steinfath *et al.*, 2010). In a recent study into drought tolerance in rainforest trees, the abundance of nonstructural carbohydrates (NSC) was found to positively correlate with drought tolerance in trees showing natural variation and in those that had been manipulated (O'Brien *et al.*, 2014). These studies show that metabolites can be used as biomarkers to predict biomass, quality traits and stress responses in species as diverse as *Arabidopsis*, potato and rainforest trees. In all these studies, it was carbohydrates that were successfully used as markers.

We recently showed that two fast-growing and high-yielding genotypes of the perennial bioenergy grass, *Miscanthus*, displayed a distinctive NSC profile compared to two slower growing genotypes and that this phenotype was consistent across 2 years and different environments (Purdy *et al.*, 2015). However, the limited number of genotypes and hybrids used in this study were insufficient to unequivocally determine whether the carbohydrate metabolic profile ('glycome') could be used as a biomarker of productivity. The phenotypic attribute so far shown to most strongly correlate with final yield is (log-transformed) maximum canopy height ($R^2 = 0.55$; Robson *et al.*, 2013). Therefore, our primary aim with this study was to identify single or multiple metabolic biomarkers that could predict yield in *Miscanthus* and to determine how the strength of the correlations compared with height as a predictor. *Miscanthus* is usually harvested at the end of winter when senescence is complete, but we sampled carbohydrates in stems in the middle of UK summer when growth was most rapid. The summer carbohydrate metabolic profile was then used to predict winter yields harvested the following year.

The choice of nonstructural carbohydrates to profile was based on previously observed genotypic differences

in abundance and partitioning in four genotypes (Purdy *et al.*, 2014, 2015). Sucrose is the most abundant soluble sugar in *Miscanthus*, and owing to the close phylogenetic relationship between *Miscanthus* and sugarcane (Hodkinson *et al.*, 2002), it was an obvious candidate for study in diverse genotypes. Sucrose is formed of a molecule each of glucose and fructose, and relationships between the hexoses and biomass traits had previously been observed (Purdy *et al.*, 2015). Unlike many C3 temperate grasses, C4 species such as *Miscanthus* do not accumulate fructans (Muguerza *et al.*, 2013) but instead accumulate starch as a transient form of storage carbohydrate (de Souza *et al.*, 2013; Purdy *et al.*, 2014).

To grow, plants must accumulate structural mass, predominantly cellulose and the cell wall hemicellulose polysaccharides. Both starch and cellulose are polymers of glucose, and we hypothesized that rapidly growing genotypes of *Miscanthus* may be accumulating cellulose more rapidly at the expense of starch biosynthesis, thus explaining the negative relationship between starch and growth observed in our previous study and that of others (Rocher, 1988; Sulpice *et al.*, 2009; Purdy *et al.*, 2014, 2015). Therefore, starch, cellulose and the hemicelluloses were also assayed to assess a potential role as yield biomarkers.

Materials and methods

Mixed population

A total of 244 *Miscanthus* genotypes were collected and planted as described previously (Allison *et al.*, 2011; Jensen *et al.*, 2011; Robson *et al.*, 2012). From this population, a selection of seven short and 11 tall plants were used in the experiment. A description of the different species is provided in Table 1. Three biological replicates per genotype were harvested from blocks 1, 2 and 3 of the trial.

Mapping family

A total of 102 genotypes from a paired cross between a diploid *M. sinensis* and a diploid *M. sacchariflorus* were sown from seed in trays in a glasshouse in 2009. In 2010, individual plants were

Table 1 The species and experimental structure of the mixed population and the mapping family

Species	*n* Total	*n* Tall	*n* Short
Miscanthus mixed population			
M. sinensis	10	5	5
Hybrid	4	4	0
M. sacchariflorus	4	2	2
Miscanthus mapping family			
M. sinensis	1	1	0
Hybrid	19	9	10

split to form three replicates of each genotype and then planted out into the field in a spaced-plant randomized block design comprising three replicate blocks. The field site is located 300 m to the south from the mixed population (described above), and therefore, stone content and soil types are as described previously (Allison et al., 2011); however, the field containing the mapping family is on a gentler slope than the mixed population.

Biomass trait measurements

Growth rate: Canopy heights of the selected plants were measured weekly. The values presented are for the 2-week period surrounding the harvests to give a value of growth rate cm day^{-1}.

Stem height: A single stem that was representative of canopy height was selected for destructive harvest and its height (cm) measured on the day of harvest.

Destructive harvests

A single stem that was representative of canopy height was selected from each plant, cut at a height of 10 cm from the base, measured then flash-frozen before freeze-drying. As NSC show diurnal fluctuations in Miscanthus (Purdy et al., 2013), the two sets of plants were harvested on different days so that each harvest could be completed within a 2-h window at the same time of day (Zt 8–10 of a 16-h photoperiod). The mixed population was harvested on 04 July 2013, and the mapping family was harvested on 19 July 2013. For the harvesting of the entire mapping family in 2014, each of the three blocks were harvested on consecutive days in July to stay within the 2-h time window specified above.

Annual yield harvest: The mixed population and mapping family were destructively harvested for yield in March 2014 (following the 2013 growing season), and the mapping family was harvested in Feb 2015 following the 2014 growing season. Biomass was dried to a constant weight, and then, the average DW weight per plant (kg) was calculated.

Nonstructural carbohydrate (NSC) compositional analyses

Soluble sugars and starch were analysed as previously described (Purdy et al., 2014, 2015). Soluble sugar extraction: approximately 20 mg (actual weight recorded) of each cryomilled (6870 Freezer Mill, Spex, Sampleprep, Stanmore, UK) plant tissue sample was weighed into 2-mL screwcap microcentrifuge tubes. Sugars were extracted four times with 1 mL of 80% (v/v) ethanol and the resulting supernatants pooled; two extractions were at 80 °C for 20 min and 10 min, respectively, and the remaining two at room temperature. A 0.5 mL aliquot of soluble sugar extract and the remaining pellet containing the insoluble fraction (including starch) were dried down in a centrifugal evaporator (Jouan RC 1022, Saint-Nazaire, France) until all the solvent had evaporated. The dried down residue from the soluble fraction was then resuspended in 0.5 mL of distilled water. Samples were stored at −20 °C for analysis.

Soluble sugar analysis: Soluble sugars of samples extracted in the previous step were quantified enzymatically by the stepwise addition of hexokinase, phosphoglucose isomerase and invertase (Jones et al., 1977). Samples were quantified photometrically (Ultraspec 4000; Pharmacia Biotech, Uppsala, Sweden) by measuring the change in wavelength at 340 nm for 20 min after the addition of each enzyme. Sucrose, glucose and fructose were then quantified from standard curves included on each 96-well plate.

Starch quantification: Starch was quantified using a modified Megazyme protocol (Megazyme Total Starch Assay Procedure, AOAC method 996.11; Megazyme International, Wicklow, Ireland). Briefly, the dried pellet was resuspended in 0.4 mL of 0.2 M KOH, vortexed vigorously and heated to 90 °C in a water bath for 15 min to facilitate gelatinization of the starch. A total of 1.28 mL of 0.15 M NaOAc (pH 3.8) was added to each tube (to neutralize the sample) before the addition of 20 µL α-amylase and 20 µL amyloglucosidase (Megazyme International). After incubation at 50 °C for 30 min and centrifugation for 5 min, a 0.02 mL aliquot was combined with 0.6 mL of GOPOD reagent (Megazyme). A total of 0.2 mL of this reaction was assayed photometrically (Ultraspec 4000; Pharmacia Biotech) on a 96-well microplate at 510 nm against a water-only blank. Starch was quantified from known standard curves on the same plate. Each sample and standard was tested in duplicate. Each plate contained a Miscanthus control sample of known concentration for both soluble sugars and starch analysis.

Cell wall carbohydrates and lignin

Lignin and matrix polysaccharides were analysed as described by Foster et al. (2010a,b). To quantify matrix polysaccharides, a Dionex ICS-5000DC (Thermo Scientific, Loughborough, UK) was used. Each chromatographic run contained sets of standards and a dilution series. Lignin quantification followed the method described by Foster et al. (2010a). Crystalline cellulose was analysed by Seaman hydrolysis and subsequent quantification of glucose (Purdy et al., 2014).

Crystalline cellulose

Approximately 60 mg (actual weight recorded) of purified cell wall was hydrolysed with 0.6 mL of 72% H_2SO_4, vortexed and left to incubate whilst shaking at 200 rpm for 1 h at 30 °C. After incubation, samples were diluted with 16.8 mL of deionized H_2O. Tubes were then capped and autoclaved at 121 °C for 1 h. Once cooled, an aliquot of 0.65 mL was neutralized with 30 mg $CaCO_3$ and centrifuged to pellet the $CaCO_3$ and the supernatant was removed to a fresh tube. Glucose was quantified enzymatically as previously described (Purdy et al., 2014). Standards of glucose were treated alongside experimental samples and included on each plate with a duplicated check sample.

The amount of glucose that was derived from hemicellulose (as described above) was subtracted from the total to give a value derived just from the Seaman hydrolysis of crystalline cellulose.

Data modelling

Principal component analysis (PCA) was performed in SIMCA-P v.11 (Umetrics AB, Malmo, Sweden) on values averaged across all biological replicates (usually three) of each genotype in each sampled population. Data were mean-centred and scaled to unit variance, and the reported PC was 'significant' in the default SIMCA-P cross-validation procedure. Machine learning employed the *AttributeSelectedClassifier* in the Explorer interface of WEKA v.3.6 (Frank *et al.*, 2004). This WEKA 'metaclassifier' firstly sought an optimal subset of biomass predictors from the individual carbohydrate levels and their derivative metrics (i.e. sums and ratios). This involved testing all potential predictor combinations by the *ExhaustiveSearch* algorithm with evaluation by *CfsSubsetEval*. The metaclassifier then trained a support vector regression algorithm, *SMOreg*, to relate the selected carbohydrate predictors to their associated biomass trait data. The resultant model was evaluated for its accuracy in predicting the relevant biomass values from the carbohydrate metrics of a 'test' set of plants. Test data were always completely excluded from model training. Default parameters were used for each algorithm.

$^{13}CO_2$ pulse labelling

Our chamber design and ^{13}C pulse-labelling approach were similar to previous methods (Hogberg *et al.*, 2008; Subke *et al.*, 2009; Biasi *et al.*, 2012) and applied to a field trial planted in 2010 with triplicate plots of a *M. sinensis*, a *M. sinensis* × *M. sacchariflorus* hybrid (*M. x giganteus*) and a *M. sacchariflorus* in a randomized block design. In each replicate plot, square ^{13}C pulse chambers were erected (2 m *l*, 2 m *w*, 3 m *h*) above the crop, resulting in a total tent volume of 12 m³. Aluminium scaffold was used to support plastic polythene film, which allowed 90% of photosynthetically active radiation to enter the chamber. During the ^{13}C pulse, the chamber was sealed at the base. To counter ambient air temperature increases within the chambers, each was cooled using a water cooled, split air conditioner (Andrew Sykes, Wolverhampton, UK).

The ^{13}C pulse labelling was carried out on 23 July 2013 at ca. 08:20 by introducing ca. 6 L of 99% ^{13}C-atom enriched pure CO_2 (CK Gases, Ibstock, UK) in sequential batches after sealing the tent.

^{13}C Harvesting and sample preparation

Pulsed samples were harvested 30 h after labelling. A single marked stem was harvested as previously described. The level of ^{13}C enrichment above natural background ^{13}C levels was determined in the soluble, starch and structural mass, which was extracted as previously described.

Solid sample analysis was performed on a Costech EC S4010 Elemental Analyser (Costech Analytical Technologies Inc, Valencia, CA, USA) coupled to a Picarro G-2131i Series CRDS analyzer (Picarro Inc, Santa Clara, CA, USA) via a split-flow interface using a method similar to (2013). Cryomilled samples of ~2 mg were weighed into ultraclean, 6 × 4 mm pressed tin cups (Elemental MicroAnalysis Ltd, Okehampton, UK),

crimped and loaded into a Zero N-Blank, 50 position carousel, autosampler. From the autosampler, samples were dropped at a throughput of 1 every 15 min into a combustion reactor, maintained at a constant 980 °C. Samples undergo flash combustion and thermally decompose. Evolved CO_2 was passed through a thermal conductivity detector (TCD) for C detection and then vented through a split-flow interface to the Picarro CRDS analyzer for ^{13}C analysis. Standard materials covering a representative range of C and $\delta^{13}C$ values were run during each analysis batch, and results were calibrated against these.

^{13}C Calculations

Stable isotope notation. Studies of this kind have generally either expressed ^{13}C enrichment values in $\delta^{13}C$, a measure of the ratio of ^{13}C and ^{12}C, reported in parts per thousand (‰) relative to a standard value (Pee Dee Belemnite – PDB) or atom %. Outputs from the Picarro $^{13}CO_2$ analyzer were in standard delta (δ) value notation ($\delta^{13}C$). $\delta^{13}C$ values are calculated using the following equation:

$$\delta^{13}C\,sample = \frac{^{13}C/^{12}C\,sample}{^{13}C/^{12}C\,PDB} - 1 \times 1000, \qquad (1)$$

where $^{13}C/^{12}CPDB$ is the isotopic ratio of the standard material PDB given as 0.0112372. Results were converted to atom % and then mg g^{-1} using the following equations:

$$Atom\% = \frac{100 \times AR \times (\delta^{13}C/1000 + 1)}{1 + AR \times (\delta^{13}C/1000 + 1)}, \qquad (2)$$

where AR = 0.011237. The absolute ratio of standard material (PDB) and $\delta^{13}C$ = standard delta value of sample.

$$mg\,g^{-1} = Atom\% \times 10. \qquad (3)$$

Statistical analyses

Differences between genotypes for biomass traits, NSC, structural carbohydrates and block effects were determined from one-way ANOVA using genotype or block as the treatment factor ($P = \le 0.05$). Genotypic differences in the deposition of ^{13}C were determined by one-way ANOVA using genotype as the treatment factor and an associated Tukey's HSD test. ANOVA, Tukey and Wilcoxon tests were performed using GENSTAT (13th Edition). Differences between genotypes grouped as 'tall' or 'short' were determined from Student's two-tailed *t*-tests (assuming unequal variance; $P = \le 0.05$) using Microsoft Excel. To compare biomass measures and NSC between across years (2013 and 2014), a Pearson correlation was performed to determine similarities in absolute values and a Spearman rank correlation analysis to compare the ordering of genotypes. Both analyses were carried out in SIGMAPLOT 12 (Systat Software, Inc, San Jose, CA, USA).

Results

Two sets of field-grown plants were studied (Table 1). The 'mixed population', from which 18 plants were selected for study in 2013 (their eighth growing season), was comprised of *M. sinensis*, *M. sinensis* × *M. sacchariflorus* hybrids and *M. sacchariflorus* genotypes. The 'mapping family', from which 20 plants were selected for study in 2013 (their fourth growing season), was comprised of *M. sinensis* × *M. sacchariflorus* hybrids, plus a single, tall, *M. sinensis* genotype. Nonstructural and structural carbohydrates and lignin were sampled in July, during the summer growing season. Measures of biomass traits obtained during the summer sampling were stem height, and growth rate over the surrounding two-week period, whereas annual yield was obtained at harvest after the following winter. Carbohydrate and biomass data for all genotypes are in Tables S1–S3.

Height has been shown to be the trait that best correlates with final yield in *Miscanthus* (Robson *et al.*, 2013). Therefore, each set of plants was divided into 'tall' and 'short' classes for comparison of carbohydrate contents. The average heights of plants grouped as short or tall from the mixed population were 79 cm and 151 cm, respectively, and in the mapping family, the average heights of the short and tall classes were 56 cm and 120 cm, respectively (Table S1).

In both sets of plants, the abundance of fructose was significantly greater in the tall plants compared to the short plants, whereas the opposite was true for starch, which was significantly more abundant in the short plants (Fig. 1a and b; Table S2). Glucose, total hexose and total soluble carbohydrates were significantly different between tall and short plants only in the mixed population.

The ratios between different NSC were examined in the short and tall groups, to further investigate the contrasting relationships of fructose and starch with plant height. The greatest difference between tall and short plants was the starch/fructose ratio, which was >fourfold greater in the short plants of the mixed population and >twofold greater in the short plants of the mapping family (Fig. 1c and d). The glucose/fructose and sucrose/fructose ratios were also significantly negatively associated with height in both populations. Differences in the starch/glucose and sucrose/starch ratios of tall and short plants were significant only in the mixed population (Fig. 1c and d).

Significant differences in hemicellulosic glucan were observed between the short and tall plants in both populations, with short plants exhibiting higher glucan levels (Fig. 2a and b). Other significant differences, seen only in the mapping family, were arabinose, galactose and mannose, which were more abundant in short plants, whereas crystalline cellulose and lignin were more abundant in tall plants (Fig. 2b; Table S3b). In the mixed population, these last cell wall components did not show significant height-associated differences (Fig. 2a; Table S3a), possibly due to a significant effect

of the replicate field blocks at the relevant trial site (Table S4). These replicate blocks were arranged parallel to the slope of the hill on which the mixed population was grown (see *Materials and Methods*). A block effect was previously reported in this population at the end of the growing season and attributed to differences in wind exposure and water dynamics (Allison *et al.*, 2011). No significant block effects were detected for any of the NSC from either set of plants, or in the cell wall composition of the mapping family (Table S4).

Having found differences in carbohydrates of height classes of plants selected in 2013, we examined the potential for using carbohydrates to model biomass traits (summer height, summer growth rate and annual yield). To provide more data for this purpose, a much larger selection (102 genotypes) was made in 2014 from the mapping family and similarly analysed for NSC and biomass traits. As an exploratory stage for modelling, we first sought to correlate carbohydrate contents with biomass traits in each set of plants, using Spearman's rank coefficients (R_S), which are robust to potential nonlinearity. Of the NSC in the mixed population, both glucose and fructose produced positive correlations with biomass traits, with the strongest correlation being between fructose and stem height (0.91; Fig. 3a). In the mapping family, biomass traits were significantly positively correlated with fructose (correlation coefficients of ~0.5), though not with glucose (Fig. 3b and c). No strong correlation between sucrose and biomass traits was detected in either set (Fig. 3a–c). In contrast, starch negatively correlated with all biomass traits (and with fructose) in both sets of plants (Table S3a–c), the strongest relationship being with growth rate in the mixed population (−0.76). The relationships observed in the mapping family in 2013 and in the extended number of genotypes in 2014 were largely consistent, both showing strong positive correlations of fructose with stem height and yield (~0.5) and strong negative relationships between starch and yield (Fig. 3b and c). Correlations with growth rate, however, were considerably lower in 2014 than 2013 (Fig. 3b and c).

Ratios between NSC produced stronger correlations with biomass traits than did the individual components (Fig. 3a–c, middle row). In both the mixed population and mapping family, significant negative correlations between biomass traits and the starch/fructose and sucrose/fructose ratios were observed. Conversely, positive correlations between the sucrose/starch ratio and biomass traits were observed in both sets of plants (Fig. 3a–c).

In both sets of plants, the matrix polysaccharide components arabinose, galactose and glucan were negatively correlated with biomass traits (Fig. 3a and b, bottom row). The strongest negative correlation in the

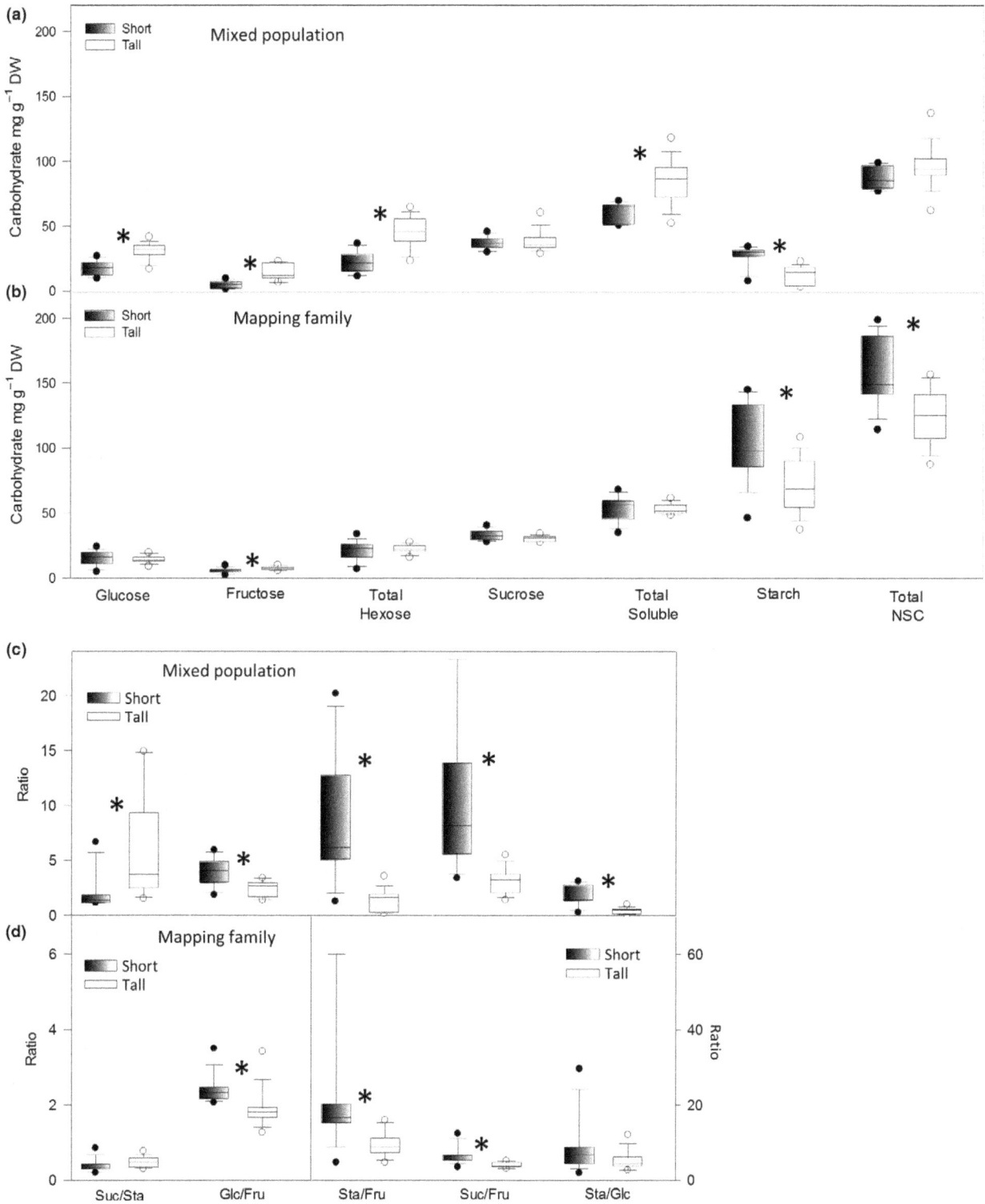

Fig. 1 Quantification of NSC in short and tall genotypes of the mixed population (a) and mapping family (b), and ratios between carbohydrate levels in short and tall genotypes of the mixed population (c) and mapping family (d). Circles show outliers. Significant differences between tall and short plants are shown by an asterisk (Student's t-test assuming unequal variances, $P =\leq 0.05$). Key: NSC = total nonstructural carbohydrate, Suc/Sta = sucrose-to-starch ratio, Glc/Fru = glucose-to-fructose ratio, Sta/Fru = starch-to-fructose ratio, Suc/Fru = sucrose-to-fructose ratio, Sta/Glc = starch-to-glucose ratio.

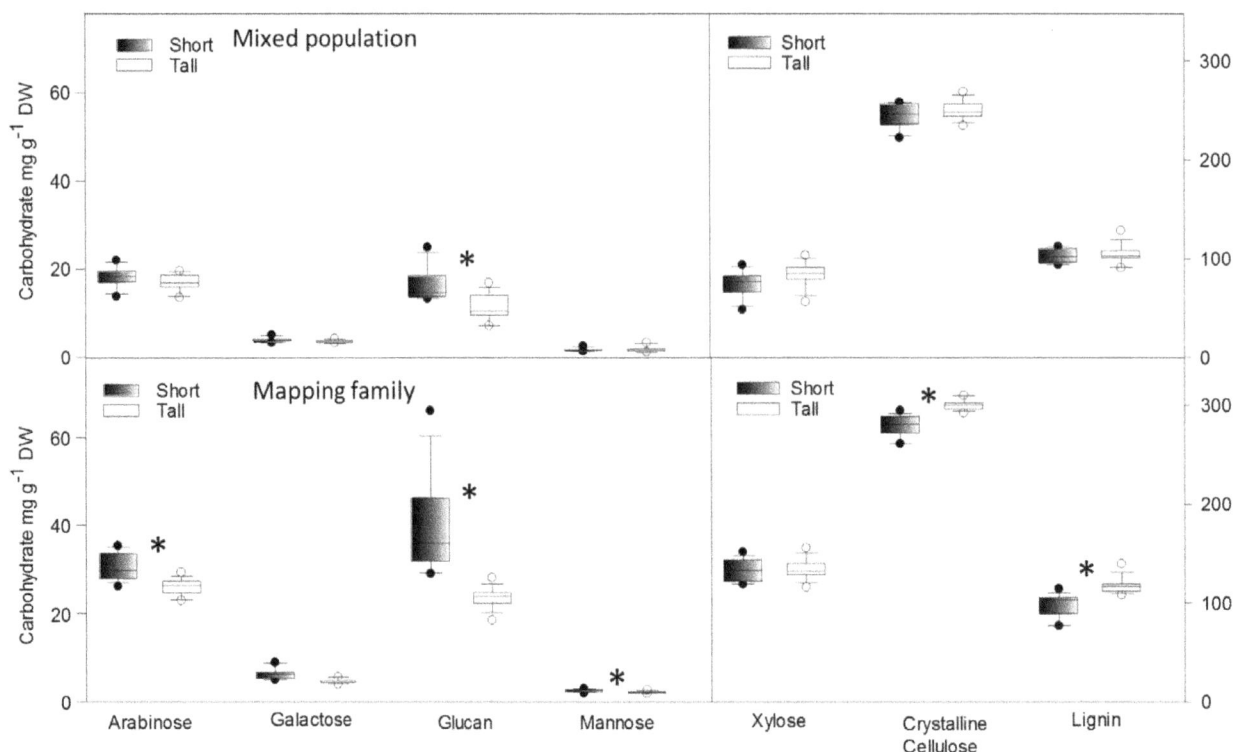

Fig. 2 Quantification of structural carbohydrates and lignin in short and tall genotypes of the mixed population (a) and mapping family (b). Circles show outliers. Significant differences between tall and short plants are shown by an asterisk (Student's *t*-test assuming unequal variances, $P \leq 0.05$).

mixed population was between arabinose and growth rate (−0.76), whilst in the mapping family, matrix glucan correlated negatively with yield (−0.92). In the mapping family, crystalline cellulose and lignin showed positive correlations with biomass traits of 0.5–0.8, whereas these relationships were not observed in the mixed population (Fig. 3a and b).

As similar correlations manifested in the various genotypes, in two different field trial sites and two different years (Fig. 3), it was pertinent to ask whether a 'carbohydrate phenotype' related to biomass traits was sufficiently robust to be identifiable across all sampled populations. When the NSC levels (glucose, fructose, sucrose, starch) measured in all populations were subjected to PCA in a unified data set, the first component, PC[1], accounted for 58.3% of overall variance (Fig. 4). The loadings for each NSC on PC[1] identified opposite variations in starch on the one hand, and the hexoses fructose and glucose on the other (Fig. 4a). Thus, PC[1] could be regarded as a composite index of NSC status, such that negative scores on PC[1] were indicative of high starch and low hexoses, and the converse for positive PC[1] scores. Moreover, PC[1] scores of the sampled genotypes showed significant correlations to biomass traits. As seen in Fig. 4b, genotypes with negative PC[1] scores (high starch, low hexoses) tended to be

characterized by shorter stature than genotypes with positive scores (low starch, high hexoses). Pearson correlations with PC[1] scores were highly significant ($P < 10^{-12}$) for height (R, 0.65), yield (R, 0.56) and growth rate (R, 0.62).

Considering the carbohydrate/biomass correlations in different *Miscanthus* populations (Fig. 3), and evidence for a biomass-correlated multivariate 'carbohydrate phenotype' across populations (Fig. 4), we sought to test the predictive power of the glycome for biomass traits by multivariate modelling. A machine-learning approach was chosen to address the complexity of the potentially relevant carbohydrate and genetic data, in the context of the small number and variability of biological replicates typical of a screening trial in the field.

As evident from Fig. 3, a number of potential carbohydrate metrics, including sums and ratios of individual NSC, would be available for inclusion in such a model. However, use of too many of these metrics would be redundant. We therefore used machine learning to identify parsimonious subsets of effective predictors from the following list of 20: glucose, fructose, total hexose, sucrose, total soluble carbohydrates, starch, total NSC, arabinose, galactose, glucan, xylose, mannose, crystalline cellulose, lignin and the ratios sucrose/starch, glucose/fructose, starch/fructose, sucrose/fructose,

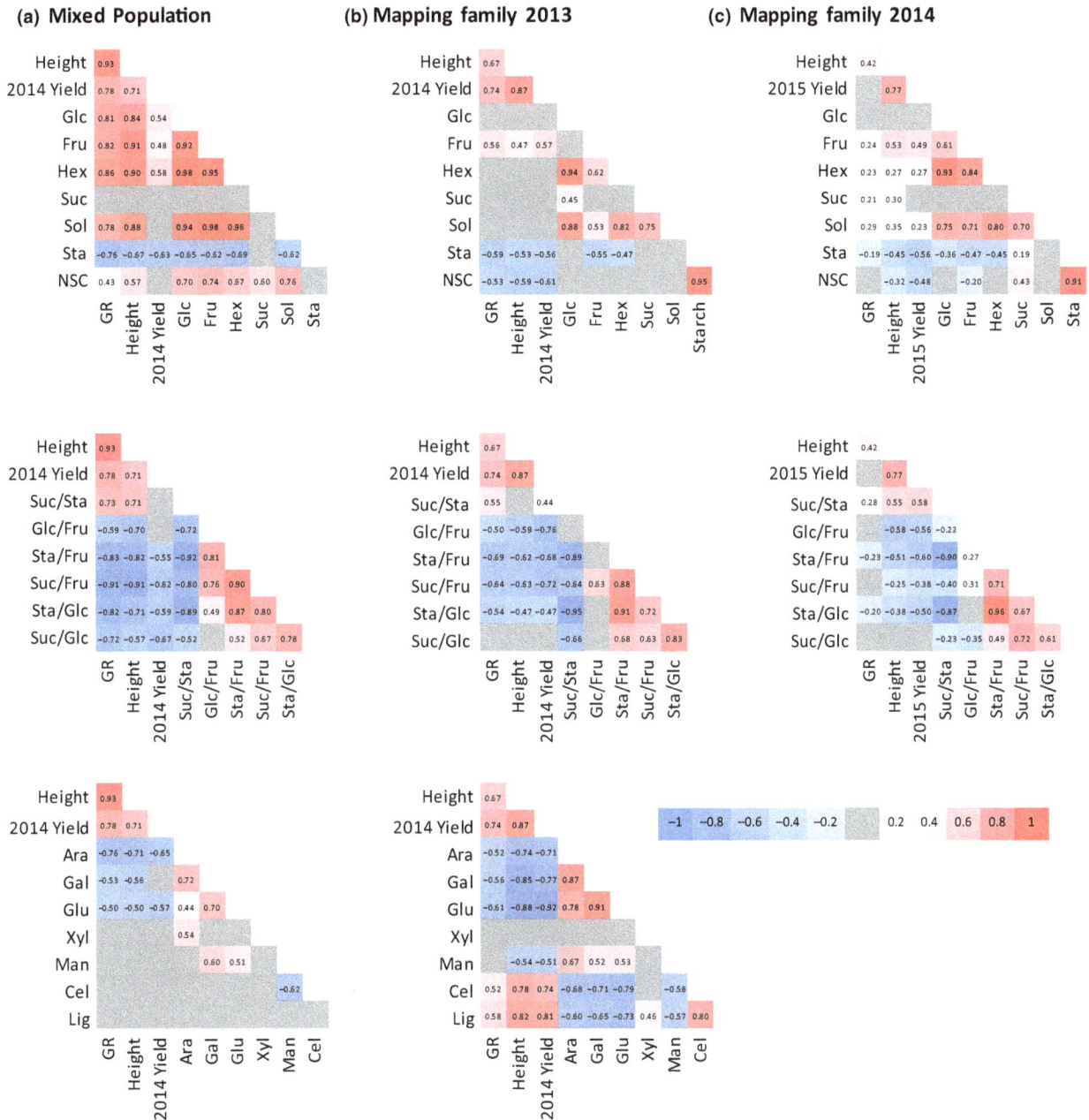

Fig. 3 Spearman's rank correlations between biomass parameters, NSC levels (top) and ratios (middle), and cell wall components (bottom) in the Miscanthus mixed population in 2013 (a) and the mapping family in 2013 (b) and 2014 (c). Significant positive correlations are coloured red, and significant negative correlations are coloured blue ($P =\leq 0.05$). Nonsignificant correlations are coloured grey. Key: Glc = glucose, Fru = fructose, Hex = total hexose, NSC = total nonstructural carbohydrate, Suc = sucrose, Sta = starch, Glc/Fru = glucose-to-fructose ratio, Suc/Glc = sucrose-to-glucose ratio, Suc/Fru = sucrose-to-fructose ratio, Suc/Sta = sucrose-to-starch ratio, Sta/Glc = starch-to-glucose ratio, Ara = arabinose, Cel = cellulose, Glu = glucan, Gal = galactose, Lig = lignin.

starch/glucose and sucrose/glucose. For the full list of 20 carbohydrate predictors, 1 048 576 possible combinations needed evaluation. We also fitted models using only the NSC predictors (8192 combinations), and excluding starch, only soluble sugar predictors (256 combinations). Evaluations were performed using a 'correlation-based feature selection' algorithm, *CfsSubsetEval*, which prefers sets of predictors that have low correlation amongst themselves, but each has high predictive worth (Wang *et al.*, 2005).

A support vector regression algorithm, *SMOreg*, was 'trained' to fit regression models of a given biomass trait

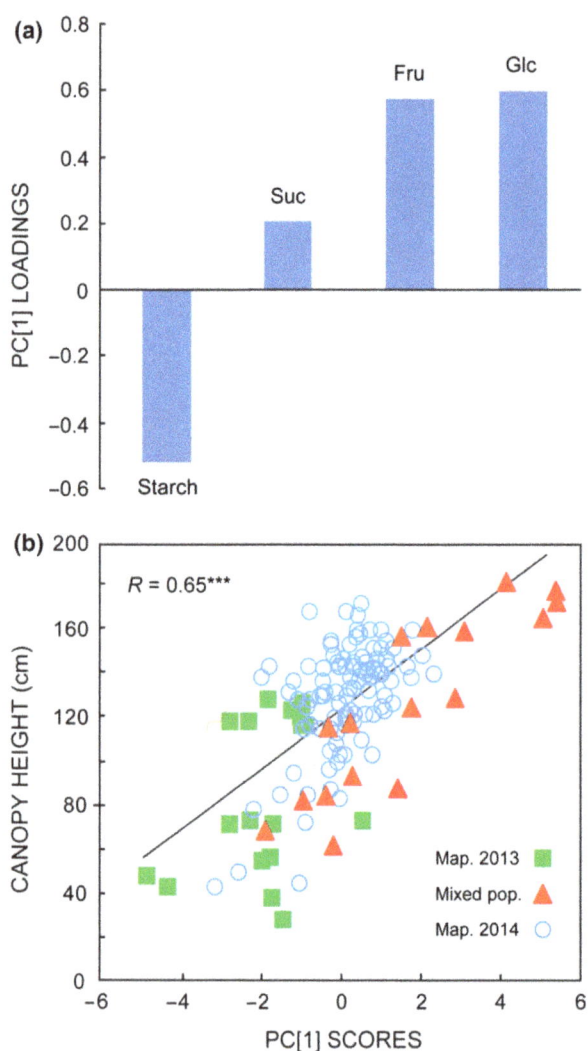

(a) PC[1] LOADINGS axis with bars: Starch (negative ~−0.5), Suc (~0.2), Fru (~0.58), Glc (~0.6)

(b) CANOPY HEIGHT (cm) vs PC[1] SCORES, R = 0.65***
Legend: Map. 2013 (squares), Mixed pop. (triangles), Map. 2014 (open circles)

Fig. 4 Correlation of multivariate carbohydrate phenotypes and biomass across all experimental populations. Horizontal axis shows scores on PC[1], the major component (58% of variance) from PCA of NSC levels (Glc, Fru, Suc, starch) of each genotype in the mixed population (triangles), the mapping population selected in 2013 (squares) and the extended mapping family analysed in 2014 (open circles). Vertical axis shows mean canopy height of each sampled genotype. R indicates Pearson correlation (***, $P < 0.001$) between heights and PC[1] scores.

using the NSC predictors selected in the above process. Support vector machines were originally developed for classification and may be conceived as mapping to higher dimensional space of a 'hyperplane' between two data classes, the 'support vectors' being the least-separated members of the opposite classes. To perform regression, the task is converted to classification by duplicating each y-axis value by addition and subtraction of a new parameter ε. The hyperplane between the plus-ε and minus-ε 'classes' is equivalent to a regression

function (Li *et al.*, 2009). Regression modelling was investigated for each *Miscanthus* population by cross-validation, in which model performance was averaged over nine successive random partitions of the data into training and validation subsets. For each partition, both predictor selection and *SMOreg* modelling used the training data, and models were tested on the held-out validation data (Table 2).

As each experimental population comprised up to 102 genotypes, usually in biological replicates of only three, and distributed in field plots potentially subject to environmental gradients (Table S4), we were interested if genetic structure was detectable in the carbohydrate data. The effects on the *SMOreg* models of including or excluding genotype information for each replicate plant were therefore examined. Models of height (Table 2a) or harvest yield (Table 2b) were significantly improved ($P < 0.05$, Wilcoxon tests) by appending genotype information to the carbohydrate predictors. Mean R of predicted vs. actual biomass values improved from 0.59 to 0.89 for height, and from 0.61 to 0.80 for yield, when models had prior knowledge of genotypes. This evidence for genetically conditioned 'carbohydrate phenotypes' was supported by the significantly better performance ($P < 0.05$, Wilcoxon tests) of models constructed on averaged replicates of each genotype (mean R values: 0.73 for height; 0.71 for yield), relative to those based on all replicates (Table 2). This was presumably due to improved signal to noise for each genotype.

Models based on all carbohydrates including cell wall constituents, or on NSC, or on soluble carbohydrates, were all statistically significant (Table 2) and suggested that the more extensive analytical procedures were unlikely to prove essential in screening of *Miscanthus* populations for biomass potential. The numbers of predictors selected in the machine-learning models ranged from 4 to 8 for the 'all carbohydrates models', 3 to 5 for the 'NSC models' and 2 to 4 for the 'soluble carbohydrates models', and detailed lists are available in Table S5. For each category of model, Table 2 highlights the carbohydrates that featured in every predictor list, whether as an individual metabolite or in a sum or ratio metric. The single metabolite that featured in every model in all categories was fructose. The other NSCs were prominent in some model categories, but ignored in others, with sucrose featuring the least frequently. Amongst cell wall constituents, glucan was ubiquitous in the all-carbohydrate models of height (Table 2a), but not yield (Table 2b).

In modelling yield for the mixed population, one *M. sacchariflorus* genotype, Sac-2, proved particularly detrimental (without genotype information) and was omitted for Table 2b. Amongst over four hundred plants sampled for this study, the two tallest individuals belonged to Sac-2, but its yields per unit height were by far the lowest in

Table 2 Machine-learning models of biomass traits using carbohydrate data

A. Canopy height		Information included in models		
		Genotype, carbohydrates	Carbohydrates only	
		All genotype replicates	All genotype replicates	Averaged by genotype
Plants	Carbohydrate fractions	R values		
Mixed population	All	0.92***	0.44***	0.76***
	Nonstructural	0.81***	0.70***	0.81***
	Soluble	0.84***	0.52***	0.70**
Mapping family 2013	All carbohydrates	0.94***	0.72***	0.77***
	Nonstructural	0.92***	0.63***	0.68***
	Soluble	0.93***	0.68***	0.70***
Mapping family 2014	Nonstructural	0.88***	0.61***	0.76***
	Soluble	0.86***	0.44***	0.66***
Carbohydrate fractions	Constituents common to predictors of all models*			
All	**Glucan**; **Fructose** (as Fru, Glc/Fru, Hex or Suc/Fru); **Glucose** (as Glc/Fru, Hex or Sta/Glc); **Starch** (as Sta, Sta/Glc or Suc/Sta)			
Nonstructural	**Fructose** (as Fru, Glc/Fru, Hex, NSC or Suc/Fru); **Glucose** (as Glc/Fru, Hex, NSC, Sta/Glc or Suc/Glc); **Starch** (as NSC, Sta, Sta/Fru, Sta/Glc or Suc/Sta)			
Soluble	**Fructose** (as Fru, Glc/Fru, Hex or Suc/Fru)			

B. Harvest yield		Information included in models		
		Genotype, carbohydrates	Carbohydrates only	
		All genotype replicates	All genotype replicates	Averaged by genotype
Plants	Carbohydrate fractions	R values		
Mixed population†	All	0.79***	0.61***	0.81***
	Nonstructural	0.79***	0.62***	0.75***
	Soluble	0.79***	0.62***	0.75***
Mapping family 2013	All carbohydrates	0.85***	0.77***	0.70***
	Nonstructural	0.86***	0.65***	0.72***
	Soluble	0.86***	0.68***	0.72***
Mapping family 2014	Nonstructural	0.75***	0.56***	0.68***
	Soluble	0.74***	0.40***	0.58***
Carbohydrate fractions	Constituents common to predictors of all models*			
All	**Fructose** (as Fru, Glc/Fru or NSC)			
Nonstructural	**Fructose** (as Fru, Glc/Fru, NSC or Suc/Fru); **Glucose** (as Glc/Fru, NSC or Suc/Glc); **Sucrose** (as NSC, Suc/Fru, Suc/Glc or Suc/Sta); **Starch** (as NSC, Sta or Suc/Sta)			
Soluble	**Fructose** (as Fru, Glc/Fru or Suc/Fru)			

Support vector regression (*SMOreg*) models were trained using subsets of predictors selected (*CfsSubsetEval*) for individual correlation with trait but low correlation with each other. Models were evaluated in ninefold cross-validations. R values indicate Pearson correlation between actual and predicted biomass data.
*Full lists in Supplementary Information.
†*M. sacchariflorus* genotype Mb306 was excluded from Table B.

the mixed population. Inclusion of this outlier genotype in the yield models saw R values for the mixed population (genotypes averaged) fall to 0.62, 0.36 and 0.52 for the all-carbohydrate, NSC and soluble carbohydrate models, respectively. It was concluded that the traits responsible for the particular morphology of Sac-2 were not accessible to modelling from carbohydrates.

Biomass and NSC data for the same genotypes in 2013 and 2014 showed high absolute (R) and rank (R_S) interyear correlation (Table 3). We therefore investigated whether predictive models relating carbohydrates to biomass traits could be applicable from 1 year to another. Figure 5 shows the application of the machine-learning regression method to prediction of (a) height, and (b) yield of the mapping family genotypes analysed in 2014, based purely on the measurements of their NSC levels (Table S6). The regression models were pretrained on the NSC and biomass data of the smaller number of mapping genotypes analysed in 2013. Correlations between predicted and actual biomass trait

Table 3 Correlations between 2013 and 2014 for biomass traits and nonstructural carbohydrate composition. For biomass traits, $N = 102$, and for carbohydrates and ratios, $N = 20$, $*P = \leq 0.05 \; **P = \leq 0.01$

Trait	Pearson's (R)	Spearman's Rank (R)	P
Canopy Height	0.8	0.8	**
Yield	0.9	0.9	**
Glucose	0.6	0.6	**
Fructose	0.8	0.7	**
Hexose	0.7	0.4	*
Sucrose	0.7	0.6	**
Starch	0.8	0.8	**
Total NSC	0.8	0.8	**
Suc/Sta	0.7	0.7	**
Glc/Fru	0.7	0.8	**
Sta/Fru	0.9	0.7	**

values were highly significant ($P < 10^{-14}$) for height and yield, with Pearson correlation values of 0.67–0.68 (Fig. 5). Predictions for growth rate data were weaker (R, 0.25; $P < 0.05$).

To investigate our hypothesis of a competitive relationship between starch and cellulose biosynthesis, we conducted a ^{13}C-labelling experiment in the field. We used a *M. sinensis* genotype ('Goliath'), which was comparatively slow-growing, a fast-growing hybrid genotype phylogenetically similar to Hyb 2 of the mixed population (*M. x giganteus*), and a *M. sacchariflorus* genotype phylogenetically similar to Sac-2. Stems were harvested 30 h after labelling. The slower growing genotype, Goliath, partitioned significantly more pulse-derived ^{13}C into starch than the fast-growing hybrid and *M. sacchariflorus* (Table 4). When analysed by ANOVA, there appeared to be no difference between the genotypes in % deposition into the cell wall even though the mean values were quite different. We considered that the analysis was being skewed by a large amount of variation between replicate in the *M. sacchariflorus* genotype (Table 4). Therefore, we also performed *t*-tests between the three genotypes, which showed that *M. x giganteus* had deposited significantly more pulse-derived ^{13}C into the insoluble fraction which would be mainly comprised of cellulose. This approach was also applied on other measurements (such as % soluble), and no additional significant differences between genotypes were observed. The *M. sacchariflorus* genotype, whilst taller than *M. x giganteus*, had a slower growth rate at the time of labelling and was not statistically distinct from either of the other genotypes in its pulse-derived ^{13}C deposition. All three genotypes had accumulated the same total amount of pulse-derived ^{13}C. Therefore, the observed differences between the hybrid and Goliath were in carbon partitioning rather than capture (Table 4).

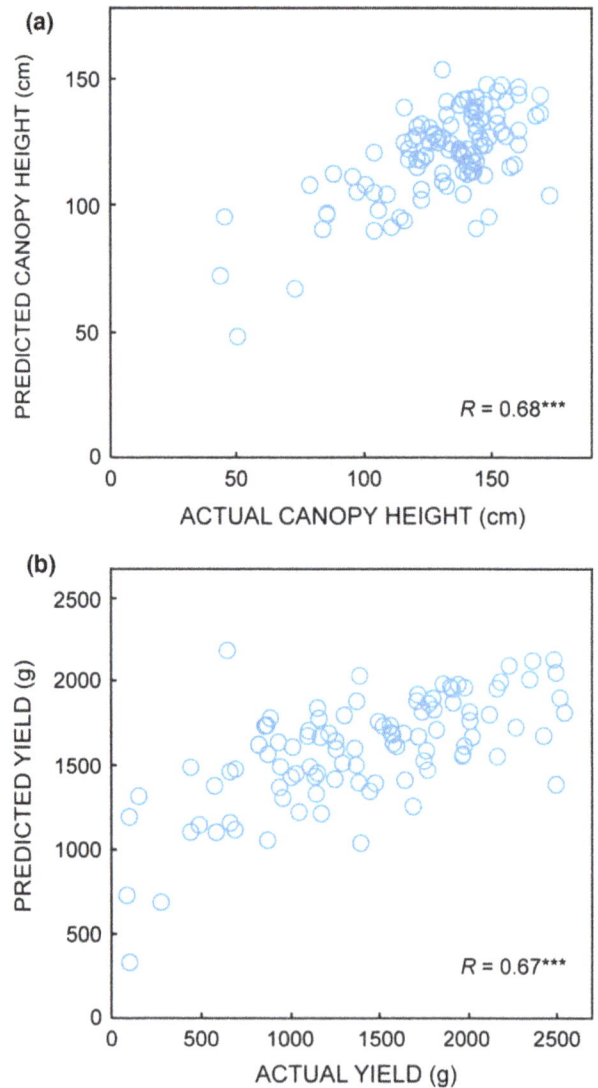

Fig. 5 Interyear prediction of biomass traits from carbohydrates. From NSC levels in the 20 genotypes selected in 2013 from the mapping population, machine learning was used to select minimal lists of predictors to model biomass traits of the sampled plants. The machine-learning models thereby 'trained' on the 2013 data were subsequently provided with the NSC data (only) of the 102 mapping family genotypes analysed in 2014 and 'tested' for prediction of (a) canopy height, and (b) yield, of the 2014 plants. For both models, the predictor metrics selected by machine learning were as follows: total NSC, and the Glc/Fru and Suc/Fru ratios. R indicates Pearson correlation (***, $P < 0.001$) between actual and predicted values.

In the mapping family in 2013, there was a positive correlation between concentrations of cellulose and fructose, and negative relationships between cellulose and the ratios of glucose/fructose and sucrose/fructose (Fig. 6). However, no negative relationship between starch and cellulose was observed, which did not support our starting hypothesis that a competitive

Table 4 Biomass traits and % ^{13}C recovered from the stem 30 h after labelling in three genotypes. $N = 3 \pm$ SE. Different letters show significant differences between genotypes according to an ANOVA and associated t-test. Values in parentheses in the % insoluble column show the additional results of t-tests between the three genotypes ($P = \leq 0.05$)

| | Stem weight (g DW) | Stem height (cm) | Growth rate (cm day^{-1}) | Pulse derived ^{13}C (mg g^{-1} DW) | ^{13}C recovered in each fraction as a % of the total | | |
					Soluble	Starch	Insoluble
Hybrid	30.0 ± 3.5 a	241.0 ± 11.6 a	2.8 ± 0.4 a	2.0 ± 0.2 a	19.0 ± 2.7 a	18.2 ± 2.8 a	62.8 ± 0.8 a (a)
Goliath	20.0 ± 1.1 ab	180.3 ± 2.0 b	1.0 ± 0.5 b	2.2 ± 0.4 a	17.4 ± 2.5 a	38.2 ± 4.8 b	44.4 ± 3.2 a (b)
M. sacchariflorus	54.1 ± 10.7 b	274.3 ± 3.5 c	1.3 ± 0.3 ab	1.5 ± 0.2 a	39.0 ± 12.3 a	26.3 ± 1.5 a	34.8 ± 13.8 a (ab)

	Cellulose	P
Glucose	−0.04	0.87
Fructose	0.50	0.02
Hexose	0.15	0.52
Sucrose	−0.05	0.83
Soluble	−0.04	0.87
Starch	−0.23	0.32
NSC	−0.26	0.27
Suc/Sta	0.08	0.73
Glc/Fru	−0.64	0.00
Sta/Fru	−0.43	0.06
Suc/Fru	−0.58	0.01
Sta/Glc	−0.21	0.37
Suc/Glc	−0.24	0.30

Fig. 6 Spearman's rank correlation between cellulose and the NSC levels and ratios in the mapping family in 2013. Significant positive correlations are coloured red, and significant negative correlations are coloured blue ($P = \leq 0.05$). Nonsignificant correlations are coloured grey.

relationship exists between these two polymers of glucose (Fig. 6).

Discussion

We have demonstrated that the abundance and partitioning of carbohydrates, particularly the NSC, can act as metabolic biomarkers of productivity in diverse genotypes of field-grown *Miscanthus*. These findings thereby support other studies that have modelled biomass from metabolites using the model species *A. thaliana* in controlled environments (Meyer *et al.*, 2007; Sulpice *et al.*, 2009, 2010, 2013; Scott *et al.*, 2010). The present article, by contrast, takes the important step of applying this approach to the field environment using a bioenergy crop species with a limited breeding

history. Despite the likely complex impacts of sunlight, temperature and rainfall on the carbohydrate status of the field plants, informative and significant biomass models could be generated.

In our field-grown Miscanthus, we observed that fructose consistently, positively correlated with yield traits. Fructose is produced exclusively from the metabolism of sucrose by the action of sucrose synthases (SuSy) and invertases, whereas glucose is produced both by the action of invertases (but not SuSy) and the metabolism of starch (Koch, 2004; Smith *et al.*, 2005; Ruan, 2014). Therefore, fructose is a direct indication of sucrose metabolism, whereas glucose provides information about both sucrose and starch metabolism. Whilst our observations are strictly correlative (there is no evidence from our results that a high abundance of fructose specifically *causes* plants to be taller or higher yielding), there is evidence in the literature that altered partitioning in the ratio of sucrose/starch can follow increases in biomass in other species (Foyer & Ferrario, 1994; Laporte *et al.*, 1997; Signora *et al.*, 1998). For example, sucrose phosphate synthase (SPS) catalyses the penultimate step in the synthesis of sucrose in the cytosol (Signora *et al.*, 1998). In tomato plants engineered to constitutively overexpress maize SPS, biomass was increased up to 100% compared to wild-type controls (Foyer & Ferrario, 1994; Laporte *et al.*, 1997). Furthermore, when the experiment was replicated in *Arabidopsis*, the ratio of sucrose to starch increased due to a reduction in starch in the transgenic plants (changes in biomass were not observed in this study; Signora *et al.*, 1998). More recently, it has been found that transgenic expression of *Arabidopsis* SPS and sucrose phosphate phosphatase also enhanced growth and biomass accumulation in hybrid poplar (Maloney *et al.*, 2015). Conversely, an *Arabidopsis* double knockout mutant of the major SPS leaf isoforms was strongly impaired in growth and accumulated high levels of starch (Volkert *et al.*, 2014). In the Poaceae, SPS has been of interest as a biochemical marker for complex agronomic traits in several species (Castleden *et al.*, 2004). In rice, an SPS gene coincides with a quantitative trait locus for plant height

(Ishimaru *et al.*, 2004; Venu *et al.*, 2014), whilst plants with a maize SPS transgene grew taller (Ishimaru *et al.*, 2004).

Sucrose-metabolizing enzymes are candidates for explaining the correlation between hexose and biomass. The overexpression of SuSy and UDP-glucose pyrophosphorylase (either as individual or as double mutants) in transgenic tobacco resulted in an increase in the abundance of hexoses, particularly fructose, a decrease in the glucose/fructose ratio and a concurrent increase in plant biomass (Coleman *et al.*, 2006). Poovaiah *et al.* (Poovaiah *et al.*, 2015) overexpressed a SuSy transgene in the biofuel feedstock switchgrass (*Panicum virgatum*), achieving increased height and biomass in some transformants. In *Populus alba* x *grandidentata* (hybrid poplar), overexpression of SuSy caused increases in soluble carbohydrates and cellulose, and a decrease in cell wall-derived arabinose (Coleman *et al.*, 2009). A plasma membrane-bound isoform of SuSy is thought to transfer UDP-glucose units directly to the extending glucan chains of cellulose (McFarlane *et al.*, 2014), providing a link between the metabolism of the nonstructural pool and the formation of structural biomass. Some of these results therefore complement our findings using natural variation, in that the highest yielding plants had high hexoses, a low glucose/fructose ratio, higher cellulose and lower arabinose. Therefore, the abundance or activity of enzymes involved in sucrose biosynthesis or metabolism are candidates for the causal basis of our observations on hexoses and biomass.

No negative correlation between cellulose and starch was observed, which did not support our starting hypothesis that a competitive relationship exists. However, significant, negative correlations were observed between cellulose and the ratios of glucose/fructose and sucrose/fructose and a positive relationship between fructose and cellulose. This suggests that when the proportion of fructose is higher (relative to glucose or sucrose), cellulose is also in greater abundance. Cellulose biosynthesis is dependent on sucrose metabolism (Amor *et al.*, 1995; Coleman *et al.*, 2009; Baroja-Fernandez *et al.*, 2012), and the positive correlation between an increased proportion of fructose and cellulose could be a demonstration of this; as UDP-glucose units, cleaved from sucrose through the action of SuSy, are transported across the plasma membrane to the extending cellulose chain, an increasing pool of fructose is left behind. Our findings from the ^{13}C-labelling experiment support this, as the differences in partitioning were observed between the fast- and slow-growing genotype. Therefore, a more rapid rate of growth depends upon a greater accumulation of cellulose, via sucrose metabolism, rather than transient storage as starch.

The strength of the significant correlations found in our study is within the same range as those reported in pairwise analyses of molecular markers and traits. Correlations between short nucleotide polymorphisms (SNPs) and starch quality showed significant correlations (R^2) of 0.17–0.67 in rice. In a human asthma, study correlations between SNPs and lung physiology ranged from (R^2) ~0.3 to 0.9 (Kim & Xing, 2009; Kharabian-Masouleh *et al.*, 2012). This demonstrates that the strength of the correlation between the starch/fructose ratio (for example) of $R = 0.6$–0.8 is within a range comparable to molecular markers. Height is the trait that best correlates with yield ($R^2 = $~0.55; Robson *et al.*, 2013). Using modelling to combine the strongest biomarkers, the predicted and actual yields of the mapping family in 2014 produced correlations of $R = 0.67$ ($R^2 = 0.44$), which is a weaker predictor than height (Robson *et al.*, 2013). However, whilst metabolic profiling was not a stronger predictor of final yield than height alone in *Miscanthus*, detailed knowledge of the relations of metabolism and biomass accumulation can be expected to yield powerful novel tools to accelerate and enhance energy plant breeding programmes (Robson *et al.*, 2013). For example, in *Miscanthus*, the juvenile phase severely hinders early phenotypic selection (Robson *et al.*, 2013), but if the metabolic profile could predict mature height in juvenile plants, metabolic biomarkers could then be used in a similar way to molecular markers. To address this hypothesis, the next stage of our experimentation is to screen first-year seedlings and 1-year-old and 2-year-old plants to discover at what stage in development the glycome can be used as a biomarker for yield in mature plants. As *Miscanthus* takes 4 years from sowing seed to reach maturity, if the markers could only be used in the 2nd year of growth, this could reduce screening time by 50%. An alternative scenario in which yield prediction through biomarkers could be highly beneficial is in species such as trees where physical phenotyping is particularly challenging, or in screening for abiotic or biotic stress tolerances. It is also possible that metabolic and molecular markers could be used synergistically in breeding programmes to improve selection.

A concern about the use of metabolic biomarkers is their reliability, given that metabolites are dynamic and their absolute abundances will vary. However, a number of studies have demonstrated that the abundance of NSC and the ratios between different pools is under genetic control (Calenge *et al.*, 2006; Purdy *et al.*, 2015). Furthermore, it is generally accepted (and experimentally demonstrated, e.g. Table 3) that a high-yielding genotype will consistently produce high yields compared to a low-yielding type, even though climatic conditions over the course of a whole growing season may vary tremendously from year to year. As yield is implicitly dependent upon the NSC pool to form the structural biomass, it is logical that the NSC composition must also be genetically controlled and similar enough between years and within genotypes to produce consistent results in yield and quality traits. Many of the metabolites measured, such as glucose and most cell wall components, were found to be unreliable for predicting yield when used alone, as they only produced significant relationships with biomass in one or other of the field sites. In contrast, fructose, starch and several of the ratios were found to be consistent indicators of biomass traits in both sets of plants and in both years of study. Therefore, as with molecular markers, it is important to choose robust markers to produce reliable results.

The model generated for the mapping family was based on data from plants in their 3rd and 4th complete growing season in the field. Whilst this is considered a mature crop, it has been shown that yields continue to increase at least until the 5th year of growth (Robson et al., 2013). Therefore, the current model may underestimate yields in subsequent years and have to be reparameterized once the annual increase in yields has plateaued.

In conclusion, our study has shown that fructose and starch positively and negatively correlate with yield traits, respectively. The glycome in the summer growing season can be used as a biomarker to predict future harvest yields in the following year. Plants that partitioned a greater proportion of captured carbon into cellulose rather than starch attained greater biomass. Metabolic biomarker identification may also be an approach that could be adapted for other agronomic traits such as stress tolerance or disease resistance.

Acknowledgements

This work was supported by the BBSRC Sustainable Bioenergy Centre (BSBEC) grant (BB/G016216/1) working within the BSBEC BioMASS (http://www.bsbec-biomass.org.uk/) Programme of the centre and Ceres inc. and the BBSRC Energy Grasses & Biorefining Institute Strategic Programme (BBS/E/W/00003134) at IBERS and the Ecosystem Land Use Modelling & Soil C Flux Trial (ELUM) at CEH Lancaster and IBERS (ET/I000100/1). The funders had no role in study design, data collection and analysis, decision to publish or preparation of the manuscript.

References

Allison GG, Morris C, Clifton-Brown J, Lister SJ, Donnison IS (2011) Genotypic variation in cell wall composition in a diverse set of 244 accessions of Miscanthus. Biomass and Bioenergy, 35, 4740–4747.

Amor Y, Haigler CH, Johnson S, Wainscott M, Delmer DP (1995) A membrane-associated form of sucrose synthase and its potential role in synthesis of cellulose and callose in plants. Proceedings of the National Academy of Sciences of the United States of America, 92, 9353–9357.

Baroja-Fernandez E, Munoz FJ, Li J et al. (2012) Sucrose synthase activity in the sus1/sus2/sus3/sus4 Arabidopsis mutant is sufficient to support normal cellulose and starch production. Proceedings of the National Academy of Sciences of the United States of America, 109, 321–326.

Biasi C, Pitkamaki AS, Tavi NM, Koponen HT, Martikainen PJ (2012) An isotope approach based on 13C pulse-chase labelling vs. the root trenching method to separate heterotrophic and autotrophic respiration in cultivated peatlands. Boreal Environment Research, 17, 184–192.

Calenge F, Saliba-Colombani V, Mahieu S, Loudet O, Daniel-Vedele F, Krapp A (2006) Natural variation for carbohydrate content in Arabidopsis. Interaction with complex traits dissected by quantitative genetics. Plant Physiology, 141, 1630–1643.

Castleden CK, Aoki N, Gillespie VJ et al. (2004) Evolution and function of the sucrose-phosphate synthase gene families in wheat and other grasses. Plant Physiology, 135, 1753–1764.

Coleman HD, Ellis DD, Gilbert M, Mansfield SD (2006) Up-regulation of sucrose synthase and UDP-glucose pyrophosphorylase impacts plant growth and metabolism. Plant Biotechnology Journal, 4, 87–101.

Coleman HD, Yan J, Mansfield SD (2009) Sucrose synthase affects carbon partitioning to increase cellulose production and altered cell wall ultrastructure. Proceedings of the National Academy of Sciences of the United States of America, 106, 13118–13123.

Foster CE, Martin TM, Pauly M (2010a) Comprehensive compositional analysis of plant cell walls (Lignocellulosic biomass) part I: lignin. Journal of Visualized Experiments, 1745, doi: 10.3791/1745.

Foster CE, Martin TM, Pauly M (2010b) Comprehensive compositional analysis of plant cell walls (lignocellulosic biomass) part II: carbohydrates. Journal of Visualized Experiments, 1837, doi: 10.3791/1837.

Foyer CH, Ferrario S (1994) Modulation of carbon and nitrogen-metabolism in transgenic plants with a view to improved biomass production. Biochemical Society Transactions, 22, 909–915.

Frank E, Hall M, Trigg L, Holmes G, Witten IH (2004) Data mining in bioinformatics using Weka. Bioinformatics, 20, 2479–2481.

Hodkinson TR, Chase MW, Lledo MD, Salamin N, Renvoize SA (2002) Phylogenetics of Miscanthus, Saccharum and related genera (Saccharinae, Andropogoneae, Poaceae) based on DNA sequences from ITS nuclear ribosomal DNA and plastid trnL intron and trnL-F intergenic spacers. Journal of Plant Research, 115, 381–392.

Hogberg P, Hogberg MN, Gottlicher SG et al. (2008) High temporal resolution tracing of photosynthate carbon from the tree canopy to forest soil microorganisms. New Phytologist, 177, 220–228.

Ishimaru K, Ono K, Kashiwagi T (2004) Identification of a new gene controlling plant height in rice using the candidate-gene strategy. Planta, 218, 388–395.

Jensen E, Farrar K, Thomas-Jones S, Hastings A, Donnison I, Clifton-Brown J (2011) Characterization of flowering time diversity in Miscanthus species. GCB Bioenergy, 3, 387–400.

Jones MG, Outlaw WH, Lowry OH (1977) Enzymic assay of 10 to 10 moles of sucrose in plant tissues. Plant Physiology, 60, 379–383.

Kharabian-Masouleh A, Waters DLE, Reinke RF, Ward R, Henry RJ (2012) SNP in starch biosynthesis genes associated with nutritional and functional properties of rice. Scientific Reports, 2, 557.

Kim S, Xing EP (2009) Statistical estimation of correlated genome associations to a quantitative trait network. Plos Genetics, 5, e1000587.

Koch K (2004) Sucrose metabolism: regulatory mechanisms and pivotal roles in sugar sensing and plant development. Current Opinion in Plant Biology, 7, 235–246.

Laporte MM, Galagan JA, Shapiro JA, Boersig MR, Shewmaker CK, Sharkey TD (1997) Sucrose-phosphate synthase activity and yield analysis of tomato plants transformed with maize sucrose-phosphate synthase. Planta, 203, 253–259.

Li HD, Liang YZ, Xu QS (2009) Support vector machines and its applications in chemistry. Chemometrics and Intelligent Laboratory Systems, 95, 188–198.

Maloney VJ, Park JY, Unda F, Mansfield SD (2015) Sucrose phosphate synthase and sucrose phosphate phosphatase interact in planta and promote plant growth and biomass accumulation. Journal of Experimental Botany, 66, 4383–4394.

McFarlane HE, Doring A, Persson S (2014) The cell biology of cellulose synthesis. Annual Review of Plant Biology, 65, 69.

Meyer RC, Steinfath M, Lisec J et al. (2007) The metabolic signature related to high plant growth rate in Arabidopsis thaliana. Proceedings of the National Academy of Sciences of the United States of America, 104, 4759–4764.

Muguerza M, Gondo T, Yoshida M, Kawakami A, Terami F, Yamada T, Akashi R (2013) Modification of the total soluble sugar content of the C4 grass *Paspalum notatum* expressing the wheat-derived sucrose: sucrose 1-fructosyltransferase and sucrose: fructan 6-fructosyltransferase genes. *Grassland Science*, **59**, 196–204.

O'Brien MJ, Leuzinger S, Philipson CD, Tay J, Hector A (2014) Drought survival of tropical tree seedlings enhanced by non-structural carbohydrate levels. *Nature Climate Change*, **4**, 710–714.

Poovaiah CR, Mazarei M, Decker SR, Turner GB, Sykes RW, Davis MF, Stewart CN Jr (2015) Transgenic switchgrass (*Panicum virgatum* L.) biomass is increased by overexpression of switchgrass sucrose synthase (PvSUS1). *Biotechnology Journal*, **10**, 552–563.

Purdy SJ, Maddison AL, Jones LE, Webster RJ, Andralojc J, Donnison I, Clifton-Brown J (2013) Characterization of chilling-shock responses in four genotypes of *Miscanthus* reveals the superior tolerance of *M. x giganteus* compared with *M. sinensis* and *M. sacchariflorus*. *Annals of Botany*, **111**, 999–1013.

Purdy SJ, Cunniff J, Maddison AL et al. (2014) Seasonal carbohydrate dynamics and climatic regulation of senescence in the perennial grass, Miscanthus. *Bioenergy Research*, **8**, 28–41.

Purdy SJ, Maddison AL, Cunniff J, Donnison I, Clifton-Brown J (2015) Non-structural carbohydrate profiles and ratios between soluble sugars and starch serve as indicators of productivity for a bioenergy grass. *AoB Plants*, **7**, plv032, doi: 10.1093/aobpla/plv032.

Robson P, Mos M, Clifton-Brown J, Donnison I (2012) Phenotypic variation in senescence in Miscanthus: towards optimising biomass quality and quantity. *Bioenergy Research*, **5**, 95–105.

Robson P, Jensen E, Hawkins S et al. (2013) Accelerating the domestication of a bioenergy crop: identifying and modelling morphological targets for sustainable yield increase in Miscanthus. *Journal of Experimental Botany*, **64**, 4143–4155.

Rocher JP (1988) Comparison of carbohydrate compartmentation in relation to photosynthesis, assimilate export and growth in a range of maize genotypes. *Australian Journal of Plant Physiology*, **15**, 677–686.

Ruan YL (2014) Sucrose metabolism: gateway to diverse carbon use and sugar signaling. *Annual Review of Plant Biology*, Vol 61, **65**, 33–67.

Scott IM, Vermeer CP, Liakata M et al. (2010) Enhancement of plant metabolite fingerprinting by machine learning. *Plant Physiology*, **153**, 1506–1520.

Scott IM, Ward JL, Miller SJ, Beale MH (2014) Opposite variations in fumarate and malate dominate metabolic phenotypes of Arabidopsis salicylate mutants with abnormal biomass under chilling. *Physiologia Plantarum*, **152**, 660–674.

Signora L, Galtier N, Skot L, Lucas H, Foyer CH (1998) Over-expression of sucrose phosphate synthase in *Arabidopsis thaliana* results in increased foliar sucrose/ starch ratios and favours decreased foliar carbohydrate accumulation in plants after prolonged growth with CO_2 enrichment. *Journal of Experimental Botany*, **49**, 669–680.

Smith AM, Zeeman SC, Smith SM (2005) Starch degradation. *Annual Review of Plant Biology*, **56**, 73–98.

Somerville C, Youngs H, Taylor C, Davis SC, Long SP (2010) Feedstocks for Lignocellulosic Biofuels. *Science*, **329**, 790–792.

de Souza AP, Arundale RA, Dohleman FG, Long SP, Buckeridge MS (2013) Will the exceptional productivity of Miscanthus x giganteus increase further under rising atmospheric CO_2? *Agricultural and Forest Meteorology*, **171**, 82–92.

Steinfath M, Strehmel N, Peters R et al. (2010) Discovering plant metabolic biomarkers for phenotype prediction using an untargeted approach. *Plant Biotechnology Journal*, **8**, 900–911.

Subke JA, Vallack HW, Magnusson T, Keel SG, Metcalfe DB, Hogberg P, Ineson P (2009) Short-term dynamics of abiotic and biotic soil $^{13}CO_2$ effluxes after *in situ* $^{13}CO_2$ pulse labelling of a boreal pine forest. *New Phytologist*, **183**, 349–357.

Sulpice R, Pyl ET, Ishihara H et al. (2009) Starch as a major integrator in the regulation of plant growth. *Proceedings of the National Academy of Sciences of the United States of America*, **106**, 10348–10353.

Sulpice R, Trenkamp S, Steinfath M et al. (2010) Network analysis of enzyme activities and metabolite levels and their relationship to biomass in a large panel of Arabidopsis accessions. *Plant Cell*, **22**, 2872–2893.

Sulpice R, Nikoloski Z, Tschoep H et al. (2013) Impact of the carbon and nitrogen supply on relationships and connectivity between metabolism and biomass in a broad panel of Arabidopsis accessions. *Plant Physiology*, **162**, 347–363.

Venu RC, Ma J, Jia Y et al. (2014) Identification of candidate genes associated with positive and negative heterosis in rice. *PLoS ONE*, **9**, e95178.

Visser P, Pignatelli V (2001) Utilization of Miscanthus. In: *Miscanthus: For Energy and Fibre* (ed. Walsh M), pp. 109–154. Earthscan, London.

Volkert K, Debast S, Voll LM et al. (2014) Loss of the two major leaf isoforms of sucrose-phosphate synthase in *Arabidopsis thaliana* limits sucrose synthesis and nocturnal starch degradation but does not alter carbon partitioning during photosynthesis. *Journal of Experimental Botany*, **65**, 5217–5229.

Wang L, Wu JQ (2013) The essential role of Jasmonic acid in plant-herbivore interactions – using the wild tobacco *Nicotiana attenuata* as a model. *Journal of Genetics and Genomics*, **40**, 597–606.

Wang Y, Tetko IV, Hall MA, Frank E, Facius A, Mayer KFX, Mewes HW (2005) Gene selection from microarray data for cancer classification – a machine learning approach. *Computational Biology and Chemistry*, **29**, 37–46.

Wanner LA, Junttila O (1999) Cold-induced freezing tolerance in Arabidopsis. *Plant Physiology*, **120**, 391–399.

11

A realistic meteorological assessment of perennial biofuel crop deployment: a Southern Great Plains perspective

MELISSA WAGNER[1], MENG WANG[2], GONZALO MIGUEZ-MACHO[3], JESSE MILLER[4], ANDY VANLOOCKE[5], JUSTIN E. BAGLEY[6], CARL J. BERNACCHI[4,7] and MATEI GEORGESCU[1,2]

[1]School of Geographical Sciences and Urban Planning, Arizona State University, Tempe, AZ 85287-5302, USA, [2]School of Mathematical and Statistical Sciences, Arizona State University, Tempe, AZ, USA, [3]Universidade de Santiago de Compostela, Galicia, Spain, [4]Department of Plant Biology, University of Illinois, Urbana, IL, USA, [5]Department of Agronomy, Iowa State University, Ames, IA, USA, [6]Climate and Ecosystems Science Division, Lawrence Berkeley National Laboratory, Berkeley, CA, USA, [7]Global Change and Photosynthesis Research Unit, United States Department of Agriculture Agricultural Research Service, Urbana, IL 61801, USA

Abstract

Utility of perennial bioenergy crops (e.g., switchgrass and miscanthus) offers unique opportunities to transition toward a more sustainable energy pathway due to their reduced carbon footprint, averted competition with food crops, and ability to grow on abandoned and degraded farmlands. Studies that have examined biogeophysical impacts of these crops noted a positive feedback between near-surface cooling and enhanced evapotranspiration (ET), but also potential unintended consequences of soil moisture and groundwater depletion. To better understand hydrometeorological effects of perennial bioenergy crop expansion, this study conducted high-resolution (2-km grid spacing) simulations with a state-of-the-art atmospheric model (Weather Research and Forecasting system) dynamically coupled to a land surface model. We applied the modeling system over the Southern Plains of the United States during a normal precipitation year (2007) and a drought year (2011). By focusing the deployment of bioenergy cropping systems on marginal and abandoned farmland areas (to reduce the potential conflict with food systems), the research presented here is the first realistic examination of hydrometeorological impacts associated with perennial bioenergy crop expansion. Our results illustrate that the deployment of perennial bioenergy crops leads to widespread cooling (1–2 °C) that is largely driven by an enhanced reflection of shortwave radiation and, secondarily, due to an enhanced ET. Bioenergy crop deployment was shown to reduce the impacts of drought through simultaneous moistening and cooling of the near-surface environment. However, simulated impacts on near-surface cooling and ET were reduced during the drought relative to a normal precipitation year, revealing differential effects based on background environmental conditions. This study serves as a key step toward the assessment of hydroclimatic sustainability associated with perennial bioenergy crop expansion under diverse hydrometeorological conditions by highlighting the driving mechanisms and processes associated with this energy pathway.

Keywords: biofuel crops, drought, hydroclimate, land-use change, modeling, renewable energy

Introduction

The U.S. Energy Independence and Security Act of 2007 (RFA, 2010) mandates the production of 80 gigaliters of ethanol from nongrain sources by 2022 (Gelfand *et al.*, 2013; Oikawa *et al.*, 2015). Perennial bioenergy crops have the potential to contribute an important relative share of U.S. ethanol demand, thereby helping to meet established mandates, while simultaneously decreasing reliance on fossil fuels. Reduced carbon emissions, increased energy security, and stabilization in energy pricing serve as key elements driving continued interest in the generation of biomass-derived energy (Haberl *et al.*, 2010; López-Bellido *et al.*, 2014; Abraha *et al.*, 2015; Miller *et al.*, 2015; Hudiburg *et al.*, 2016). Nongrain perennial biofuel crops such as the perennial grasses switchgrass (*Panicum virgatum*) and miscanthus (e.g., *Miscanthus × giganteus*) are particularly appealing due to their reduced carbon footprint, averted competition

Correspondence: Matei Georgescu
e-mail: Matei.Georgescu@asu.edu

with food crops, minimal fertilizer usage and upkeep, and ability to grow on abandoned and degraded farmlands (Foley *et al.*, 2005; Davis *et al.*, 2012; Campbell *et al.*, 2013; Bagley *et al.*, 2014; Hudiburg *et al.*, 2015; Vanloocke *et al.*, 2010). By reducing greenhouse gas emissions, the deployment of perennial bioenergy crops could play an important role in mitigating anthropogenic climate change (Anderson-Teixeira *et al.*, 2012).

In addition to biogeochemical impacts (e.g., Melillo *et al.*, 2009; Gopalakrishnan *et al.*, 2012), large-scale cultivation of perennial bioenergy crops will result in land use and land cover change (LULCC) impacts in which biogeophysical implications must be considered. Conversion of existing landscapes to perennial cropping systems will produce surface energy balance changes, affecting atmospheric boundary-layer dynamics (Weaver & Avissar, 2001), with implications for regional hydroclimate modification (Vanloocke *et al.*, 2010; Georgescu and Lobell 2010; Georgescu *et al.*, 2011; Georgescu *et al.*, 2013; Pielke, 2001, 2005; Le *et al.*, 2011; Xu *et al.*, 2013; Bagley *et al.*, 2014; Goldstein *et al.*, 2014; Devaraju *et al.*, 2015). The few studies that examined biogeophysical consequences of perennial biofuel crops used regional climate (e.g., Georgescu *et al.*, 2009, 2011; Anderson *et al.*, 2013; Khanal *et al.*, 2013), ecosystem (e.g., Vanloocke *et al.*, 2010; Le *et al.*, 2011), and watershed-scale (Wagle & Kakani, 2014) models as well as micrometeorological assessments (e.g., Hickman *et al.*, 2010; Abraha *et al.*, 2015; Miller *et al.*, 2015). These studies highlight the importance of time-varying representation that characterizes the distinct biofuel cropping systems appropriately (Bright *et al.*, 2012, 2016). For example, recent field-scale plantation measurements of perennial biofuel crops illustrate greater albedo values during the growing season when compared with those of annual crops (e.g., maize and soybean) (Miller *et al.*, 2015). Higher albedo values can lead to a decrease in net surface radiation and have been shown to regionally cool temperatures in climate modeling studies that replaced annual with perennial biofuel crops (Georgescu *et al.*, 2009, 2011; Vanloocke *et al.*, 2010; Bagley *et al.*, 2015; Miller *et al.*, 2015). This regional cooling was associated with both changes in the net surface radiation as well as the increased evapotranspiration (ET) resulting from the relatively denser and deeper rooting systems of perennial, relative to annual bioenergy crops, drawing down soil moisture from deeper soil depths (Vanloocke *et al.*, 2010; Georgescu *et al.*, 2011; Anderson *et al.*, 2013; Hallgren *et al.*, 2013; Ferchaud *et al.*, 2015).

Enhanced ET and soil moisture depletion at deeper rooting depths could have important hydroclimate implications. Researchers have concluded that reduced surface runoff (McIsaac *et al.*, 2010; Vanloocke *et al.*, 2010; Chen *et al.*, 2015) and the subsequent reductions in streamflow (Anderson *et al.*, 2013; Goldstein *et al.*, 2014; Khanal *et al.*, 2014; Feng *et al.*, 2015) are direct consequences of the diminished soil water volume. This reduction, according to Ferchaud *et al.* (2015) and Feng *et al.* (2015), could be compensated by lower soil moisture depletion near the surface, as a result of the denser canopy cover reducing soil evaporation. These suggestions are in agreement with recent modeling work that focused on the Cornbelt region (Georgescu *et al.*, 2011) and Central Plains (Anderson *et al.*, 2013) demonstrating increasing depletion of deep soil water volume for the Midwestern United States after the deployment of perennial bioenergy cropping systems. The aforementioned findings illustrate the potential for unintended consequences on the coupled atmosphere–hydrologic system and suggest that perennial biofuel expansion could lead to water stress and consequently lower yields (Vanloocke *et al.*, 2010; Le *et al.*, 2011; Anderson *et al.*, 2013; Xu *et al.*, 2013; Goldstein *et al.*, 2014). Therefore, hydrometeorological assessments (e.g., near-surface temperature, surface energy balance, and soil moisture effects) are necessary to assess biophysical implications associated with large-scale deployment of perennial biomass energy crops.

The access to deeper soil moisture and enhanced ET relative to existing vegetation may reduce the need for irrigation for perennial biofuel crops in many regions (Miller *et al.*, 2015) and could make these crops more resilient to drought. Under these conditions, the enhanced ET decreases atmospheric demand for water (Seneviratne *et al.*, 2010), whereas access to deeper soil moisture mitigates plant stress due to water availability (Wu *et al.*, 2002). Field measurements under drought conditions have reported the highest vapor deficits, highest ET-to-precipitation ratios, and lowest water-use efficiencies for perennial relative to annual cropping systems (Abraha *et al.*, 2015). However, micrometeorological assessments of perennial biofuel crops have observed only small changes in albedo, ET, and yields when contrasting drought years to nondrought years (Abraha *et al.*, 2015; Miller *et al.*, 2015; Yimam *et al.*, 2015), illustrating the enhanced resiliency of perennial systems (Joo *et al.*, 2016). In fact, perennial crops outperformed annual biofuel crops with significantly higher leaf area index (LAI) and minimal crop damage during adverse hydrometeorological conditions (Miller *et al.*, 2015). Early growing season precipitation proved an important factor in the availability of soil moisture (Wagle & Kakani, 2014) and crop yield (Yimam *et al.*, 2015). Perennial biofuel crops, under a no-irrigation scenario, could reduce stress on already overburdened water resources and mitigate the effects of drought, especially in areas that have historically dealt with drought, such as the Southern High Plains.

Given the concerns associated with the diminished water resources, marginal lands are receiving additional recognition as a key component of a sustainable approach to biofuel crop expansion (Campbell *et al.*, 2008; Cai *et al.*, 2010; Zumkehr & Campbell, 2013). Although important in terms of improving process-level understanding, the majority of the aforementioned work has not provided a realistic representation of perennial biofuel crop expansion because these studies have generally focused on the replacement of existing annual with perennial biofuel crops. Planting perennial biofuel crops on marginal lands (i.e., land that would not be useful for annual crops) is a more sustainable approach because it averts competition with food production (Campbell *et al.*, 2008; Bagley *et al.*, 2014). Using marginal lands for biofuel production could decrease or negate the potential negative impacts on water resources (Feng *et al.*, 2015). However, the net effect of this pathway remains uncertain (Gelfand *et al.*, 2013; Rahman *et al.*, 2014) and requires site-specific examination of physical changes associated with varying deployment pathways.

Here, we perform the first realistic examination of hydrometeorological impacts associated with perennial bioenergy crop expansion and focus on land that would not useful for annual crops. Using a suite of simulations with a coupled land surface–atmosphere model, applied at a high resolution, we focus on the quantification of meteorological effects during a normal hydrologic year (2007) and a drought year (2011) for the Southern Plains of the United States (Hoerling *et al.*, 2013; United States Drought Monitor, 2014; Tadesse *et al.*, 2015). The historically drought-prone nature of this region serves as an ideal test-bed to examine the region's drought resiliency under an alternate land-use scenario. This study quantifies hydrometeorological impacts for varying scenarios of perennial biofuel crop expansion under diverse hydrometeorological conditions and will fill a critical void associated with bioenergy-based land-use conversion. This work sheds light on the sustainability of perennial bioenergy crop deployment over marginal and abandoned farmland areas and therefore contributes to the emerging body of literature characterizing the potential contribution of perennial biomass energy crops to the portfolio of renewable energy options.

Materials and methods

Regional climate model design

We used the Weather Research and Forecasting (WRF) model version 3.6.1 coupled with the Noah land surface model (LSM) to examine the potential effects of perennial biofuel crop expansion under normal climate and drought conditions for the years 2007 and 2011, respectively. WRF is a nonhydrostatic mesoscale model, commonly utilized in climate research and operational forecasting (Skamarock *et al.*, 2005). The Noah LSM calculates time-varying soil temperature and moisture, horizontally and vertically through a multilayer soil column extending down to 2 m, and specifies surface energy partitioning of available radiant energy to drive atmospheric processes and characterize the meteorological response to vegetation forcing (Ek *et al.* 2003). The LSM used in this work has been used to address a spectrum of topics ranging from the assessment of hypothetical bioenergy expansion (e.g., Georgescu *et al.*, 2011) to improving predictive capabilities through the strategic use of remote sensing data (e.g., Cao *et al.*, 2015). We used a triple nested grid configuration with the outermost and innermost domains utilizing 32- and 2-km grid spacing, respectively (see Fig. 1a). Centered over the northeastern part of Oklahoma, the innermost domain defines the boundaries of our study area and is discretized by 149 and 113 grid points in the east–west and north–south directions, respectively (see Fig. 1b). The city of Tulsa, Oklahoma, is located in the center of the study area and is surrounded predominantly by grasslands (see Fig. 1b).

We acquired Final Operational Global Analysis (FNL) data from the National Centers for Environmental Prediction for the years 2007 and 2011. FNL data are reanalysis data derived from observational weather data, Global Forecast System (GFS) model runs, and other analyses. These data are available from the Research Data Archive (http://rda.ucar.edu), are provided at 1-degree by 1-degree resolution on a global domain, and were used to initialize and force the lateral boundaries for the outermost domain for all WRF simulations. All simulations were initialized on December 1st of the preceding year and a one-month spin-up was performed to allow the system to reach a state of equilibrium. This initial month, for all experiments, was discarded, and only the 12 months pertaining to each year was used for the subsequent analysis.

Data used for model evaluation

We conducted an ensemble of model simulations by varying a suite of select physics schemes (microphysics and radiation) aimed at selecting an optimal control experiment (see Table 1). The convective scheme was not varied in this study, because convection is explicitly resolved in the innermost domain. Model performance metrics for temperature and precipitation were evaluated for each year using Taylor diagrams. The utility of Taylor diagrams stems from their simultaneous capability to summarize statistical measures of centered root mean square error (RMSE), normalized standard deviation, and correlation coefficient (Taylor, 2001). Metrics were based on the comparison between model simulations and observed meteorological data, which were obtained from seven stations available from the Global Historical Climate Network (see Table 2). Mean daily temperatures and daily precipitation totals were averaged over the seven stations and compared to the corresponding modeled grid cells.

In addition to evaluation against the observed station values, annual spatial patterns of total precipitation and average temperature were calculated. Model simulations were first

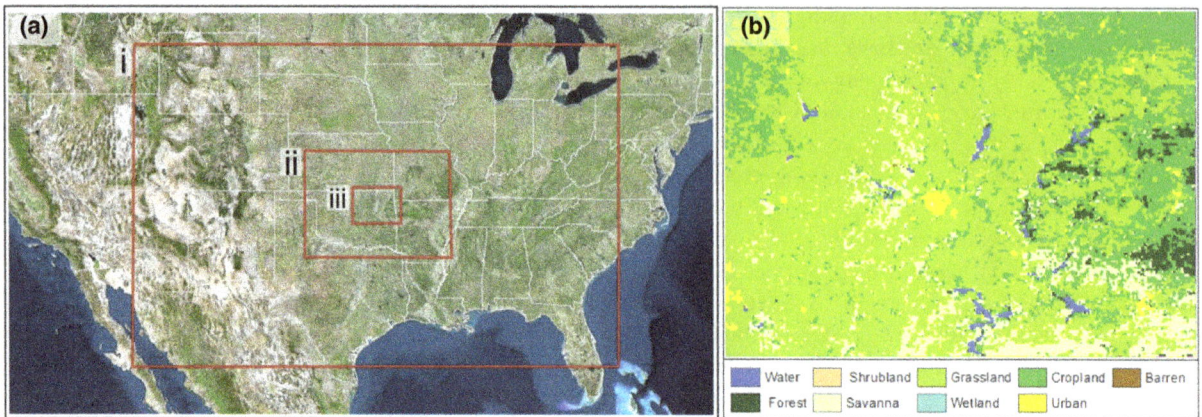

Fig. 1 (a) Study area showing triple nested configuration with (i) outermost (32-km grid spacing), (ii) middle (8-km grid spacing), and (iii) innermost (2-km grid spacing) domain resolution, respectively. (b) Innermost domain land use/land cover.

Table 1 List of model simulations and the configuration (microphysics, shortwave and longwave radiation schemes used to select an optimal control simulation)

Simulation	Microphysics	Shortwave radiation	Longwave radiation
1	WSM 3	Dudhia	RRTM
2	Thompson *et al.*	Dudhia	RRTM
3	Morrison 2-moment	Dudhia	RRTM
4	WDM 6-class	Dudhia	RRTM
5	Thompson *et al.*	RRTMG	RRTM
6	WDM 6-class	RRTMG	RRTM
7	Thompson *et al.*	RRTMG	RRTMG
8	WDM 6-class	RRTMG	RRTMG

Table 2 List of Global Historical Climatological Network stations by name and their location (latitude and longitude)

Station name	Latitude	Longitude
Mannford	36.17472	−96.44333
Spavinaw	36.38944	−95.05972
Bartlesville	36.76833	−96.02611
Pawhuska	36.66917	−96.34722
Tulsa Intl Airport	36.19944	−95.88722
Muskogee	35.65667	−95.36139
Ponca	36.73056	−97.09972

resampled from 2- to 4-km resolution to match the observed spatial gridded datasets obtained from the University of Idaho at 4-km grid spacing (Abatzoglou, 2013). For the simulated annual total precipitation, yearly model-aggregated values were divided by the observed yearly total precipitation to produce an annual precipitation ratio. We also calculated the average annual temperatures for the observed and simulated experiments to spatially assess the model performance. Simulated annual average temperature was subtracted from the observed annual average temperature, yielding average annual temperature differences.

Perennial biofuel crop suitability

To determine the locations for perennial biofuel deployment, we utilized bioenergy crop suitability data at 1-km resolution based on Cai *et al.* (2010). These data identify the marginal and abandoned lands suitable for biofuel crop deployment globally based on soil productivity, slope, soil temperature, humidity index, and land-use information. We chose the most realistic scenario, which included low-productivity grasslands, savanna, and shrublands with regular or marginal productivity, while excluding the total current pasture land and regions of crop production (see Fig. 2a). Extracting only information within the innermost domain, these data were resampled to 2-km resolution to match the gridded specification of our innermost domain. To ensure deployment over only marginal or abandoned farmlands, suitable locations were compared with the WRF land cover data [MODIS Modified International Geosphere-Biosphere Program (IGBP)] using a GIS platform. Any pixel coincident with agriculture was deemed unsuitable for deployment. Lastly, the suitability data were reclassified using quantile classification method into four suitability classes ranging from low to very high suitability.

We assume two deployment scenarios to examine the range of potential hydrometeorological impacts and associated uncertainty given a range of perennial bioenergy crop distribution. The first scenario employs the full perennial biofuel deployment as previously discussed and covers 56 667.2 square kilometers (see Fig. 2b). The second scenario limits the potential perennial biofuel deployment to the upper 25% of suitable areas, covering only 14 376.3 square kilometers (see Fig. 2c).

Biophysical representation of bioenergy crops and simulations performed

Perennial biofuel crop deployment was represented via the modification of biophysical parameters characteristic of perennial grasses: albedo, LAI, and vegetation fraction, or the fraction of vegetative cover within a 2-km grid cell, under a no-irrigation scenario. Using the observed field-scale values obtained from Miller *et al.* (2015), daily albedo data were

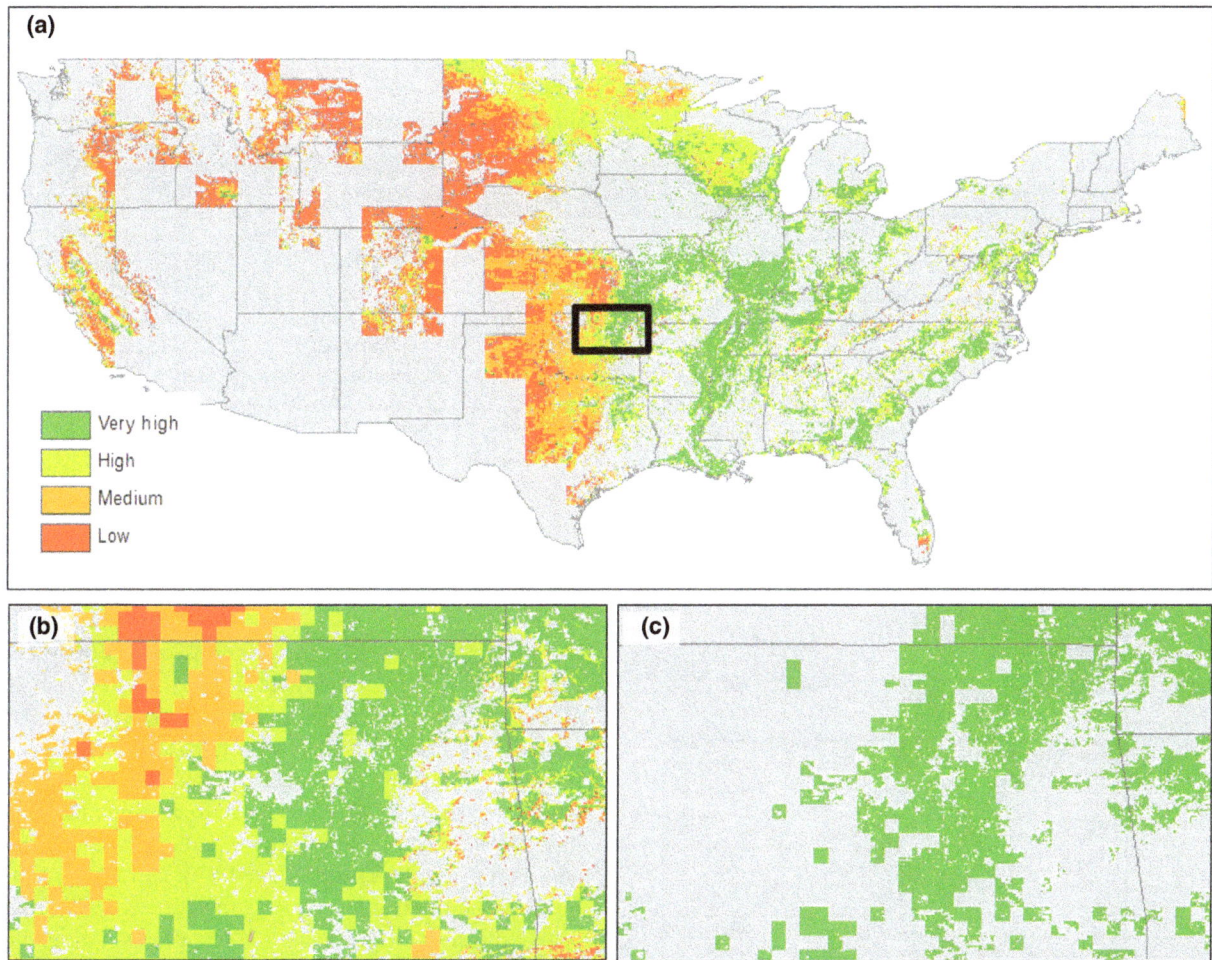

Fig. 2 Biofuel crop suitability for (a) contiguous United States using a quantile classification. Unsuitable locations are displayed in gray, while low and very high suitability locations are displayed in red and green, respectively. Potential perennial biofuel deployment for (b) 100% deployment scenario [all suitable classes presented in panel (a)] and (c) 25% deployment scenario (only very high suitability shown in green). The black box shown in (a) is the domain illustrated in (b) and (c).

averaged across the available sites for two plant types: switchgrass and miscanthus. A linear interpolation was applied to fill gaps separately for each year (2010 and 2011). These values were then averaged across plant types and over the two years, yielding mean daily albedo values for perennial biofuel crops (Fig. 3a). Mean daily albedo values were interpolated to three-hourly values (corresponding to the temporal frequency of WRF output) and ingested into WRF for both simulated years (2007 and 2011). When compared with albedo values of the existing land cover utilized for the control experiments (Fig. 3a), mean daily albedo values of perennial biofuel crops are higher. This difference is especially evident during the growing season from May to October.

LAI and vegetation fraction values were also modified to further characterize the realistic phenological evolution of perennial bioenergy cropping systems. These values were scaled to follow the changes in observed albedo values using the known maximum and minimum values (see Fig. 3b, c). From May to October, LAI values for perennial biofuel crops are 58% higher than LAI values of existing land cover. Unlike

LAI, the differences in vegetation fraction between perennial biofuel crops and existing land cover are smaller, with vegetation fraction values approximately 20% higher than existing grasslands from May to October.

After the modification of albedo, LAI, and vegetation fraction, two sets of simulations were performed to examine the effect of perennial bioenergy crop deployment (see Table 3). The first set of simulations aimed to establish a baseline using existing land cover from the best (hereafter *Exp1*) and least (hereafter *Exp2*) skilled model configurations as compared to suitable station and gridded observational data (see Data used for model evaluation). This set was determined from an ensemble of eight simulations with varying physics options that regional climate models such as WRF are well known to be highly sensitive to (Table 1). The second set of simulations represented 100% and 25% perennial biofuel crop deployment scenarios in the respective suitable areas by modifying biophysical parameters of albedo, LAI, and vegetation fraction values. Additionally, each of the perennial bioenergy cropping system simulations (i.e., 100% and 25%) was conducted for Exp1 (i.e., most skilled

(a)

(b)

(c)

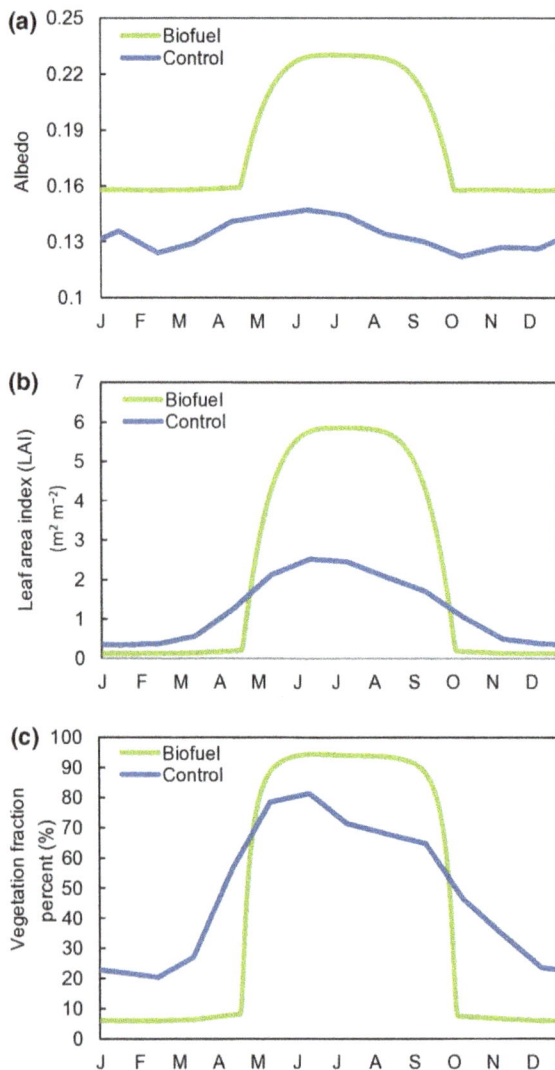

Fig. 3 Phenological evolution of biophysical parameters for existing grasslands and perennial biofuel representation: (a) albedo, (b) leaf area index (LAI), and (c) vegetation fraction. Default biophysical parameter representation (control) and perennial biofuel crop values are shown in solid blue and green lines, respectively.

control simulation: *Exp1_100* and *Exp1_25*) and Exp2 (i.e., least skilled control simulation: *Exp2_100* and *Exp2_25*). A full description of simulations performed with and without perennial biofuel crop deployment is presented in Table 3.

Results

Model evaluation

Overall, WRF simulations demonstrated greater skill with temperature than with precipitation. The suite of ensemble members exhibited high skill in simulating near-surface temperatures as indicated by the high

Table 3 List of model simulations with and without perennial biofuel crop deployment

WRF simulations	Scenario	Spin-up	Analysis time
Control	Simulation 6 (Exp1): best skill	Dec 1–31, 2006	Jan 1–Dec 31, 2007
	Simulation 1 (Exp2): least skill	Dec 1–31, 2010	Jan 1–Dec 31, 2011
Biofuel Deployment (Albedo, Veg Frac, LAI)	100% Simulation 6 (Exp1_100) Simulation 1 (Exp2_100)	Dec 1–31, 2006 Dec 1–31, 2010	Jan 1–Dec 31, 2007 Jan 1–Dec 31, 2011
	25% Simulation 6 (Exp1_25) Simulation 1 (Exp2_25)		

correlations – generally between 0.97 and 0.99 – and centered RMSE near 1 degree Celsius (see Fig. 4a, b). Modeled near-surface temperature performance was similarly skilled for both years, and the sensitivity to differing physical parameterizations was minimal. Unlike temperature, the sensitivity to the suite of examined physics parameterizations resulted in large variability in precipitation skill (Fig. 4c, d). During 2007 (see Fig. 4c), the correlation values varied between 0.30 and 0.40, indicating moderate skill. Simulation 4, employing WDM6 microphysics (i.e., single moment representation with six hydrometeor species), produced the highest correlation (0.40), lowest centered RMSE (0.97) and standard deviation (0.98), relative to the observed values. During 2011, the range of ensemble member performance indicated improved skill, with higher correlation values generally ranging between 0.55 and 0.65 (see Fig. 4d). Simulation 4, utilizing the WDM6 microphysics parameterization, exhibited one of the highest correlations and best matched the observed variance (dotted circle lines) and bias (solid blue lines), with values just shy of 1 demonstrating excellent simulation of day-to-day precipitation variability. Based on the aforementioned metrics, all ensemble members performed with high fidelity in their simulation of near-surface temperatures, while Simulation 4, making use of WDM6 microphysics, best demonstrated precipitation magnitude for both hydrometeorological years.

Examination of spatial characteristics of simulated meteorological variables can provide further confidence in simulation skill by illustrating geographically explicit performance. Spatial patterns of annually averaged temperature (Fig. 5) confirmed the previously demonstrated strong model skill for near-surface temperature. Small differences were apparent between the observed mean annual temperatures for 2007 and 2011 and the suite of ensemble member simulations produced biases generally <2 °C. The differences among individual members were mainly attributable to the selection of shortwave radiation schemes. In 2007, Simulations 5–8 (Table 1 and Fig. 5e–h), utilizing RRTMG shortwave radiation scheme, produced a warm bias of 1.5–2.5 °C, whereas in 2011, Simulations 1–4 (Fig. 5i–l), utilizing Dudhia shortwave radiation scheme, produced a cold bias of 1.5–2.5 °C. Overall, the temperature differences attributable to microphysics schemes were more localized and generally <0.5 °C.

The selection of microphysics scheme was key to model performance of precipitation despite its apparent underestimation (Fig. 6). Simulation 1 with WSM-3 microphysics (see Table 1) produced the least model skill for both years with ratio values of 0.7 or less for most of the domain (Fig. 6a, i). For 2007, Simulation 3 with Morrison microphysics (Table 1) considerably overestimated the amount of precipitation in the eastern portion of the domain (Fig. 6c). Simulation 6 with WDM6 microphysics (see Table 1) also overestimated precipitation in the southeast corner during 2007 (see Fig. 6f), but the simulated overestimation was considerably less in magnitude and extent. For 2011, Simulation

6 with WDM6 microphysics (see Table 1) best captured the spatial extent and magnitude of observed rainfall (see Fig. 6h). Of the eight different model configurations, Simulation 6 demonstrated the greatest skill in modeling precipitation for both years, while Simulation 1 produced the least skill.

Although Simulation 4 had the best model skill in simulating precipitation according to station average metrics (Taylor diagrams), Simulation 6 best captured the spatial pattern of annual precipitation, and statistical metric differences between the pair of ensemble members were small. Both Simulation 4 and Simulation 6 used the WDM-6 microphysics scheme, but had different shortwave radiation schemes. Despite a slight warm bias in 2007, RRTMG shortwave radiation scheme used in Simulation 6 improved the overall spatial pattern of precipitation with minimal differences between the simulated and observed values.

Based on the aforementioned results, Simulation 6 was selected as the best representative control experiment (Exp1), whereas Simulation 1 was selected as the least performing member (Exp2). From this point forward, we utilize both Simulation 6 (Exp1) and Simulation 1 (Exp2) with different perennial biofuel crop deployment scenarios (see Table 3) to examine the sensitivity of bioenergy crop deployment.

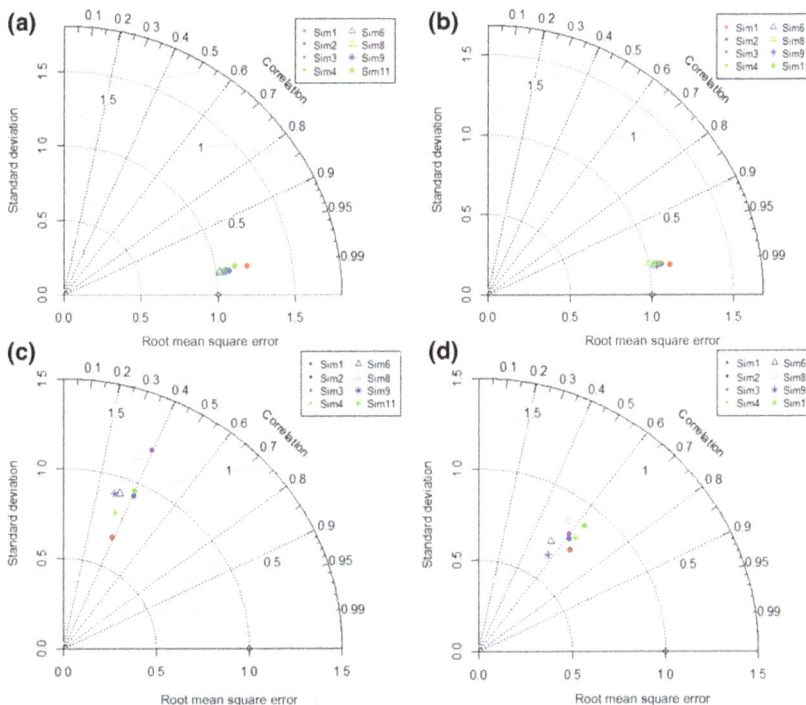

Fig. 4 Taylor diagrams of annual daily mean temperature between averaged observations and model grid points for (a) 2007 and (b) 2011 and annual daily precipitation for (c) 2007 and (d) 2011, where each symbol corresponds to model performance of the individual members presented in Table 1. Angular axis shows the spatial correlations between modeled and observed variables. Radial x-axis and y-axis show the normalized standard deviation and centered root mean square error (RMSE), respectively.

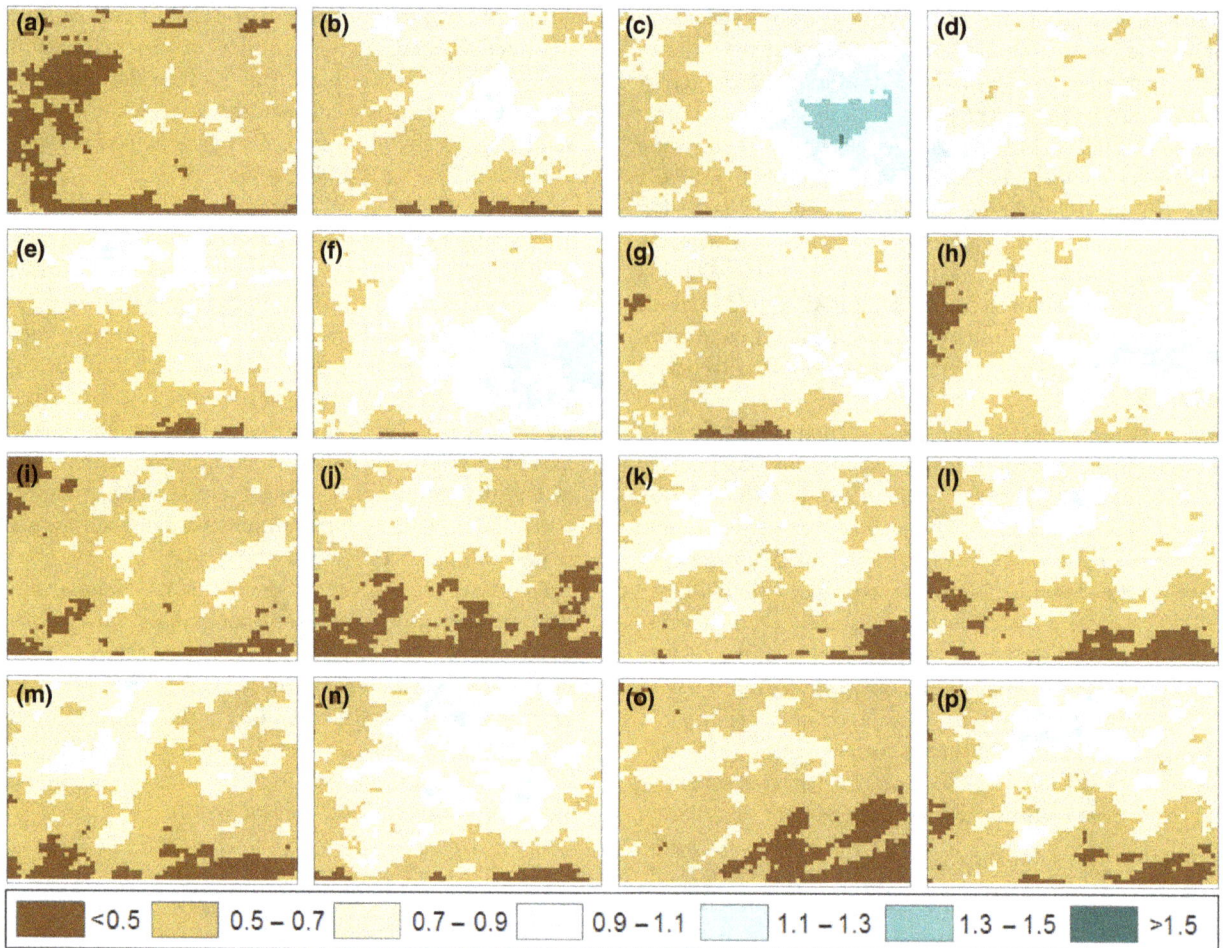

Fig. 5 Mean annual differences between the observed and modeled temperatures (°C). Figure (a–h) corresponds to the year 2007 and represents the Simulations 1–8 parameters (from left to right) as listed in Table 1. Figure (i–p) corresponds to the year 2011 and represents the Simulations 1–8 parameters (from left to right) as listed in Table 1.

Simulated impacts on temperature, ET, and soil moisture

The simulated impact of perennial bioenergy crop deployment indicates a decrease in near-surface air temperature for both normal hydrometeorological (2007) and drought (2011) years. For both years, time-series analyses of domain-averaged near-surface temperatures (i.e., 2 m temperatures; Fig. 7) show that the greatest temperature differences occurred during the growing season – from May to October – with a slight increase in temperatures from April to May, and October to November, when LAI values are lower for perennial bioenergy crops (see Fig. 3). Temperature decreases associated with perennial biofuel crop deployment were slightly more pronounced in 2007 (normal year) than in 2011 (drought year) with an average decrease of 1.0 °C compared with 0.80 °C, respectively. In 2007, the reduction in near-surface temperature associated with perennial bioenergy crop deployment frequently

exceeded 1–1.5 °C during the growing season for both deployment scenarios. In terms of scenario and experimental differences, the reduction in near-surface temperature was greater for the full deployment scenario (i.e., Exp1_100 and Exp2_100) than for the partial deployment scenario (i.e., Exp1_25 and Exp2_25), while only minimal differences were apparent between Exp1 and Exp2 (i.e., the thermal impact of bioenergy crop deployment does not appear to be a function of model performance).

Because the largest simulated effects on near-surface temperature were apparent during the growing season, we examined the spatial impacts of temperature associated with perennial bioenergy crop deployment from May to October. For both simulated years, maximum temperature differences were more pronounced for the full deployment relative to the partial deployment scenario, consistent with the fraction of crop deployment (Fig. 8). Results for full deployment scenarios revealed a

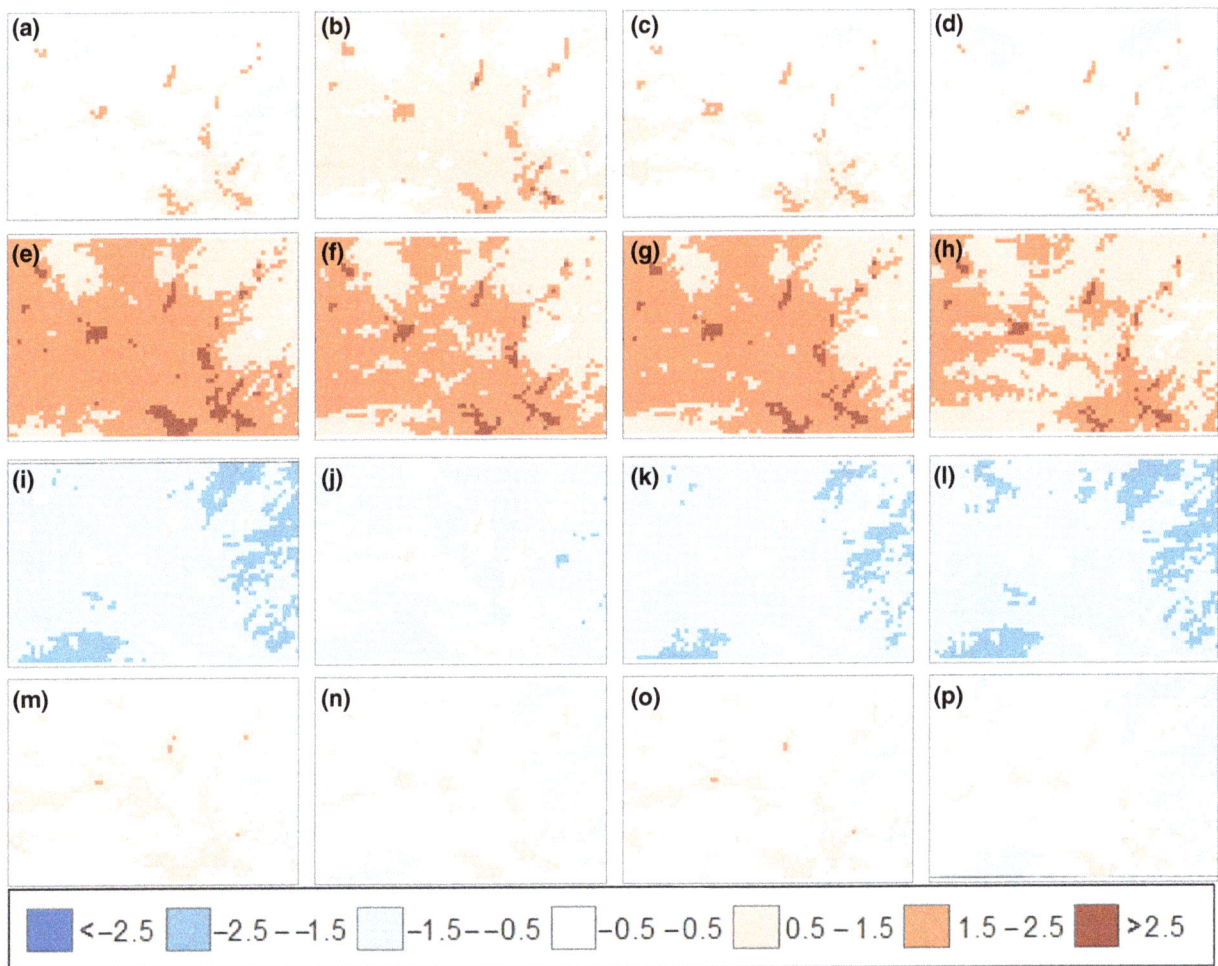

Fig. 6 Annual precipitation ratio (the total simulated precipitation divided by the total observed precipitation). Panel representation is same as in Fig. 5. Figures (a–h) correspond to the year 2007, and represent simulations 1–8 parameters (from left to right) as listed in Table 1. Figures (i–p) correspond to the year 2011, and represent simulations 1–8 parameters (from left to right) as listed in Table 1.

temperature decrease following a west–east gradient. In 2007, temperatures decreased by up to 1.8 and 0.7 °C in the west and east, respectively (Fig. 8e, f). In 2011, this gradient was slightly moderated with a corresponding 1.3 and 0.5 °C near-surface temperature decrease in the west and east, respectively (Fig. 8g, h). Temperature differences were more localized under the partial deployment compared with the full deployment scenario, despite the temperature decreases beyond deployment areas. Subtle differences were also observed between experiments, more notably with Exp1_25 and Exp2_25 in 2007 (Fig. 8a, b). Temperature decreases were slightly greater in the deployment areas for Exp1 compared with Exp2 with up to 1.3 °C of cooling compared with 1.1 °C, respectively. Moreover, Exp1 exhibited a stronger cooling signal extending well beyond the deployment area by up to 0.5 °C. Although the magnitude of cooling exhibited by Exp1 was noticeably greater than that of Exp2, the simulated spatial consistency between

Exp1 and Exp2 indicates a limited sensitivity of bioenergy crop deployment to model performance.

Unlike temperature, simulated impacts of perennial bioenergy crop deployment show diverging results for ET based on model performance and climate year. Time-series analyses of domain-averaged ET (Fig. 9) indicate that simulation results (i.e., Exp1 and Exp2) diverge from the late spring to September 2007 (Fig. 9a). During this period, ET associated with Exp1_25 and Exp1_100 increased by an average of 0.73 mm day^{-1}, while ET associated with Exp2_25 and Exp2_100 decreased by an average of 0.19 mm day^{-1} (i.e., increased ET for Exp1 simulations, but decreased ET for Exp2 simulations). Simulated impacts of precipitation showed similarly diverging results, pointing to a positive signal with Exp1 and negative signal with Exp2 (see Fig. S1 and Table S1). However, because the standard deviation explained a greater fraction of the mean, there is reduced confidence in the robustness of

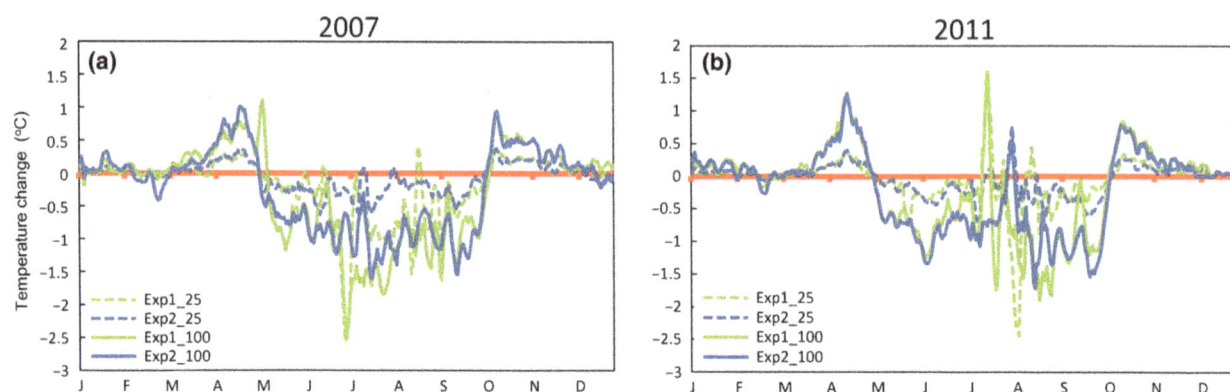

Fig. 7 Domain-averaged time series of temperature differences between perennial biofuel crop and control for (a) 2007 and (b) 2011. Green and blue lines represent Exp1 (best skilled control simulation) and Exp2 (least skilled control simulation), respectively. Dashed and solid lines represent 25% and 100% perennial biofuel deployment scenarios, respectively.

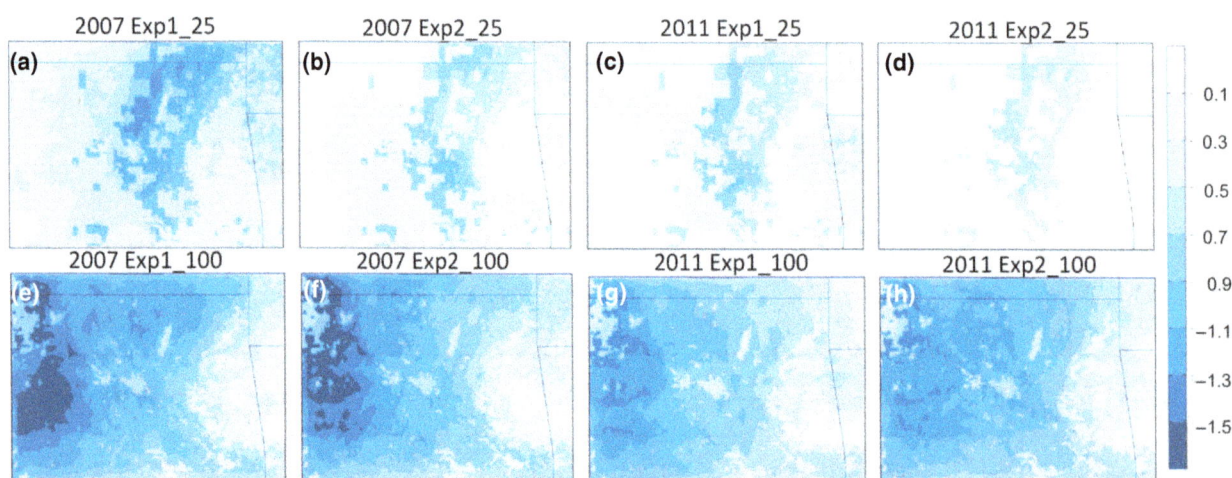

Fig. 8 Mean temperature differences (average of May-October) between perennial biofuel crop deployment experiments and control experiments for Exp1_25 and Exp2_25 for the years 2007 (a, b) and 2011 (c, d) and Exp1_100 and Exp2_100 deployment for the years 2007 (e, f) and 2011 (g, h).

precipitation results (Table S1) unlike ET. Further examination of thermodynamic impacts, as drivers of simulated precipitation, is necessary. Simulated differences were also evident in 2011, although of reduced magnitude, as ET increases from July to September varied among experiments, pointing to issues of model performance. Outside of the summer months, the simulated effects of ET associated with perennial biofuel crops generally followed similar patterns for both years: a decrease in ET of about 1 mm day^{-1} from April to May and a smaller decrease (~0.5 mm day^{-1}) from October to November (likely driven by greater LAI and vegetation fraction for simulations using default biophysical characteristics relative to bioenergy cropping simulations). The decrease from April to May was more pronounced in 2011, with ET values approximately 38% lower than those in 2007.

In addition to experimental sensitivity, annual differences in simulated ET impacts from perennial biofuel crop deployment were also apparent between normal hydrometeorological (2007) and drought (2011) years. Simulated ET impacts associated with perennial bioenergy crop deployment were moderated during the drought year relative to the normal climate year. Focusing on Exp1 (the best skilled ensemble member), ET associated with perennial biofuel crop deployment was 37% lower, on average, during the growing season in 2011 relative to 2007 (0.28 mm day^{-1} compared with 0.44 mm day^{-1}, respectively). This decrease was even more pronounced from July to September, as ET associated with perennial biofuel crop deployment was 3.4 times lower in 2011 relative to 2007 (0.21 mm day^{-1} compared with 0.73 mm day^{-1}). During the summer of 2011, ET decreased from June to July for both Exp1 and

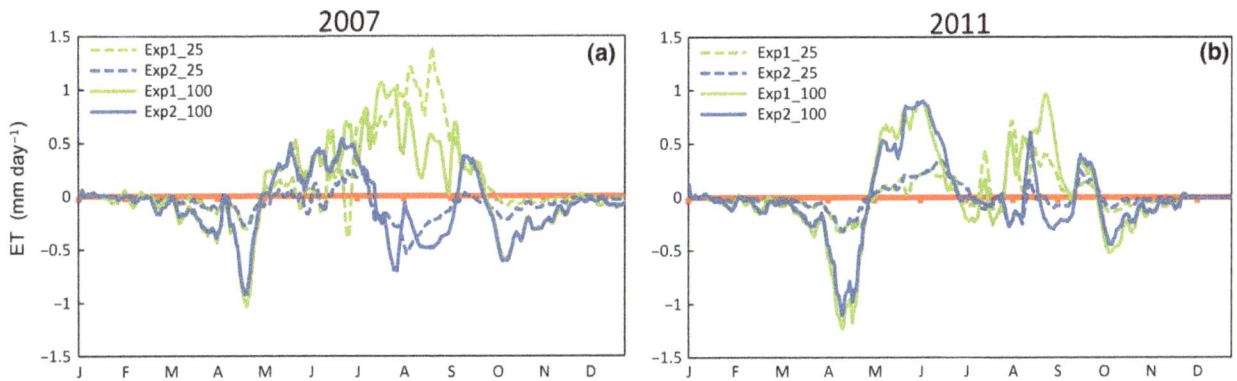

Fig. 9 Domain-averaged time series of evapotranspiration (ET) differences between perennial biofuel crop deployment experiments and control experiments for (a) 2007 and (b) 2011. Panel representation is the same as in Fig. 7.

2 and then remained just below zero for most of July to August. Despite this decrease, ET was enhanced under perennial biofuel crop deployment by an average of 0.25 mm day^{-1} and 0.13 mm day^{-1} over the growing season for Exp1 and 2, respectively, indicating the potential to mitigate the effects of drought through the enhanced ET.

Soil moisture impacts associated with perennial biofuel crop deployment followed the seasonal patterns of withdrawal and recharge based on crop phenology and atmospheric processes. The domain-averaged evolution of soil moisture associated with perennial biofuel crop deployment demonstrated the depletion from May through August, followed by a recharge period in the fall and winter due to the increased large-scale precipitation activity. After the growing season, soil moisture was restored near the surface, but remained partially depleted at rooting depths, 10–40 cm and 40 cm–1 m (Fig. 10). These deficits were most noticeable at rooting depth 40 cm–1 m, as the amount of soil moisture was 20% and 10% less than their initial state in 2007 and 2011, respectively (Fig. 10c, f). Other annual differences were also apparent from April to May, as soil moisture was approximately 29% lower in 2011 than in 2007 (Fig. 10c, f). In addition to annual differences, Exp2 resulted in lower soil moisture than Exp1 at all rooting depths with the largest disparities during 2007 (Fig. 10a–c). Moreover, at the beginning of the year, Exp1 noted almost 20% higher soil moisture in 2007 near the surface and 10–40 cm. Experimental differences in soil moisture amounts were attenuated in 2011 relative to 2007, pointing to less water uptake associated with these cropping systems under drier hydrometeorological conditions. Lastly, different deployment scenarios also affected the amount of soil moisture depletion with slightly higher withdrawals associated with full deployment scenario, especially during 2011.

Mechanisms driving simulated changes

The principle processes driving the aforementioned impacts are grounded on the changes in the radiation budget. During the growing season, the net shortwave radiation associated with perennial biofuel crop deployment decreased by an average of 15.8 W m^{-2}, while net longwave radiation associated with perennial biofuel crop deployment increased by an average of 5.6 W m^{-2}, for both years (Fig. 11). Only minor differences were discernible between years for perennial bioenergy crop deployment with slightly higher net shortwave radiation values in 2007 and slightly higher net longwave radiation in 2011. Larger differences existed between deployment scenarios. Under full deployment scenarios, the decrease in net shortwave radiation was on average 14.5 W m^{-2} greater (Fig. 11a, b), while net longwave radiation fluxes were only 4.3 W m^{-2} higher on average relative to partial deployment scenarios (Fig. 11c, d). Additionally, Fig. 11c, d shows decreases in net longwave radiation associated with perennial bioenergy crop deployment from April to May and October to November, which correlated with temperature decreases in the previously presented time-series analysis (see Fig. 7). This decrease was more pronounced in the spring of both years by an average of 10 W m^{-2}, but even more so in 2011, and is driven by the imposed, observationally based, biogeophysical representation of perennial crops.

Simulated energy fluxes associated with perennial bioenergy crop deployment illustrate the diverging results for both sensible and latent heat fluxes based on model performance and climate year, similar to ET (Fig. 12). Model simulation results for both sensible and latent heat fluxes show a diverging evolution between Exp1 and Exp2 from July to September 2007 (Fig. 12a, c). Despite the experimental differences, both Exp1 and Exp2 pointed to increasing latent heat fluxes and

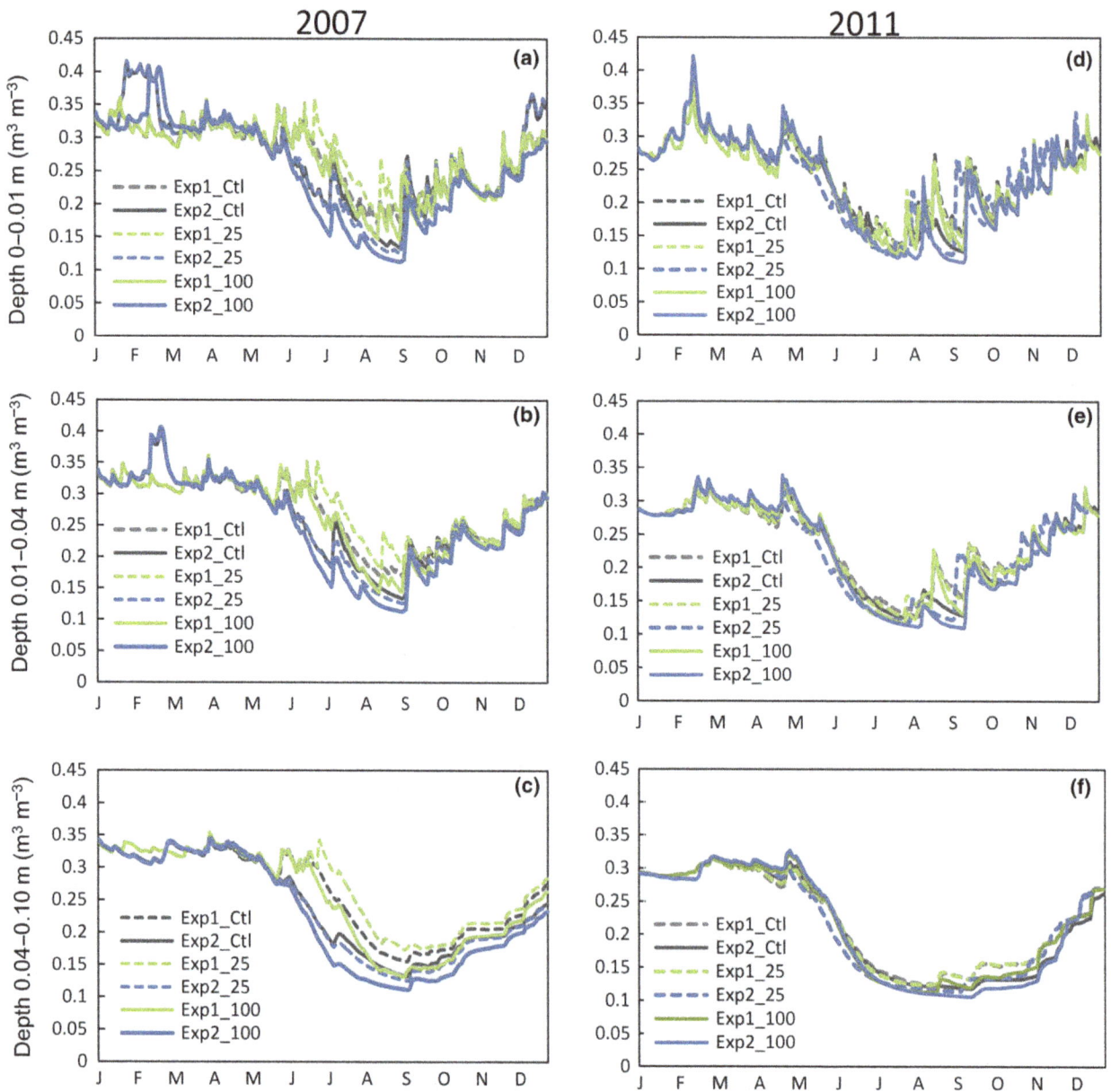

Fig. 10 Domain-averaged time series of soil moisture at model depths 0–0.01, 0.01–0.04, and 0.04–0.10 meters (from top to bottom). Figures (a–c) and (d–f) correspond to the years 2007 and 2011, respectively. Gray, green, and blue lines represent control, Exp1, and Exp2, respectively. Dashed and solid lines represent 25% and 100% perennial biofuel deployment scenarios, respectively.

decreasing sensible heat fluxes associated with perennial bioenergy crop deployment during the growing season. In 2007, energy flux differences associated with perennial bioenergy crops were most distinctive under the full deployment scenario, varying by 30–40 W m^{-2} from July to September. In 2011, these differences were moderated as both sensible and latent heat fluxes varied 10–20 W m^{-2}. Outside of the growing season, changes in energy fluxes associated with perennial bioenergy crop deployment were also apparent mainly from April to May, with spikes in sensible heat and commensurate

declines in latent heat fluxes (Fig. 12). These changes were most noticeable with latent heat fluxes especially in 2011 (Fig. 12a, b), which also correlated with temperature and radiation flux differences.

Because the largest simulated effects on energy fluxes were most apparent during the growing season, we examined the spatial effects of energy fluxes associated with perennial bioenergy crop deployment from May to October. During the growing season, latent heat fluxes increased (Fig. 13a–h), while sensible heat fluxes decreased (Fig. 13i–p). Energy flux differences were

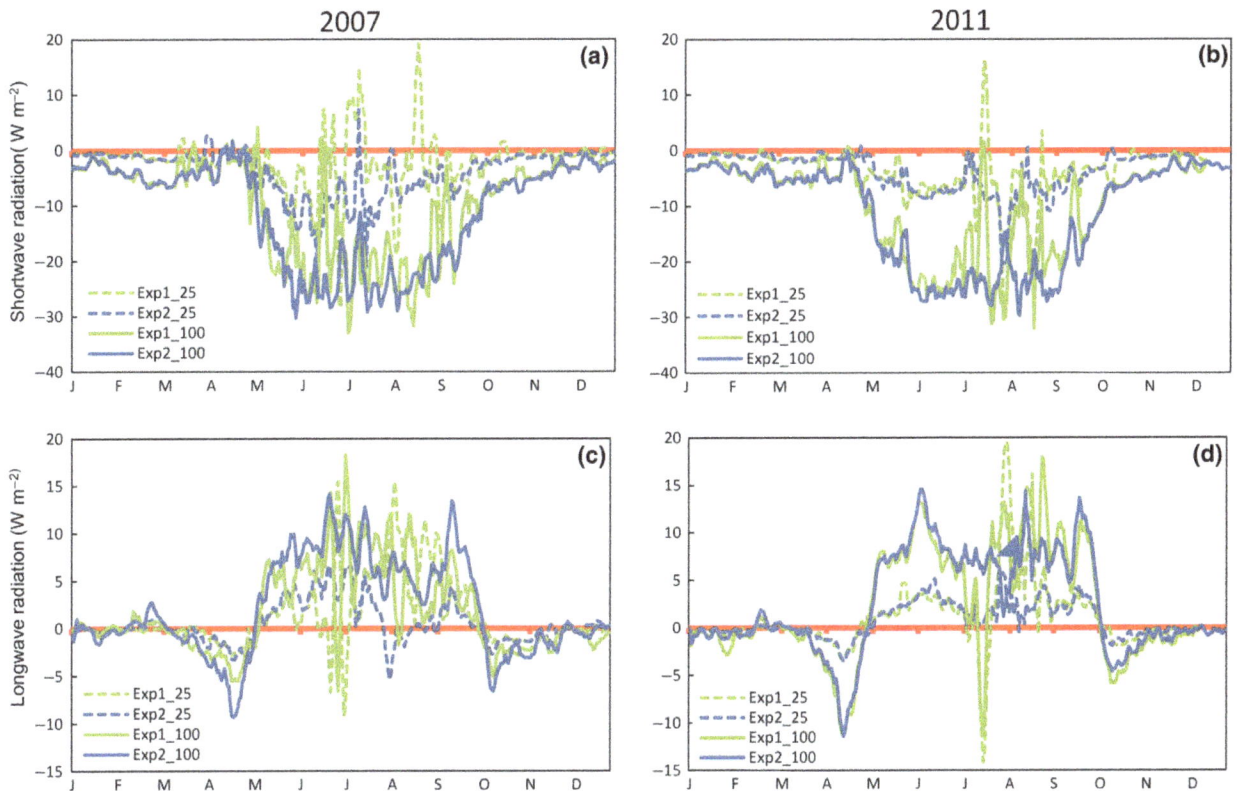

Fig. 11 Domain-averaged time series of net shortwave radiation differences between perennial biofuel crop experiments and control experiments for the years (a) 2007 and (b) 2011 and the net longwave radiation differences for the years (c) 2007 and (d) 2011. Panel representation is the same as in Fig. 7.

most noticeable under the full deployment compared with the partial deployment scenarios, with the exception of Exp1_25 and Exp1_100 in 2007 (Fig. 13b, d, j, l). For these two experiments, sensible heat fluxes increased in the eastern part of the domain (Fig. 13d), which correlated with a decrease in latent heat fluxes (Fig. 13l). Despite this anomaly, both latent and sensible heat fluxes exhibited a west-to-east gradient under full deployment scenario with the highest flux differences located in the west, coincident with the spatial pattern of temperature differences. Energy flux differences were more localized under the partial deployment compared with the full deployment scenario, despite the increases in latent heat and the decreases in sensible heat fluxes observed beyond the deployment areas.

Contrasting different hydrometeorological years, changes in sensible and latent heat fluxes associated with perennial bioenergy crop deployment were moderated in 2011 relative to 2007. In 2007, latent heat fluxes increased up to 50 W m^{-2}, while sensible heat fluxes decreased up to 50 W m^{-2}. In 2011, latent heat flux increases were approximately 10–25 W m^{-2} lower relative to 2007 and covered a smaller portion of the domain. Sensible heat fluxes followed a similar pattern with moderated decreases in 2011. This attenuation of

energy flux differences was most distinctive with Exp2 and the partial deployment scenario, indicating the sensitivity to model performance and the availability of atmospheric moisture.

Discussion and conclusions

This study examined the hydrometeorological impacts of perennial biofuel crop expansion using varying realistic deployment scenarios [i.e., partial deployment (14 376.3 square kilometers) and full deployment (56 667.2 square kilometers)] under diverse hydrometeorological conditions [i.e., drought year (2011) and normal year (2007)]. We focus bioenergy deployment only within suitable marginal and abandoned lands (see Cai et al., 2010). Our analyses show that perennial bioenergy crop deployment leads to the widespread cooling (1–2 °C) and the enhanced ET (0.5–1.0 mm day^{-1}) during the growing season – May to October (see Figs 7 and 9). In this study, soil moisture was depleted from mid-May to mid-August contributing to the enhanced ET, but nearly restored during senescence (mid-August to December) (see Fig. 10). The amount of soil moisture depletion was 20% and 10% less than the initial state in 2007 and 2011, respectively, and largely a function of

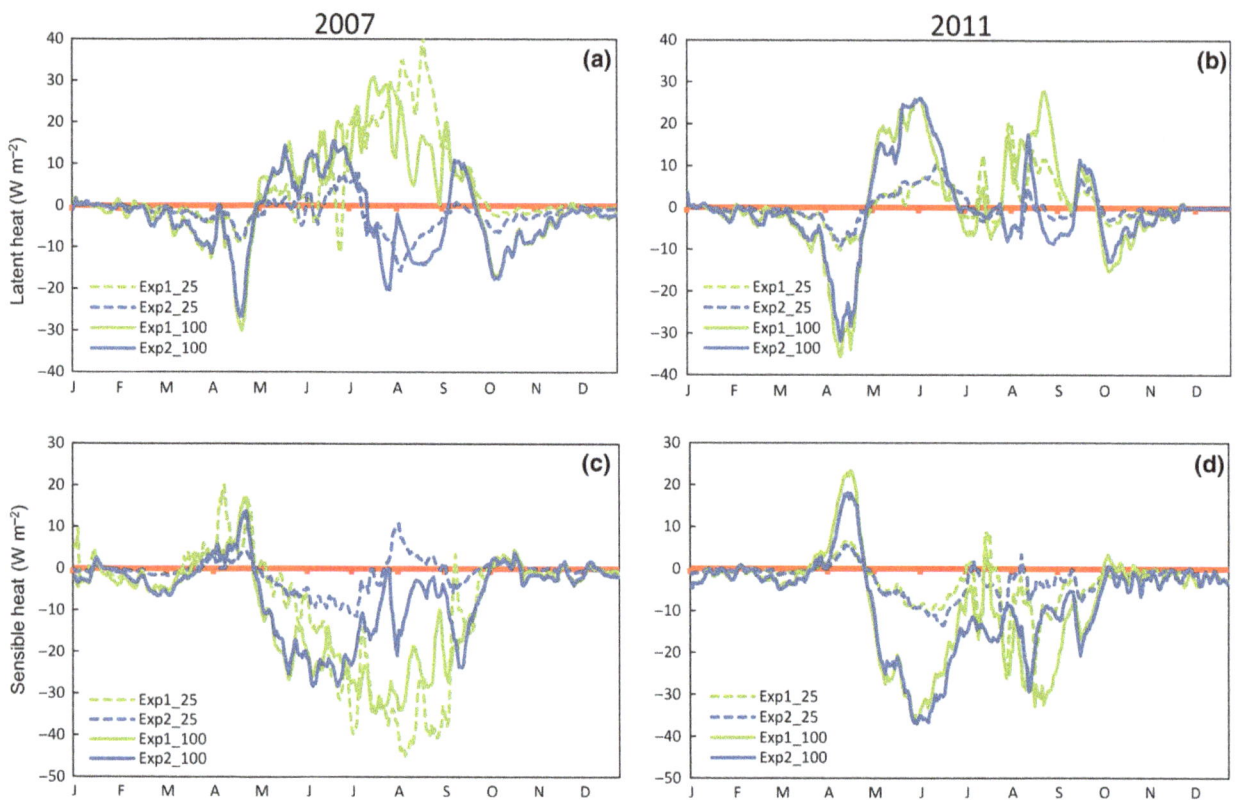

Fig. 12 Domain-averaged time series of latent heat fluxes differences between perennial biofuel crop experiments and control experiments for the years (a) 2007 and (b) 2011 and sensible heat fluxes differences for the years (c) 2007 and (d) 2011. Panel representation is the same as in Fig. 7.

moisture availability tied to hydrometeorological conditions. These hydrometeorological impacts were more evident under the full deployment, but still detectable within the deployment areas under the partial deployment scenario as well as in the surrounding environment. This finding indicates that perennial biofuel crops could still have important hydroclimate implications, even with small-scale distribution.

During the growing season, perennial bioenergy crops have higher LAI, albedo, and vegetation fraction values compared with existing grasslands (see Fig. 3). These physiological differences resulted in near-surface cooling and enhanced ET, in agreement with the findings of previous studies (Georgescu et al., 2009, 2011; Hickman et al., 2010; McIsaac et al., 2010; Vanloocke et al., 2010; Zeri et al., 2013; Miller et al., 2015; Zhu et al., 2016). Higher albedo values, however, proved to be the dominant mechanism for these simulated impacts through the subsequent decrease in absorbed surface shortwave radiation resulting in cooler temperatures. Previous studies attributed regional cooling, in part, to the enhanced ET from the dense and deep rooting systems of perennial bioenergy crops drawing down soil moisture at deeper soil depths (Clifton-Brown et al.,

2007; Vanloocke et al., 2010; Georgescu et al., 2011; Anderson et al., 2013; Hallgren et al., 2013; Ferchaud et al., 2015). While physiological or morphological factors (i.e., LAI and vegetation fraction) can alter the surface energy balance and enhance ET during green-up, hydrometeorological impacts could primarily be a function of radiative forcing, as indicated here, because the rooting systems of perennial biofuel crops and existing grasslands were characteristically similar and therefore unaltered in this study.

This study also examined whether perennial biofuel crop expansion could ameliorate drought conditions by contrasting the potential hydroclimate impacts of perennial biofuel crops during a drought year (2011) and normal year (2007). Our analyses show that perennial bioenergy crop deployment reduced the impacts of drought through the enhanced ET and simultaneously cooling of the near-surface environment, by decreasing the atmospheric demand for water (Seneviratne et al., 2010) and mitigating plant stress due to water availability (Wu et al., 2002). During the growing season, near-surface cooling and enhanced ET associated with these crops were moderated under the drought year relative to the normal climate year due to the moisture

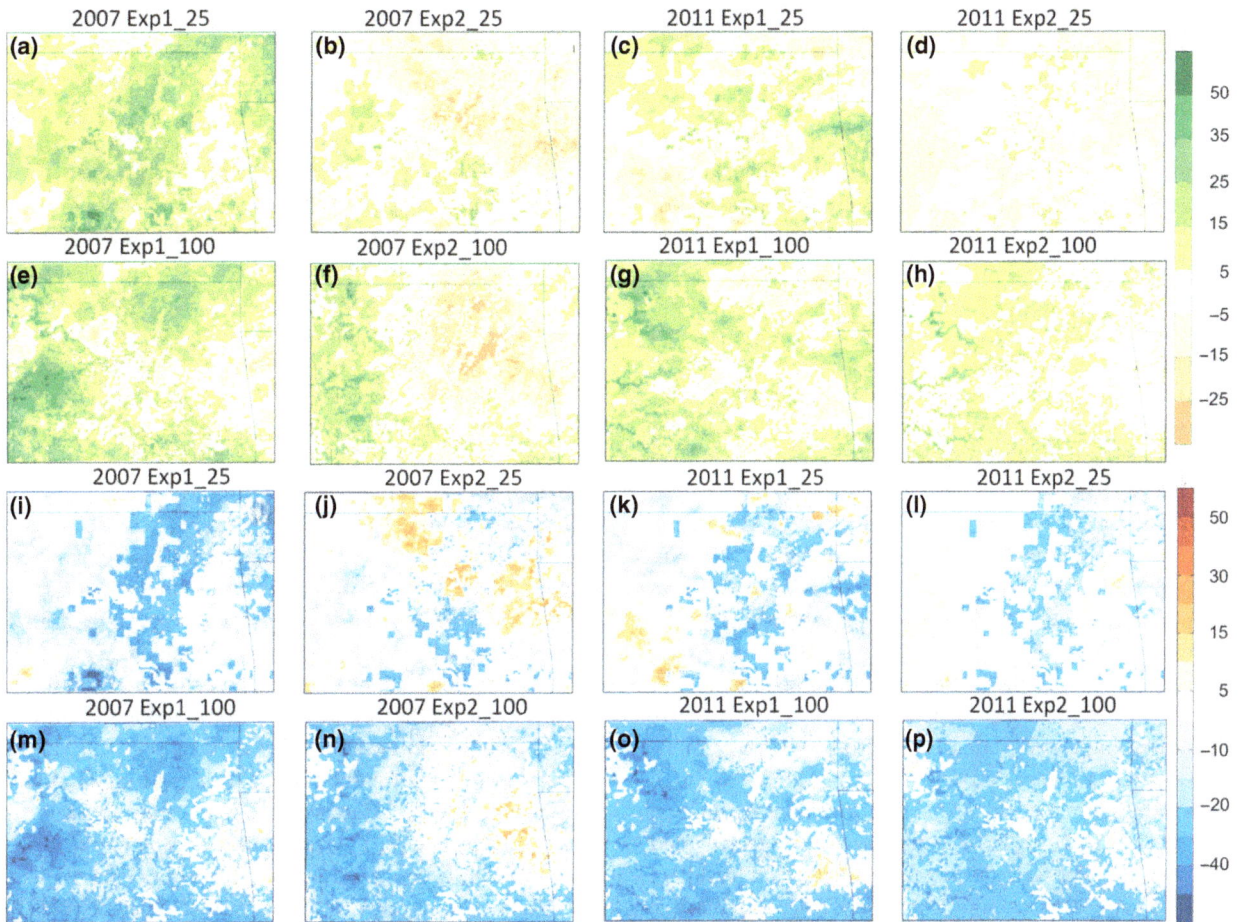

Fig. 13 Mean latent flux differences (average of May–October) between perennial biofuel crop experiments and control experiments for Exp1-2 (from left to right) 25% deployment for the year 2007. Figures (a, b) and 2011 (c, d) and 100% deployment for the years 2007 Figures (e, f) and 2011 (g, h) and sensible heat fluxes 25% deployment for the years 2007 Figures (i, j) and 2011 (k, l) and 100% deployment for the year 2007. Figures (m, n) and 2011 (o, p).

availability and the subsequent alterations in near-surface radiation and energy budgets. These changes revealed the differential effects of simulated perennial bioenergy crop impacts based on background environmental conditions. Despite the moderation of these hydrometeorological impacts, perennial biofuel crop expansion was shown to mitigate the impacts of drought (i.e., via an increased ET) and can potentially serve as a more sustainable pathway to renewable energy than current strategies given these additional unintended, but positive, consequences. The length of timescale perturbations associated with these cropping systems, however, must be considered as the small, and negative feedback of soil moisture depletion could amplify drought severity under a multiyear drought scenario. Therefore, additional multiyear high-resolution simulations are necessary to definitively conclude whether such approaches could simultaneously provide the biomass necessary for ethanol production while

ameliorating large-scale climate change through the alteration of near-surface regional climate.

It is important to note that further hydroclimate impacts were observed from April to May and October to November outside the 'active' growing season. During these periods, perennial bioenergy crop expansion points to warmer near-surface temperatures and decreased ET. These hydrometeorological impacts were more prominent during the green-up period than during the senescence. Vegetation fraction values were significantly higher than albedo and LAI for existing grasslands relative to perennial bioenergy crops during this green-up period (see Fig. 3a). This difference could be an artifact of how vegetation fraction values were scaled. It is possible that vegetation fraction evolution could progress more slowly during the spring and late fall, but requires field observation data that were unavailable (Wagle *et al.*, 2016). Other factors such as soil moisture availability and litter layer thickness could have contributed to these decreases in longwave radiation and latent heat fluxes from April to May 2011 as soil moisture levels were approximately 29% lower in 2011 relative to 2007 (see Fig. 8a–f).

Sensitivity to bioenergy crop deployment was largely independent of control simulation skill for thermal impacts, but proved to be a determining factor for simulated ET. Experimental differences were in good agreement on the simulated thermal impacts of perennial bioenergy crops as temperature decreases were generally within 0.2 °C (Fig. 7) and spatially consistent between Exp1 (the best skilled ensemble member) and Exp2 (the least skilled ensemble member). Unlike temperature, simulated ET impacts associated with perennial bioenergy crops diverged, mainly during July to September 2007. During this period, Exp1 showed that ET increased on average by 0.73 mm day^{-1}, while Exp2 showed that ET decreased by an average of 0.19 mm day^{-1}. Other experimental differences were evident in 2011 as ET increases varied according to model solution from July to September, pointing to issues of model performance and simulated atmospheric processes (see Fig. 9). During the growing season, Exp1 consistently displayed higher ET, latent heat fluxes, and soil moisture relative to Exp2, especially during 2007. Although both experiments noted a dry bias in model evaluation (see Fig. 6a, f, i, n), Exp2 significantly underestimated the amount of precipitation, most likely as a result of microphysics scheme selection (see Table 3). This dry bias associated with Exp2 could explain the significantly lower latent heat fluxes, lower soil moisture levels, and higher sensible heat fluxes, affecting simulated ET associated with perennial bioenergy crop expansion.

Simulated impacts of precipitation associated with perennial biofuel crop deployment also highlighted the sensitivity of control simulation skill, but are characterized as lower confidence relative to the aforementioned impacts. For both years, the simulated precipitation pointed to a positive signal under Exp1, while the simulated precipitation indicated a negative signal under Exp2, likely indicative of the dry bias associated with Exp2 (see Fig. S1 and Table S1). Similar to the simulated ET and latent heat flux, precipitation totals were attenuated during the drought year relative to normal climate year, pointing to moisture availability tied to hydrometeorological conditions. Unlike the aforementioned impacts, the simulated precipitation totals were also reduced under the full deployment relative to the partial deployment scenario, even during the normal climate year. This finding points to radiative forcing as the driving mechanism for simulated results, as decreased low-level heating resulting from reduced shortwave absorption lowers the available energy necessary for convection and, consequently, the amount of precipitation (Findell & Elathir, 2003; Seneviratne et al., 2010). This mechanism, however, requires further examination focusing exclusively on the untangling of disparate forcing mechanisms owing to differential background dynamics at different times of year and is beyond the scope of this manuscript.

The use of a singular LSM, rather than a suite of models with diverse physiological representations of plant functional types and physical representations of land surface flux energy partitioning, is a necessary area for future research. For example, simulated canopy temperature and associated surface fluxes, which regulate the regional hydrologic cycle, are not resolved separately in NOAH, but rather are calculated over a combined surface of vegetation and soil (Niu et al., 2011). This important point emphasizes the need for utility of a fully coupled earth system model that includes a dynamically evolving biophysical representation of perennial bioenergy cropping systems, such as the community land model (Zeng et al., 2002). Such an effort will improve upon the physiological characterization of the investigated plant type, which is dependent on ambient environmental conditions rather than on the presumed periodic depiction utilized here. In effect, imposed energy crop biophysical characteristics may be different under a drought compared with a normal precipitation year. Finally, the impacts on groundwater require examination using appropriate models that account for such hydrologic elements (e.g., NOAH-MP and LEAF-2 dynamically coupled to WRF [e.g., Miguez-Macho et al., 2007]).

Future work should also assess the hydrometeorological impacts of perennial biofuel crops using high-resolution simulations in other geographical regions to evaluate the range of potential outcomes with this energy pathway. Yield estimates should also be calculated to provide information on how diverse hydrometeorological conditions could potentially affect bioenergy crop yields by using an ecosystem model such as Agro-IBIS (e.g., Vanloocke et al., 2010). Long-term, continuous climate simulations of perennial biofuel crop expansion are also needed to understand the resiliency of these cropping systems under a variable current climate and under projected warmer climates associated with higher levels of greenhouse gas emissions. Varying aspects of this work are underway.

Acknowledgement

This work was funded by NSF Grant EAR-1204774.

References

Abatzoglou JT (2013) Development of gridded surface meteorological data for ecological applications and modelling. *International Journal of Climatology*, **33**, 121–131.

Abraha M, Chen J, Chu H et al. (2015) Evapotranspiration of annual and perennial biofuel crops in a variable climate. GCB Bioenergy, 7, 1344–1356.

Anderson CJ, Anex RP, Arritt RW, Gelder BK, Khanal S, Herzmann DE, Gassman PW (2013) Regional climate impacts of a biofuels policy projection. Geophysical Research Letters, 40, 1217–1222.

Anderson-Teixeira KJ, Snyder PK, Twine TE, Cuadra SV, Costa MH, DeLucia EH (2012) Climate-regulation services of natural and agricultural ecoregions of the Americas. Nature Climate Change, 2.3, 177–181.

Bagley JE, Davis SC, Georgescu M et al. (2014) The biophysical link between climate, water, and vegetation in bioenergy agro-ecosystems. Biomass and Bioenergy, 71, 187–201.

Bagley JE, Miller J, Bernacchi CJ (2015) Biophysical impacts of climate-smart agriculture in the Midwest United States. Plant, Cell, and Environment, 38, 1913–1930.

Bright RM, Cherubini F, Strømman AH (2012) Climate impacts of bioenergy: inclusion of carbon cycle and albedo dynamics in life cycle impact assessment. Environmental Impact Assessment Review, 37, 2–11.

Bright RM, Bogren W, Bernier P, Astrup R (2016) Carbon-equivalent metrics for albedo changes in land management contexts: relevance of the time dimension. Ecological Applications, 26, 1868–1880.

Cai X, Zhang X, Wang D (2010) Land availability for biofuel production. Environmental Science and Technology, 45, 334–339.

Campbell JE, Lobell DB, Genova RC, Field CB (2008) The global potential of bioenergy on abandoned agriculture lands. Environmental Science and Technology, 42, 5791–5794.

Campbell JE, Lobell DB, Genova RC, Zumkehr A, Field CB (2013) Seasonal energy storage using bioenergy production from abandoned croplands. Environmental Research Letters, 8, 035012.

Cao Q, Yu D, Georgescu M, Han Z, Wu J (2015) Impacts of land use and land cover change on regional climate: a case study in the agro-pastoral transitional zone of China. Environmental Research Letters, 10, 124025.

Chen G, Zhao L, Qi Y (2015) Enhancing the productivity of microalgae cultivated in wastewater toward biofuel production: a critical review. Applied Energy, 137, 282–291.

Clifton-Brown JC, Breuer J, Jones MB (2007) Carbon mitigation by the energy crop, Miscanthus. Global Change Biology, 13, 2296–2307.

Davis AS, Hill JD, Chase CA, Johanns AM, Liebman M (2012) Increasing cropping system diversity balances productivity, profitability and environmental health. PLoS ONE, 7, e47149.

Devaraju N, Bala G, Nemani R (2015) Modelling the influence of land-use changes on biophysical and biochemical interactions at regional and global scales. Plant, Cell & Environment, 38, 1931–1946.

Ek MB, Mitchell KE, Lin Y, et al. (2003) Implementation of Noah land surface model advances in the National Centers for Environmental Prediction operational mesoscale Eta model. Journal of Geophysical Research: Atmospheres, 108, (D22) 8851. doi:10.1029/2002JD003296.

Feng Q, Chaubey I, Her YG, Cibin R, Engel B, Volenec J, Wang X (2015) Hydrologic and water quality impacts and biomass production potential on marginal land. Environmental Modelling and Software, 72, 230–238.

Ferchaud F, Vitte G, Bornet F, Strullu L, Mary B (2015) Soil water uptake and root distribution of different perennial and annual bioenergy crops. Plant and Soil, 388, 307–322.

Findell KL, Eltahir EA (2003) Atmospheric controls on soil moisture-boundary layer interactions. Part I: framework development. Journal of Hydrometeorology, 4, 552–569.

Foley JA, DeFries R, Asner GP et al. (2005) Global consequences of land use. Science, 309, 570–574.

Gelfand I, Sahajpal R, Zhang X, Izaurralde RC, Gross KL, Robertson GP (2013) Sustainable bioenergy production from marginal lands in the US Midwest. Nature, 493, 514–517.

Georgescu M, Lobell DB, Field CB (2009) Potential impact of US biofuels on regional climate. Geophysical Research Letters, 36, L21806.

Georgescu M and Lobell DB (2010) Perennial questions of hydrology and climate. Science, 330(1), 33.

Georgescu M, Lobell DB, Field CB (2011) Direct climate effects of perennial bioenergy crops in the United States. Proceedings of the National Academy of Sciences, 108, 4307–4312.

Georgescu M, Lobell DB, Field CB Mahalov A(2013) Simulated hydroclimatic impacts of projected Brazilian sugarcane expansion. Geophysical Research Letters, 40, 972–977.

Goldstein JC, Tarhule A, Brauer D (2014) Simulating the hydrologic response of a semiarid watershed to switchgrass cultivation. Hydrology Research, 45, 99–114.

Gopalakrishnan G, Negri C, Salas W (2012) Modeling biogeochemical impacts of bioenergy buffers with perennial grasses for a row-crop field in Illinois. GCB Bioenergy, 4, 739–750.

Haberl H, Beringer T, Bhattacharya SC, Erb KH, Hoogwijk M (2010) The global technical potential of bio-energy in 2050 considering sustainability constraints. Current Opinion in Environmental Sustainability, 2, 394–403.

Hallgren W, Schlosser CA, Monier E, Kicklighter D, Sokolov A, Melillo J (2013) Climate impacts of a large-scale biofuels expansion. Geophysical Research Letters, 40, 1624–1630.

Hickman GC, Vanloocke A, Dohleman FG, Bernacchi CJ (2010) A comparison of canopy evapotranspiration for maize and two perennial grasses identified as potential bioenergy crops. GCB Bioenergy, 2, 157–168.

Hoerling M, Kumar A, Dole R et al. (2013) Anatomy of an extreme event. Journal of Climate, 26, 2811–2832.

Hudiburg TW, Davis SC, Parton W, Delucia EH (2015) Bioenergy crop greenhouse gas mitigation potential under a range of management practices. GCB Bioenergy, 7, 366–374.

Hudiburg TW, Wang W, Khanna M et al. (2016) Impacts of a 32-billion-gallon bioenergy landscape on land and fossil fuel use in the US. Nature Energy, 1, 15005.

Joo E, Hussain MZ, Zeri M et al. (2016) The influence of drought and heat stress on long term carbon fluxes of bioenergy crops grown in the Midwestern US. Plant, Cell and Environment, 39, 1928–1940.

Khanal S, Anex RP, Anderson CJ, Herzmann DE, Jha MK (2013) Implications of biofuel policy-driven land cover change for rainfall erosivity and soil erosion in the United States. GCB Bioenergy, 5, 713–722.

Khanal S, Anex RP, Anderson CJ, Herzmann DE (2014) Streamflow impacts of biofuel policy-driven landscape change. PLoS ONE, 9, e109129.

Le PV, Kumar P, Drewry DT (2011) Implications for the hydrologic cycle under climate change due to the expansion of bioenergy crops in the Midwestern United States. Proceedings of the National Academy of Sciences, 108, 15085–15090.

López-Bellido L, Wery J, López-Bellido RJ (2014) Energy crops: prospects in the context of sustainable agriculture. European Journal of Agronomy, 60, 1–12.

McIsaac GF, David MB, Mitchell CA (2010) Miscanthus and switchgrass production in Central Illinois: impacts on hydrology and inorganic nitrogen leaching. Journal of Environmental Quality, 39, 1790–1799.

Melillo JM, Reilly JM, Kicklighter DW et al. (2009) Indirect emissions from biofuels: how important?. Science, 326, 1397–1399.

Miguez-Macho G, Fan Y, Weaver CP, Walko R, Robock A (2007) Incorporating water table dynamics in climate modeling: 2. Formulation, validation, and soil moisture simulation. Journal of Geophysical Research: Atmospheres, 112, D13108. doi:10.1029/2006JD008112

Miller JN, VanLoocke A, Gomez-Casanovas N, Bernacchi CJ (2015) Candidate perennial bioenergy grasses have a higher albedo than annual row crops in the Midwestern US. Global Change Biology, 21, 4237–4249.

Niu GY, Yang ZL, Mitchell KE et al. (2011) The community Noah land surface model with multiparameterization options (Noah MP): 1. Model description and evaluation with local scale measurements. Journal of Geophysical Research: Atmospheres, 116, D12109. doi:10.1029/2010JD015139

Oikawa PY, Jenerette GD, Grantz DA (2015) Offsetting high water demands with high productivity: sorghum as a biofuel crop in a high irradiance arid ecosystem. GCB Bioenergy, 7, 974–983.

Pielke RA (2001) Influence of the spatial distribution of vegetation and soils on the prediction of cumulus convective rainfall. Reviews of Geophysics, 39, 151–177.

Pielke RA (2005) Land use and climate change. Science, 310, 1625–1626.

Rahman MM, Mostafiz SB, Paatero JV, Lahdelma R (2014) Extension of energy crops on surplus agricultural lands: a potentially viable option in developing countries while fossil fuel reserves are diminishing. Renewable and Sustainable Energy Reviews, 29, 108–119.

RFA (Renewable Fuel Association) (2010) Statistics, Annual World Ethanol Production by Country (based on F.O. Licht estimates). Available at: http://www.ethanolrfa.org/pages/statistics#E (accessed 26 October 2010).

Seneviratne SI, Corti T, Davin EL et al. (2010) Investigating soil moisture–climate interactions in a changing climate: a review. Earth-Science Reviews, 99, 125–161.

Skamarock WC, Klemp JB, Dudhia J, Gill DO, Barker DM, Wang W, Powers JG (2005) A description of the advanced research WRF version 2 (No. NCAR/TN-468+STR). National Center For Atmospheric Research Boulder Co Mesoscale and Microscale Meteorology Div.

Tadesse T, Wardlow BD, Brown JF, Svoboda MD, Hayes MJ, Fuchs B, Gutzmer D (2015) Assessing the vegetation condition impacts of the 2011 drought across the US southern great plains using the vegetation drought response index (VegDRI). Journal of Applied Meteorology and Climatology, 54, 153–169.

Taylor KE (2001) Summarizing multiple aspects of model performance in a single diagram. *Journal of Geophysical Research: Atmospheres*, **106**, 7183–7192.

United States Drought Monitor (2014). Available at: http://droughtmonitor.unl.edu/MapsAndData/GISData.aspx (accessed 16 October 2014)

Vanloocke A, Bernacchi CJ, Twine TE (2010) The impacts of Miscanthus× giganteus production on the Midwest US hydrologic cycle. *GCB Bioenergy*, **2**, 180–191.

VanLoocke A, Twine TE, Kucharik CJ, Bernacchi CJ (2016) Assessing the potential to decrease the Gulf of Mexico hypoxic zone with Midwest US perennial cellulosic feedstock production. *GCB Bioenergy*, **9**, 858–875.

Wagle P, Kakani VG (2014) Seasonal variability in net ecosystem carbon dioxide exchange over a young Switchgrass stand. *GCB Bioenergy*, **6**, 339–350.

Wagle P, Kakani VG, Huhnke RL (2016) Evapotranspiration and ecosystem water use efficiency of switchgrass and high biomass sorghum. *Agronomy Journal*, **108**, 1007–1019.

Weaver CP, Avissar R (2001) Atmospheric disturbances caused by human modification of the landscape. *Bulletin of the American Meteorological Society*, **82**, 269–281.

Wu W, Geller MA, Dickinson RE (2002) The response of soil moisture to long-term variability of precipitation. *Journal of Hydrometeorology*, **3**, 604–613.

Xu X, Scanlon BR, Schilling K, Sun A (2013) Relative importance of climate and land surface changes on hydrologic changes in the US Midwest since the 1930s: implications for biofuel production. *Journal of Hydrology*, **497**, 110–120.

Yimam YT, Ochsner TE, Kakani VG (2015) Evapotranspiration partitioning and water use efficiency of switchgrass and biomass sorghum managed for biofuel. *Agricultural Water Management*, **155**, 40–47.

Zeng X, Shaikh M, Dai Y, Dickinson RE, Myneni R (2002) Coupling of the common land model to the NCAR community climate model. *Journal of Climate*, **15**, 1832–1854.

Zeri M, Hussain MZ, Anderson-Teixeira KJ, DeLucia E, Bernacchi CJ (2013) Water use efficiency of perennial and annual bioenergy crops in central Illinois. *Journal of Geophysical Research: Biogeosciences*, **118**, 581–589.

Zhu P, Zhuang Q, Joo E, Bernacchi C (2016) Importance of biophysical effects on climate warming mitigation potential of biofuel crops over the conterminous United States. *GCB Bioenergy*, **9**, 577–590.

Zumkehr A, Campbell JE (2013) Historical US cropland areas and the potential for bioenergy production on abandoned croplands. *Environmental Science and Technology*, **47**, 3840–3847.

12

Impact of drought stress on growth and quality of miscanthus for biofuel production

TIM VAN DER WEIJDE[1,2,*], LAURIE M. HUXLEY[3,*], SARAH HAWKINS[3], EBEN HAESER SEMBIRING[1], KERRIE FARRAR[3], OENE DOLSTRA[1], RICHARD G. F. VISSER[1] and LUISA M. TRINDADE[1]

[1]Wageningen UR Plant Breeding, Wageningen University and Research, PO Box 386, 6700 AJ Wageningen, The Netherlands, [2]Graduate School Experimental Plant Sciences, Wageningen University, Droevendaalsesteeg 1, 6708 PB Wageningen, The Netherlands, [3]Institute of Biological, Environmental & Rural Sciences (IBERS), Aberystwyth University, Plas Gogerddan, Aberystwyth, Ceredigion SY23 3EE, UK

Abstract

Miscanthus has a high potential as a biomass feedstock for biofuel production. Drought tolerance is an important breeding goal in miscanthus as water deficit is a common abiotic stress and crop irrigation is in most cases uneconomical. Drought may not only severely reduce biomass yields, but also affect biomass quality for biofuel production as cell wall remodeling is a common plant response to abiotic stresses. The quality and plant weight of 50 diverse miscanthus genotypes were evaluated under control and drought conditions (28 days no water) in a glasshouse experiment. Overall, drought treatment decreased plant weight by 45%. Drought tolerance – as defined by maintenance of plant weight – varied extensively among the tested miscanthus genotypes and ranged from 30% to 110%. Biomass composition was drastically altered due to drought stress, with large reductions in cell wall and cellulose content and a substantial increase in hemicellulosic polysaccharides. Stress had only a small effect on lignin content. Cell wall structural rigidity was also affected by drought conditions; substantially higher cellulose conversion rates were observed upon enzymatic saccharification of drought-treated samples with respect to controls. Both cell wall composition and the extent of cell wall plasticity under drought varied extensively among all genotypes, but only weak correlations were found with the level of drought tolerance, suggesting their independent genetic control. High drought tolerance and biomass quality can thus potentially be advanced simultaneously. The extensive genotypic variation found for most traits in the evaluated miscanthus germplasm provides ample scope for breeding of drought-tolerant varieties that are able to produce substantial yields of high-quality biomass under water deficit conditions. The higher degradability of drought-treated samples makes miscanthus an interesting crop for the production of second-generation biofuels in marginal soils.

Keywords: cell wall composition, cellulose, drought tolerance, hemicellulose, lignin, miscanthus, saccharification efficiency

Introduction

Perennial biomass crops, such as miscanthus, are being developed for the production of biofuels to replace our fossil fuel-based energy supply chain with a renewable and more sustainable biomass-based alternative. Miscanthus is a leading candidate crop for biomass production owing to its rapid biomass accumulation and high nutrient and water-use efficiencies (Jones & Walsh, 2001; Heaton et al., 2010; Van Der Weijde et al., 2013). In addition, miscanthus biomass typically has a high quality for biofuel production as it is characterized by low

moisture and high cell wall and carbohydrate contents, which are traits that contribute favorably to the yield of fermentable sugars to be used for the production of cellulosic ethanol (Wyman, 2007; Himmel & Picataggio, 2008; Zhao et al., 2012).

A consistent and predictable supply of high-quality lignocellulosic feedstocks is crucial to the success of cellulosic biorefineries (Perlack et al., 2005; Van Der Weijde et al., 2013). To achieve this, crops must be high yielding and have stable performance across diverse environments. Drought is one of the most widespread abiotic stresses (Chaves et al., 2003; Farooq et al., 2009), and the incidence of local and regional drought events is increasing worldwide due to climate change (Sheffield & Wood, 2008; Dai, 2013). In addition, miscanthus is seen as a crop with a high potential for production on

*Authors contributed equally to this work.

Correspondence: Luisa M. Trindade
e-mail: luisa.trindade@wur.nl

marginal land, minimizing competition with food crops for arable land (Quinn et al., 2015). Plants growing on marginal soils, such as eroded soils or bare lands, will regularly encounter periods with a water deficit. Unlike most food products, lignocellulosic feedstocks are considered low-value, high-volume commodities and crops such as miscanthus must be produced under low-input regimes. Under these provisions, irrigation is likely to be uneconomical and/or unsustainable for the production of miscanthus biomass (Bullard, 2001). In most miscanthus crop production scenarios, particularly those involving the production of biomass on marginal soils, periods of drought stress may regularly occur (Quinn et al., 2015).

Attractive characteristics of miscanthus with regard to drought tolerance include (i) that its C4 photosynthesis system is characterized by a higher water-use efficiency compared to C3 photosynthetic plants and (ii) that its perennial growth habit and extensive root system enable better exploitation of soil water reserves present in deeper soil layers than annual plants (Heaton et al., 2010; Byrt et al., 2011; Van Der Weijde et al., 2013). Moreover, the genus Miscanthus harbors extensive genetic diversity as it is adapted to a wide range of geographical conditions across East Asia (Clifton-Brown et al., 2002, 2008). These features provide scope for the selection and breeding of stress-tolerant miscanthus varieties.

Aside from the adverse effects of drought on plant growth, drought influences virtually all plant physiological processes, including cell wall biosynthesis. These effects are important if miscanthus grown on marginal soils are to be used for biofuel production, as the composition and structural rigidity of the cell wall are key factors determining the techno-economic efficiency of biofuel production (Wyman, 2007; Himmel & Picataggio, 2008; Zhao et al., 2012; Torres et al., 2016). The contents of the two main cell wall polysaccharides, cellulose, and hemicellulose determine the maximum theoretically extractable content of fermentable sugars. The relative contents of the major cell wall components – particularly the content of lignin – as well as the extent of cross-linking between them are important parameters determining the efficiency of converting cell wall polysaccharides into fermentable sugars (Wyman, 2007; Himmel & Picataggio, 2008; Zhao et al., 2012). One of the consequences of drought is a loss of cell turgor (Farooq et al., 2009). A primary plant response to the loss of turgor is stiffening of cell walls to provide structural resistance and arrest cell extension (Moore et al., 2008; Tenhaken, 2015). Longer exposure to drought stress challenges plants to modify their cell walls to sustain growth under conditions with reduced water potential (Moore et al., 2008). Drought stress is thus likely to affect the biomass quality of the feedstock (Iraki et al., 1989a; Moore et al.,

2008; Moura et al., 2010; Pauly & Keegstra, 2010; Frei, 2013; Emerson et al., 2014; Tenhaken, 2015).

Although the cell wall is clearly affected by drought stress, surprisingly little is known about drought-induced changes in cell wall composition (Tenhaken, 2015). Transcriptome studies often report cell wall-related genes to be differentially expressed upon drought stress, but actual biochemical changes in cell wall components are sparsely investigated. Studies that have investigated biochemical changes in cell wall composition consistently report a decrease in cellulose content upon drought stress (Frei, 2013). However, both increases and decreases in contents of lignin and hemicellulosic polysaccharides upon drought stress are reported in different crops and plant tissues (Guenni et al., 2002; Vincent et al., 2005; Al-Hakimi, 2006; Moore et al., 2008; Hu et al., 2009; Jiang et al., 2012; Meibaum et al., 2012; Emerson et al., 2014; Rakszegi et al., 2014). Therefore, it is yet largely unknown how water deficits affect biomass quality of bioenergy crops.

Increasing our understanding of drought-induced cell wall modifications and their impact on biomass quality is of major importance for developing miscanthus varieties for biomass production under low-input conditions and/or on marginal soils. In this study, plant growth and the compositional quality of stem and leaf material were analyzed in 50 diverse miscanthus genotypes, comprising Miscanthus sinensis, Miscanthus sacchariflorus, and interspecific hybrids, cultivated under drought and control growing conditions. To our knowledge, this is the first study to explore the magnitude of available variation in plant growth and biomass quality under drought stress in the germplasm pool of bioenergy feedstock miscanthus.

Materials and methods

Plant material

The experiment comprised 50 miscanthus genotypes including 35 M. sinensis, 8 M. sacchariflorus, and 7 M. sinensis × M. sacchariflorus species. All genotypes used in this study were supplied by Wageningen University and Aberystwyth University, in a collaboration that is part of the EU Seventh Framework Programme OPTIMISC (www.optimisc-project.eu). Like-sized tillers were split from clonal stock plants into eight separate parts and transferred to prelined 1-meter pipes filled with John Innes number-3 soil (Fig. 1a, b). Plants were allowed to grow with sufficient watering for 84 days prior to the start of treatment.

Drought experiment

The experiment was designed to evaluate genotypic responses to total water withdrawal. A total of 50 miscanthus genotypes

Fig. 1   Establishment of 50 miscanthus genotypes ($n = 500$) in 1-m pipes between March (a) and June 2014 (b) prior to screening.

were planted in a randomized split-plot block design with four blocks. Each block was randomly split in two segments, each containing the full set of genotypes, which received one of two treatments: well-watered vs. complete water withdrawal for 28 days, commencing June 2014. In total, four replicates per genotype per treatment were evaluated ($n = 400$). The experiment was conducted in a glasshouse at IBERS (52°43′N, 04°02′W).

After 28 days of treatment, all five replicate plants per genotype per treatment were harvested. Using secateurs, plant tillers were cut just above soil level. Stem (with panicle) and leaf material were separated and oven-dried to a constant dry weight (DW) at 60 °C for 72 h to determine stem, leaf, and plant weights in gram dry matter per plant, as well as the stem: leaf ratio (g g^{-1}). Plant weight as defined here refers to the aboveground biomass (stem + leaf) of the plants. Drought tolerance was calculated as the percentage of maintained biomass under water stress [average plant weight under drought stress ($n = 4$)/average plant weight under control treatment ($n = 4$) × 100%]. One genotype, OPM-17, yielded insufficient material for analysis and was excluded from the study. To obtain enough material for biochemical analyses, the samples were pooled for stem and leaf samples independently, by randomly combining two of the four replicate samples per genotype per treatment into one of two pools. All pooled leaf and stem samples [$n = 400$ (50 genotypes × 2 treatments × 2 pools × 2 tissue types)] were ground using a hammer mill with a 1-mm screen prior to biochemical analysis.

Biochemical analysis of the cell wall

Contents of neutral detergent fiber (NDF), acid detergent fiber (ADF), and acid detergent lignin contents (ADL) of stem and leaf dry matter were determined according to protocols developed by ANKOM Technology that are essentially based on the work of Goering and Van Soest (Van Soest, 1967; Goering & Van Soest, 1970). Neutral and acid detergent extractions were performed using an ANKOM 2000 Fiber Analyzer (ANKOM Technology Corporation, Fairpoint, NY, USA). Acid detergent lignin was determined after 3-h hydrolysis of the ADF residue in 72% H_2SO_4 with continuous shaking. All analyses were performed in triplicate. The weight fractions of detergent fiber

residues in dry matter were subsequently used to estimate the content of cell wall in dry matter (NDF% dm) and to obtain the contents of cellulose [(ADF% dm − ADL% dm)/NDF% dm × 100%], hemicellulosic polysaccharides [(NDF% dm − ADF% dm)/NDF% dm × 100%], and lignin (ADL% dm/NDF% dm × 100%) relative to the cell wall content.

Analysis of saccharification efficiency

Saccharification efficiency of the samples was assessed by the conversion of cellulose into glucose by mild alkaline pretreatment and enzymatic saccharification reactions. Reactions were carried out in triplicate using 500 mg subsamples per stem or leaf sample. All subsamples were incubated for 13 min with thermostable α-amylase (ANKOM Technology Corporation), followed by three-five-minute incubations with warm deionized water (60 °C) to remove interfering soluble sugars. The remaining biomass was then subjected to a mild alkaline pretreatment, carried out in 50-ml plastic centrifuge tubes with 15 ml 2% NaOH at 50 °C with constant shaking (160 RPM) for two hours in an incubator shaker (Innova 42; New Brunswick Scientific, Enfield, CT, USA). In this study, the objective of the pretreatment was not to maximize cellulose conversion but to treat samples to better allow discrimination of genotypic differences in cellulose conversion efficiency. Pretreated samples were washed twice with deionized water (5 min, 50 °C) and once with 0.1 M sodium citrate buffer (pH 4.6, 5 min, 50 °C).

Saccharification reactions were subsequently carried out according to the NREL Laboratory Analytical Procedure 'Enzymatic saccharification of lignocellulosic biomass' (Selig et al., 2008). Pretreated samples were hydrolyzed for 48 h with 300 μl (25.80 mg of enzyme) of the commercial enzyme cocktail Accellerase 1500 (DuPont Industrial Biosciences, Leiden, the Netherlands) supplemented with 15 μl (0.12 mg of enzyme) endo-1,4-β-xylanase M1 (EC 3.2.1.8; Megazyme International Ireland, Bray, Ireland) in an incubator shaker (Innova 42; New Brunswick Scientific) set at 50 °C and constant shaking (160 RPM). This enzyme mixture has the following reported specific activities: endoglucanase 2200–2800 CMC U g^{-1}, beta-glucosidase 450–775 pNPG U g^{-1}, and endoxylanase 230 U mg^{-1}. Reactions were carried out in 44 ml 0.1 M sodium citrate buffer (pH

4.6), containing 0.4 ml 2% sodium azide to prevent microbial contamination.

Glucose contents in the enzymatic saccharification liquors were determined in duplicate using the enzyme-linked D-glucose assay kit (R-Biopharm, Darmstadt, Germany). This assay was adapted to a 96-well microplate format, and the increases in sample absorption following enzyme-mediated conversion reactions were spectrophotometrically determined at 340 nm using a Bio-Rad Microplate Reader (Bio-Rad, Richmond, CA, USA). All sample absorbance measurements were corrected using blanks, containing water instead of sample solution. Glucose release was determined by calculating the glucose content in the saccharification liquor from absorbance measurements using Eqn (1).

$$\text{Glucose release (mg)} = \frac{V \times \text{MW}}{\varepsilon \times d \times v \times 1000} \times \text{df} \times \Delta\text{Abs} \quad (1)$$

where V = final well volume (= 3.02 ml); MW = molecular weight of glucose (= 180.16 g mol^{-1}); ε = the molar extinction coefficient of NADPH (= 6.3 l × mol^{-1} × cm^{-1}); d = light path-length (= 1.016 cm); v = sample volume (ml); df = dilution factor (= 10); and ΔAbs = increase in sample absorbance, corrected for the increase in blank absorbance. Cellulose conversion was calculated from the release of glucose relative to the cellulose content in the sample, as detailed in Eqn (2).

$$\text{Cellulose conversion (\%)} = \frac{\text{Glucose release (mg)}}{\text{CC} \times 1.111} \times 100\% \quad (2)$$

where CC = cellulose content in the sample (in mg), and 1.111 = the mass conversion factor that converts cellulose to equivalent glucose (the molecular weight ratio of 180.16–162.16 g mol^{-1} for glucose and anhydro-glucose, Dien, 2010).

Analysis of miscanthus biomass using near-infrared spectroscopy (NIRS)

Multivariate prediction models based on near-infrared (NIR) spectral data were developed to allow high-throughput prediction of biomass quality traits. Near-infrared absorbance spectra of stem and leaf samples were obtained using a Foss DS2500 near-infrared spectrometer (Foss, Hillerød, Denmark). Averaged spectra were obtained consisting of eight consecutive scans from 400 to 2500 nm with an interval of 2 nm using ISI-Scan software (Foss). Obtained spectra were further processed by weighted multiplicative scatter correction and mathematical derivatization and smoothing treatments using WINISI 4.9

statistical software (Foss). These statistical transformations of spectra help to minimize effects resulting from light scatter and differences in particle size. Parameters for derivatization and smoothing were set at 2-6-4-1, in which the first number of this mathematical procedure refers to order of derivatization, the second number to the gap in the data points over which the derivation is applied, and the third number and fourth number refer to the number of data points used in the smoothing of the first and second derivative.

For the creation of prediction models, a calibration set of 110 samples was randomly selected from the complete set of samples, but with an approximate 1 : 1 representation of leaf and stem samples. The biochemical reference data and near-infrared spectra of the calibration samples were used for the development and validation of prediction models using WINISI version 4.9 (Foss). The prediction equations were generated using modified partial least-squares regression analyses (Shenk & Westerhaus, 1991), and obtained calibration statistics are reported in Table 1. Another 20 of the remaining samples were randomly selected as an external validation set to evaluate the quality of the generated prediction models. The prediction models were validated using the squared Pearson coefficient of correlation (r^2) between predicted and biochemical data of the external validation set ($n = 20$) and by evaluating for these samples the standard error of prediction (SEP) and its comparison to the standard error of laboratory (SEL) for each of the traits (Table 2). The prediction models were used to determine the cell wall, cellulose, hemicelluloses, and lignin contents, as well as the cellulose conversion rate of all leaf and stem samples.

Statistical analysis

General analyses of variance (ANOVA) were performed to determine the significance ($P < 0.05$) of genotype, treatment, and interaction sources of variation. For growth-related traits, ANOVAS were performed following the completely randomized split-plot block design of the experiment. The four original biological replicates per genotype per treatment were used as a fixed block effect with a nested split-plot on which treatment was applied. Variance analyses for biomass quality-related traits were performed considering that the four biological replicates were combined into two pools. For variance analyses, these two pools were considered as two independent replicates per genotype per treatment and used as a block effect. As these pools were not actual blocks in the original experimental

Table 1 Summary of calibration statistics of mPLS models used for the prediction of biomass quality traits

Trait	Samples	Mean chemical analysis	Mean NIRS prediction	r^2	SEC	SECV
Cell wall content (% dm)	104	67.38	67.14	0.99	0.56	1.25
Cellulose (% ndf)	106	45.82	45.75	0.96	0.77	1.15
Hemicellulose (% ndf)	107	47.28	47.33	0.96	0.86	1.40
Lignin (% ndf)	105	6.89	6.85	0.81	0.40	0.59
Cellulose conversion (%)	103	49.99	49.34	0.61	4.42	4.68

r^2, coefficient of determination; SEC, standard error of calibration; SECV, standard error of cross validation.

Table 2 Summary of validation statistics of mPLS models used for the prediction of biomass quality traits

Trait	Samples	Slope	Intercept	r^2	SEP	SEL
Cell wall content (% dm)	19	0.78	0.13	0.86	2.36	0.51
Cellulose (% ndf)	19	0.92	0.59	0.82	1.53	0.39
Hemicellulose (% ndf)	19	0.93	−0.54	0.86	1.55	0.34
Lignin (% ndf)	19	1.01	−0.09	0.74	0.43	0.26
Cellulose conversion (%)	19	1.33	−1.63	0.73	4.62	2.99

r^2, coefficient of determination; SEP, standard error of prediction; SEL, standard error of laboratory.

design, they could not be used as a fixed block effect, but instead were used as a random block effect. The analyses were performed for stem and leaf samples separately following a mixed effect model (3):

$$Y_{ijk} = \mu + G_{ij} + T_k + GT_{ijk} + B_j + e_{ij} \qquad (3)$$

where Y_{ijk} is the response variable, μ is the grand mean, G_{ij} is the genotype effect, T_k is the treatment effect, GT_{ijk} is the interaction term between genotype and treatment, B_j is the block effect, and e_{ijk} is the residual error.

Multiple comparisons analyses were performed to distinguish significant ($P < 0.05$) genotypic differences within each treatment using Fisher's protected least significant difference (LSD) test on genotype means. The significance of differences ($P = 0.05$) in trait means between two groups of genotypes that were formed based on tolerance level was evaluated using unpaired two-sample t-tests. Correlation analyses were performed on genotype means to identify the significance, strength, and direction of correlations among traits using Pearson's correlation coefficients. All statistical analyses were performed using the statistical software package GENSTAT, 16th edition (VSN International, Hemel Hempstead, UK).

Results

Drought stress affects plant weight and morphology

Growth, composition, and bioconversion efficiency of 50 miscanthus genotypes were evaluated using the leaf and stem tissues of plants grown under drought stress and control conditions. The drought treatment had a significant impact on almost all evaluated traits (Tables 3 and 4). The results showed that both final plant weight and the stem:leaf ratio were significantly affected by treatment. The set of genotypes showed significant differences in genotype performance with a low residual error (Table 3).

The mean and the range in genotype performance for plant weight and stem:leaf ratio in control and drought conditions are displayed in Table 5. Mean plant weight under control conditions was 20.10 g per plant ($n = 200$), whereas under drought stress, plant weight was on average 11.10 g per plant ($n = 200$). Drought

Table 3 Tables of analyses of variance for yield and stem-to-leaf ratio of 50 miscanthus genotypes grown under drought stress compared to control conditions

Trait	Source of variation	Degrees of freedom	Mean squares	F-prob.
Plant weight (g dm per plant)	Wplot stratum	3	115.33	
	Wplot.SplitPlot stratum			
	Treatment	1	8259.40	<0.001
	Residual	3	22.70	
	Wplot.SplitPlot.Unit stratum			
	Genotype	48	164.88	<0.001
	Genotype × treatment	48	42.48	<0.001
	Residual	257	17.92	
Stem : leaf ratio (g g^{-1})	Wplot stratum	4	0.91	
	Wplot.SplitPlot stratum			
	Treatment	1	1.53	0.135
	Residual	3	0.37	
	Wplot.SplitPlot.Unit stratum			
	Genotype	48	0.40	<0.001
	Genotype × treatment	48	0.06	0.393
	Residual	257	0.06	

Wplot, whole blocks in the experiment containing two split-plots to which treatment was applied; SplitPlot, split-plots in the experiment containing all genotypes.

Table 4 Tables of analyses of variance for stem and leaf biomass quality traits of 50 miscanthus genotypes grown under drought stress compared to control conditions

Trait	Source of variation	Stem			Leaf		
		df	m.s.	F-prob.	df	m.s.	F-prob.
Cell wall content (% dm)	Treatment	1	3603.55	0.004	1	2608.88	<0.001
	Residual	2	14.94		2	2.58	
	Genotype	48	82.06	<0.001	48	30.57	<0.001
	Treatment × genotype	48	10.43	<0.001	48	4.42	0.036
	Residual	94	2.90		95	2.87	
Cellulose (% ndf)	Treatment	1	1154.58	0.009	1	39.82	0.033
	Residual	2	10.10		2	1.40	
	Genotype	48	9.83	<0.001	48	4.24	<0.001
	Treatment × genotype	48	3.24	0.020	48	1.81	0.009
	Residual	94	1.98		95	1.03	
Hemicellulose (% ndf)	Treatment	1	1239.44	0.009	1	81.03	0.008
	Residual	2	11.19		2	0.64	
	Genotype	48	12.85	<0.001	48	6.35	<0.001
	Treatment × genotype	48	4.17	0.018	48	2.03	0.068
	Residual	94	2.52		95	1.42	
Lignin (% ndf)	Treatment	1	8.96	0.015	1	0.01	0.522
	Residual	2	0.14		2	0.02	
	Genotype	48	1.67	<0.001	48	0.61	<0.001
	Treatment × genotype	48	0.35	0.027	48	0.22	0.002
	Residual	94	0.22		95	0.11	
Cellulose conversion (%)	Treatment	1	3486.63	0.003	1	689.99	0.001
	Residual	2	9.53		2	1.04	
	Genotype	48	54.22	<0.001	48	7.03	<0.001
	Treatment × genotype	48	7.85	<0.001	48	1.60	0.020
	Residual	94	2.66		95	0.98	

df, degrees of freedom; m.s., mean squares.

treatment in this experiment thus reduced plant weight on average by 45%. Moreover, drought-treated plants, with on average a stem : leaf ratio of 0.77, were generally more leafy than the corresponding control plants, which had on average a stem : leaf ratio of 0.91. Variation in plant weight and stem : leaf ratio among genotypes was extremely large under both stress and control conditions. Final mean plant weight of genotypes under control conditions ranged from 5.80 to 35.63 g, while the range under drought stress was from 2.78 to 20.38 g per plant (Fig. 2). Under both drought and control conditions, leaf biomass contributed on average more to total plant weight than stem biomass (Table 5), but for some genotypes, stems comprised the largest weight fraction of total plant weight.

Genotypes responded very differently to the drought treatment, as shown by the variation in plant weight and drought tolerance (Fig. 2) and by the significance of the genotype-by-treatment interaction term (Table 3). For example, two genotypes, OPM-6 and OPM-19, are both high-yielding genotypes, but differed considerably in drought tolerance. OPM-6 was the genotype with

the highest plant weight under drought stress (20.38 g per plant), which was even higher than the average plant weight (20.10 g per plant) over all genotypes under control conditions. This particular genotype had a plant weight of 26.98 g per plant under control conditions, leading to a drought tolerance of 75.53% (only a 25% reduction in plant weight due to the drought treatment). OPM-19 was the genotype with the highest plant weight under control conditions (35.63 g per plant), but was more severely affected by drought stress. Its plant weight under stress conditions was 17.75 g per plant, leading to a drought tolerance of 49.82% (a 50% reduction in plant weight due to the drought treatment). Variation in drought tolerance among all genotypes ranged from 29.60% to 109.90%. The two genotypes with a tolerance value above 100% had a higher plant weight under drought conditions than under control conditions, although the difference in mean plant weight was smaller than the variation between replicates and both genotypes were low biomass types. On the other side of the tolerance spectrum, genotypes displayed large reductions (up to

Table 5 Genotypic variation in plant weight, stem: leaf ratio, and quality traits across 50 miscanthus genotypes under control and drought conditions

	Trait	Unit	Treatment	Average	Min	Max	Range	CV (%)	LSD
Plant growth	Plant weight	g dm per plant	Control	20.10	5.80	35.63	29.83	25.12	3.19
			Drought	11.10	2.78	20.38	17.60	30.51	4.24
	Stem: leaf ratio	g g^{-1}	Control	0.91	0.55	1.49	0.94	20.35	0.12
			Drought	0.77	0.36	1.39	1.03	36.45	0.18
	Drought tolerance	%	–	57.75	29.60	109.90	80.32	–	–
Stem composition	Cell wall content	% dm	Control	73.06	62.23	78.71	16.49	1.99	2.92
			Drought	64.57	51.28	73.66	22.38	3.00	3.90
	Cellulose	% ndf	Control	51.06	47.57	53.78	6.22	2.33	2.39
			Drought	46.25	39.33	49.49	10.16	3.47	3.23
	Hemicellulose	% ndf	Control	41.57	37.97	45.33	7.36	3.18	2.66
			Drought	46.56	42.60	52.55	9.95	3.92	3.68
	Lignin	% ndf	Control	7.38	6.26	9.50	3.24	5.94	0.88
			Drought	6.93	5.67	8.67	3.00	7.15	1.00
	Cellulose conversion	%	Control	42.22	37.69	50.75	13.06	2.89	2.45
			Drought	50.57	43.45	62.18	18.73	3.91	3.98
Leaf composition	Cell wall content	% dm	Control	70.25	64.74	75.72	10.98	1.87	2.64
			Drought	62.92	57.10	70.46	13.36	3.18	4.03
	Cellulose	% ndf	Control	43.37	40.29	45.89	5.60	2.22	1.93
			Drought	42.50	40.48	44.58	4.11	2.51	2.14
	Hemicellulose	% ndf	Control	49.93	46.63	54.14	7.51	2.21	2.22
			Drought	51.18	48.64	54.06	5.42	2.50	2.57
	Lignin	% ndf	Control	6.35	5.40	7.12	1.72	5.65	0.72
			Drought	6.37	5.64	7.35	1.71	4.65	0.60
	Cellulose conversion	%	Control	50.37	46.16	52.65	6.50	1.78	1.80
			Drought	54.13	51.36	57.24	5.87	1.99	2.17

CV (%) = coefficient of variation (root-mean-squared error/average × 100%); LSD = least significant difference (0.05).

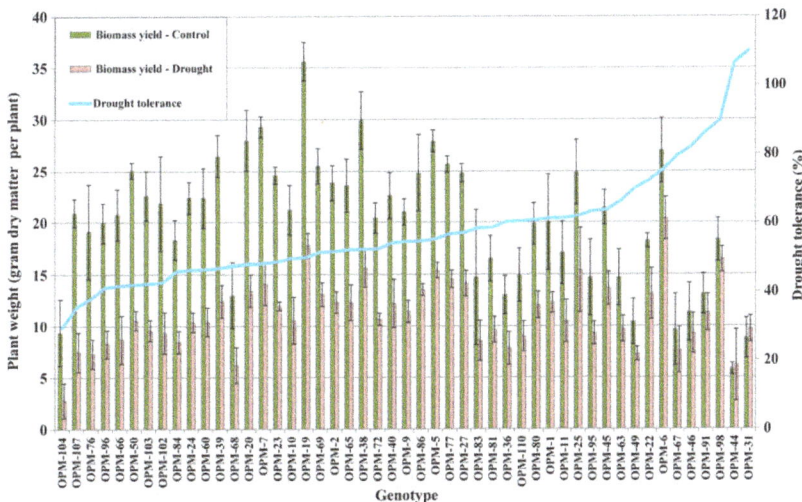

Fig. 2 Plant yield of 49 miscanthus genotypes (expressed in gram dry matter per plant) with varying drought tolerance grown under drought and control conditions. Error bars indicate the standard error of a genotype mean (average of four replicates per genotype per treatment).

70%) in plant weight due to drought treatment. Yields in 39 of the 49 genotypes were reduced by 30–60% following drought treatment, with the majority of the genotypes with a high plant weight under control conditions showing yield reductions of at least 40% under drought (Fig. 2).

Drought stress affects biomass quality in miscanthus

Drought treatment significantly affected most of the biomass quality traits of miscanthus, including cell wall content, cellulosic, and hemicellulosic polysaccharide contents and the efficiency of cellulose conversion (Tables 4 and 5). Stem lignin content, on the other hand, was only moderately affected ($P = 0.015$), and drought stress had no significant effect on lignin in leaf tissues ($P = 0.522$). Significant differences among the set of genotypes were found for all biomass quality characteristics. Furthermore, the effects of drought on biomass quality were more apparent in some genotypes than in others, as indicated by the presence of significant genotype-by-treatment interactions for most traits (Table 4).

Overall, the biochemical composition of the stem samples of drought-treated plants compared to their respective control plants was considerably changed (Table 5). Average cell wall content decreased from 73% to 65% of stem dry matter and from 70% to 63% in leaf dry matter. Average cellulose content decreased in stem tissue, from 51% to 46%, but in leaf tissue remained 43%. In contrast, average content of hemicelluloses increased, from 42% to 47% in stem and from 50% to 51% in leaf tissue. Lignin content was not substantially different between drought-treated and control plants (Table 4), remaining at 6% in leaf and 7% in stem tissue (Table 5).

Genotypic variation for cell wall composition and cellulose conversion was extensive. Generally, genotypic variation in cell wall composition was larger in drought-treated plants compared to control plants and compositional variation between genotypes larger for stem than for leaf tissue. In drought-treated plants, mean cell wall content ranged from 51% to 74% of stem dry matter and 57–70% of leaf dry matter among genotypes (Table 5). Similarly, cellulose content ranged from 39% to 49% in stem and 40% to 45% in leaf, the content of hemicellulosic polysaccharides ranged from 43% to 53% in stem and 49% to 54% in leaf, and lignin content ranged from 5.7% to 8.7% in stem and 5.6% to 7.4% in leaf materials.

Saccharification efficiency was significantly affected by drought treatment. In both stem and leaf materials, considerably higher cellulose conversion efficiencies were achieved in drought-treated plants compared to their respective control plants. Stem cellulose conversion increased from 42% (under control conditions) to 51% (under drought treatment) (Table 5). Similarly, leaf cellulose conversion increased from 50% to 54%. Extensive variation among genotypes was found for cellulose conversion efficiency in both drought-treated and control plants. Stem cellulose conversion ranged from 43% to 62% under drought and from 38% to 51% under control conditions (Table 5, Fig. 3). Less variation was observed in leaf cellulose conversion, but significant genotypic differences were detected (Tables 4 and 5).

Cell wall composition does not play a major role in drought tolerance

To evaluate whether differences existed in response to drought between tolerant genotypes and susceptible genotypes, the top six drought-tolerant (OPM-31, 44, 46, 67, 91, and 98) and top six drought-susceptible (OPM-50, 66, 76, 96, 104, and 107) genotypes were grouped together to compare changes in plant weight and

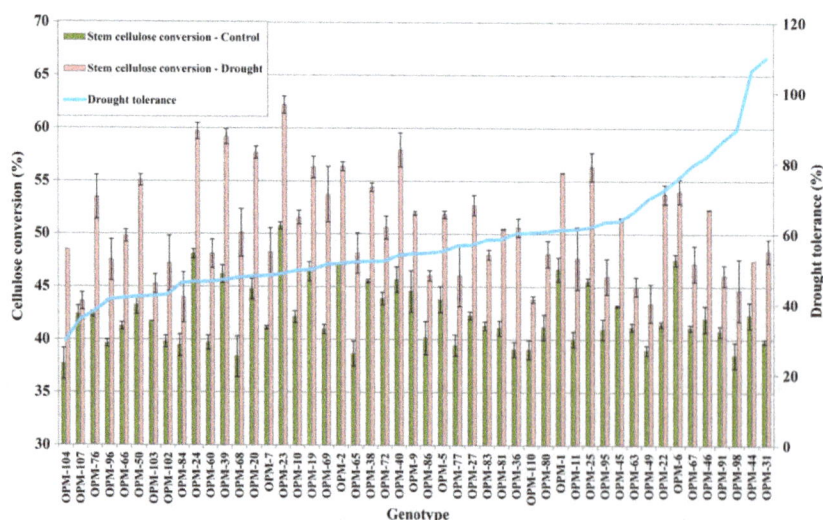

Fig. 3 Cellulose conversion of stem tissues of 49 miscanthus genotypes with varying drought tolerance grown under drought and control conditions. Error bars indicate the standard error of a genotype mean (average of two replicates per genotype per treatment).

biomass quality between the two groups (Fig. 4a, b). Difference in trait means between drought-treated plants and their corresponding control plants is presented and further referred to as the trait name preceded by a 'Δ' symbol. These differences are a measure of the plasticity in cell wall components, with a larger difference in cell wall composition between drought-treated and control plants indicative of greater plasticity. In the tolerant group, hardly any reduction in leaf and stem biomass was observed, whereas the weight of both plant fractions was highly reduced in the susceptible group. The differences in cell wall content and biomass quality between the two groups contrasting for drought tolerance were small. The only significant difference between the two groups was a significantly larger increase in leaf Δlignin (0.71 vs. 0.02) in the tolerant group compared to the susceptible group (Fig. 4a). Between these two extreme groups, cell wall plasticity was found to be highly similar (Fig. 4a, b).

To further investigate interrelations between drought tolerance and cell wall characters, a correlation analysis was performed on the whole set of genotype means of all traits. The primary objective was to investigate whether cell wall composition and cell wall plasticity play a role in tolerance to drought. A few significant trait associations (with low coefficients of determination) were observed between drought tolerance and biomass quality traits, including correlations with leaf cellulose and hemicelluloses content (Fig. 5a, $r^2 = 0.13$ and -0.11, respectively) and leaf Δcellulose, leaf Δlignin, and stem Δlignin (Fig. 5b, $r^2 = -0.08$, 0.21 and 0.10, respectively). No significant correlations were found between drought tolerance and cellulose

conversion. The increase in cellulose conversion in stems of drought-treated plants was highly correlated to Δhemicellulose (Fig. 5c) and to cell wall content (Fig. 5d).

Discussion

Variation in drought tolerance in miscanthus

The extensive variation observed among the evaluated genotypes regarding plant weight under drought stress (2.78–20.38 g plant^{-1}) indicates large genotype differences in vegetative growth vigor under dry cultivation conditions. The average loss in plant weight under drought stress compared to control conditions was considerable (45%); however, the range of variation in drought tolerance among the evaluated genotypes (30–110%) was comprehensive and is evidence of the suitability of this test panel for the experiments that were conducted. This indicates that the genotypes tested may interesting candidates for investigation of mechanisms underlying drought tolerance and could possibly be used in breeding programs.

Some plant defense strategies against the injurious effects of drought, such as dehydration avoidance, are rarely compatible with high biomass yields (Blum, 2005). Drought tolerance and plant yield of the genotypes included in this study were evaluated (Fig. 2). Plants that achieved higher plant weights in drought conditions than in control conditions (drought tolerance >100%) were quite small and had low plant weights in both control and drought conditions. The applied drought treatment was potentially less harsh for small

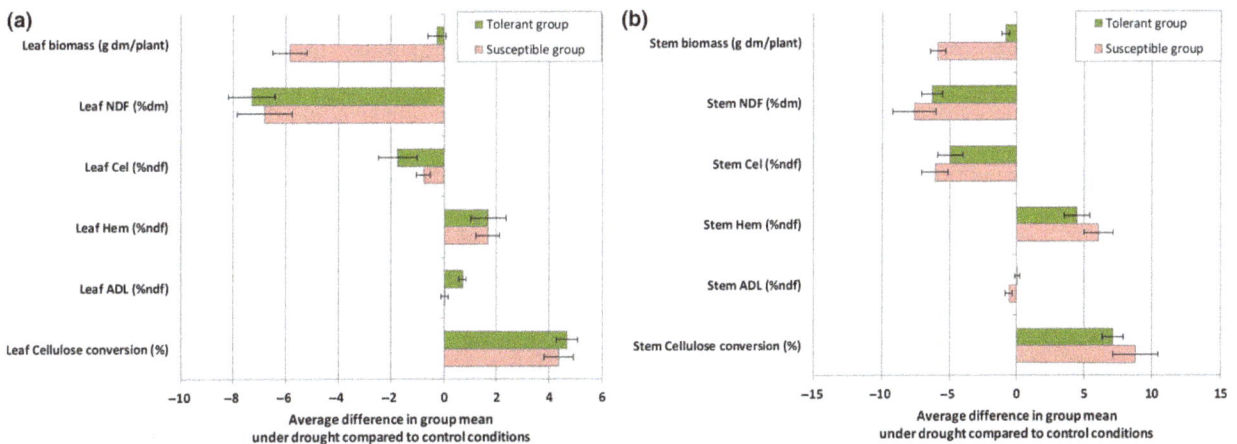

Fig. 4 Change in leaf (a) and stem (b) weight per plant and composition of drought-stressed plants relative to the control plants grouped by tolerance/susceptibility to drought. Unit on x-axis is determined by the unit of the trait on the y-axis. Error bars indicate standard errors on group means ($n = 6$ for tolerant and $n = 6$ for susceptible group). The significance of differences in group means per trait was evaluated by unpaired two-sample t-tests. Group means per trait that have a different suffix are significantly different ($P < 0.05$).

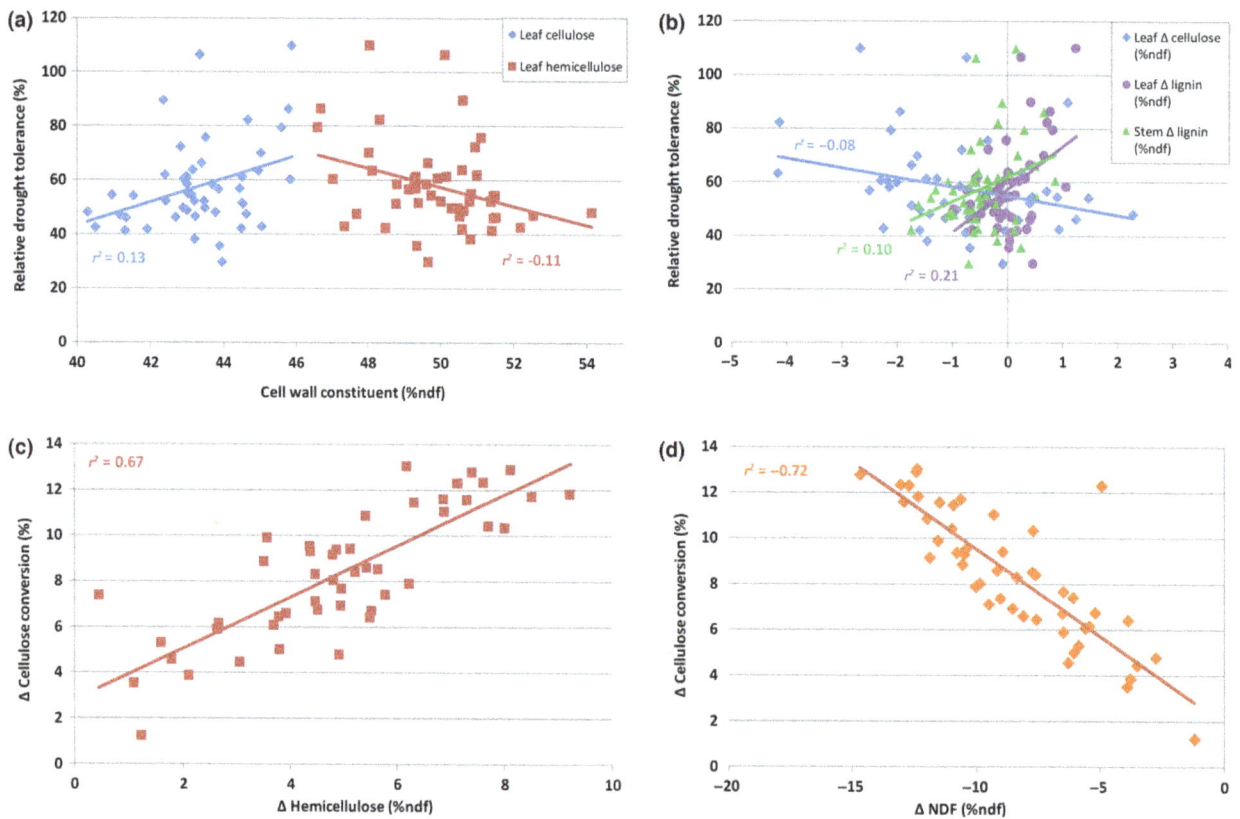

Fig. 5 Correlations between drought tolerance and cell wall composition (a), between drought tolerance and leaf Δcellulose, leaf Δlignin and stem Δlignin (b), between stem Δcellulose conversion and stem Δhemicellulose (c), and between stem Δcellulose conversion and stem ΔNDF (d), where Δ = genotype difference between drought and control conditions.

plants than for large plants; small plants need less water and are less likely to lose water due to a proportionally small leaf surface area (Blum, 2005). However, some genotypes exhibited both a relatively high plant weight and a relatively high drought tolerance, indicating that some genotypes utilize drought tolerance strategies that to some extent could be compatible with high yield. For example, one of the more drought-tolerant (75%) genotypes, OPM-6, in this experiment achieved a plant weight in drought conditions that was similar to the plant weight of *M. × giganteus* (OPM-9) in control conditions (Fig. 2). Extrapolations of the reported plant weights to estimate yield potential under field conditions should be approached with care. The current experiment is more suitable for investigating early vegetative growth than yield potential. Moreover, genotypes that had a relatively low plant weight in this experiment, might still achieve substantial yields under field conditions, perhaps by optimizing planting density. The findings reported here suggest the availability of drought-tolerant varieties in miscanthus germplasm resources that may achieve substantial biomass yields, even under dry cultivation conditions.

Drought reduces cell wall and cellulose content, while increasing hemicellulosic polysaccharides

A key objective of this study was to determine the effects of drought stress on biomass composition and conversion properties, to evaluate whether growing miscanthus under water deficit conditions affects its biomass quality for biofuel production. Biomass composition was substantially affected by drought, with significant reductions in cell wall and cellulose content and a significant increase in hemicellulosic polysaccharides observed in plants grown under drought compared to control conditions (Table 5, Fig. 4a, b). One of the most striking effects of drought was a large decrease in average cell wall content (11.62% in stem and 10.43% in leaf tissue). A drastic reduction in cell wall content was also reported in field-grown *M. × giganteus*, after evaluation of its biomass composition in a year with low precipitation as compared to a year with average precipitation (Emerson *et al.*, 2014).

It was previously shown that cultured tobacco cells subjected to osmotic stress were reduced in size and had thinner cell walls compared to untreated cells (Iraki et al., 1989b). Under normal conditions, cells expand using turgor pressure and cell walls thicken when they no longer need to be elastic to accommodate cell expansion (Lam et al., 2013; Da Costa et al., 2014). During drought, plants have to act to maintain turgor, leading to a stop or slowing down of cell division and cell expansion, which reduces average cell size (Farooq et al., 2009). However, if water would become available again, cell walls need to be able to accommodate cell expansion. Therefore, it is unlikely that the small-sized cells of drought-stressed plants will undergo extensive premature cell wall thickening. Such physiological and developmental processes could explain the lower cell wall content found in drought-stressed plants compared to the control plants.

A physiological explanation for the reduction in cellulose during drought stress may be found in the formation of osmolytes (such as soluble sugars and proline). Osmolytes are solutes formed to aid the maintenance of osmotic equilibrium in the cell under dry growing conditions, and plant stress due to water deficit is associated with a disturbance of the osmotic equilibrium of cells. The production of osmolytes at the expense of cellulose biosynthesis (or financed by cellulose deconstruction) is well reported in the literature (Guenni et al., 2002; Piro et al., 2003; Vincent et al., 2005; Al-Hakimi, 2006; Moore et al., 2008; Hu et al., 2009; Jiang et al., 2012; Meibaum et al., 2012; Emerson et al., 2014; Rakszegi et al., 2014). The fact that in stem tissue, the reduction in cellulose is much more apparent than in leaf tissue may indicate that in most plants, the production of osmolytes in stems is more associated with a concomitant decrease in cell wall cellulose than in leaves.

Lignin content in leaves of drought-treated plants was not significantly different from that of control plants and in stem tissue only a slight decrease in lignin content was observed. Previously a large reduction in lignin content was reported to be one of the side-effects of drought on biomass composition of field-grown M. × giganteus (Emerson et al., 2014). However, there is no consistency among studies in different crops and tissues regarding the effect of drought on lignin content, with some studies reporting an increase in lignin (Guenni et al., 2002; Hu et al., 2009; Jiang et al., 2012; Meibaum et al., 2012) and some reporting a decrease in lignin content (Vincent et al., 2005; Al-Hakimi, 2006). The associations between drought stress and lignin are complex and perhaps influenced by yet uncharacterized factors that may explain discrepancies between studies. The small effect of drought on cell wall lignin content and the large effect on cell wall and cellulose content reported here were consistently observed for a diverse set of genotypes comprising three miscanthus species.

Similarly, inconsistent effects of drought on hemicellulose content were previously reported (Al-Hakimi, 2006; Moore et al., 2008; Jiang et al., 2012; Emerson et al., 2014; Rakszegi et al., 2014), whereas in this study, drought-treated plants of all genotypes consistently had a significantly higher content of hemicellulosic polysaccharides compared to their respective control plants. Some of the discrepancies may also be explained by a difference in the duration of the applied drought treatment. A long-term exposure to drought, such as the treatment applied in this study, challenges plants to alter their cell wall structure to sustain cell expansion with reduced water potential. Hemicelluloses contribute to cell wall rigidity by reinforcing the cell wall matrix through cross-linking to lignin and to cellulose fibers (Le Gall et al., 2015). Lignin also provides cell wall rigidity, but is mostly deposited in mature cells that no longer require the flexibility to accommodate cell expansion (Lam et al., 2013; Da Costa et al., 2014). Compared to lignin, hemicellulose cross-links are more easily broken to ensure cell wall plasticity. An increase in the relative proportion of hemicelluloses might enable cell walls of drought-treated plants to uphold their structural rigidity without compromising plasticity (Le Gall et al., 2015; Tenhaken, 2015).

In this experiment, the effects of drought were evaluated in a controlled glasshouse environment, in which environmental factors other than those related to the drought treatment were reduced to a minimum. Compared to the often contradictory results reported in previous studies regarding the effects of drought on biomass quality, in this study the observed effects were highly consistent for a diverse set of genotypes.

Drought improves saccharification efficiency in miscanthus

In addition to cell wall composition, drought treatment was shown to significantly affect cell wall degradability. Cellulose conversion was substantially increased in biomass samples of drought-stressed plants compared to those of control plants, indicating that available cell wall polysaccharides were more easily released as fermentable sugars by pretreatment and enzymatic saccharification reactions (Fig. 3; Table 5). According to these results, the occurrence of drought during growth of bioenergy feedstocks can have highly beneficial side-effects on the processing efficiency of the biomass for the production of biofuel.

The observed increase in cellulose conversion in drought-treated plants was shown to be highly correlated to an increase in the relative proportion of hemicelluloses (Fig. 5c). It has been reported that the content of hemicelluloses is positively correlated to saccharification efficiency (Xu *et al.*, 2012; Torres *et al.*, 2014; Van Der Weijde *et al.*, 2016). The positive effect of hemicelluloses on cell wall digestibility was associated with a reduction in cellulose crystallinity (Xu *et al.*, 2012). Hemicelluloses, unlike cellulose, are highly branched polysaccharides that form an amorphous network through different types of cross-links (Hatfield *et al.*, 1999; Doblin *et al.*, 2010). Hydrolytic enzymes can more efficiently penetrate the cell wall matrix during enzymatic saccharification and have a higher affinity to the cellulose substrate when the ratio of hemicellulose to cellulose in the cell wall matrix is increased (Xu *et al.*, 2012). This explains how a relative increase in cell wall hemicelluloses in response to drought treatment resulted in a reduction in cell wall recalcitrance to deconstruction.

Saccharification efficiency is often negatively correlated to cell wall content (Jung & Casler, 2006; Torres *et al.*, 2014; Van Der Weijde *et al.*, 2016). The reduction in cell wall content observed in drought-treated plants may be another side-effect of drought that contributes positively to saccharification efficiency. As was reported previously, drought treatment of cultured tobacco cells reduced cell wall thickness (Iraki *et al.*, 1989b). Similarly, the reduction in cell wall content observed in drought-treated miscanthus plants could be due to thinner cell walls, as discussed above. Thinner cell walls, in turn, might be more easily penetrated by hydrolytic enzymes due to increased accessible surface area compared to thicker cell walls. This could provide an explanation for the negative correlation found between ΔNDF and Δcellulose conversion (Fig. 5d). However, microscopic investigations of differences in cell wall thickness were beyond the scope of this study.

Overall, growing miscanthus under drought conditions substantially affected biomass composition and saccharification efficiency, with cell walls of plants grown under drought conditions being more readily deconstructed during mild alkaline pretreatment and enzymatic saccharification. Hereby, the occurrence of drought during the growth of miscanthus biomass may contribute beneficially to its compositional quality for biofuel production, through the enhanced efficiency of releasing cell wall polysaccharides as fermentable sugars during processing. Importantly, this effect appears to occur even in genotypes that maintained high biomass yield despite drought.

Implications for breeding drought-tolerant varieties for biofuel production

These results show that genotypic variation for drought tolerance exists within miscanthus germplasm resources and that the development of drought-tolerant varieties that produce substantial biomass yields should be possible. Drought stress significantly reduced cell wall and cellulose content, which reduces the amount of structural sugars available per unit of biomass. This effect was previously reported to have a negative impact on theoretical ethanol yields of *M. × giganteus* grown during a year with limited compared to a year with average precipitation (Emerson *et al.*, 2014). However, in the current study, it was shown that drought also substantially increased cellulose conversion, which considerably enhances the techno-economic performance of bioconversion processes (Wyman, 2007; Himmel & Picataggio, 2008; Zhao *et al.*, 2012; Torres *et al.*, 2016). The occurrence of drought during the growth of miscanthus may thus beneficially affect biomass quality, through the substantial increase in cellulose conversion efficiency. The question that remains is whether in terms of the total ethanol yield per hectare, the reductions in cellulose and biomass yield associated with drought are compensated for by an increase in conversion efficiency. However, the selection of drought-tolerant high-yielding genotypes should minimize any penalty.

The absence of strong correlations among drought tolerance and compositional characters in the set of genotypes and the observation that in the tolerant group similar differences in biomass composition were observed as in the susceptible group are strong indicators that biomass quality characteristics and drought tolerance are largely under independent genetic control. Hence, drought tolerance and biomass quality are not mutually exclusive breeding goals and biomass quality can be selected for independently and simultaneously, without adversely affecting drought tolerance and vice versa. The wide range of variation for the evaluated traits observed among this set of miscanthus genotypes provides evidence of ample scope in the miscanthus germplasm pool for breeders to improve both drought tolerance and biomass composition to supply optimized varieties for the biofuel industry.

Acknowledgements

The research leading to these results has received funding from the European Union Seventh Framework Programme (FP7/2007–2013) under grant agreement no. 289159. We gratefully acknowledge the contributions of Kai Schwartz, Heike Meyer, and Claire Lessa Alvim Kamei for bringing the genotypes into *in vitro* culture. The help of Chris Maliepaard with statistical analyses and Annemarie Dechesne during laboratory analyses is gratefully acknowledged. We further acknowledge Genencor International B.V./DuPont Industrial Biosciences for kindly supplying us with their commercial Accellerase 1500 enzyme cocktail used in this study.

References

Al-Hakimi AMA (2006) Counteraction of drought stress on soybean plants by seed soaking in salicylic acid. *International Journal of Botany*, **2**, 421–426.

Blum A (2005) Drought resistance, water-use efficiency, and yield potential – are they compatible, dissonant, or mutually exclusive? *Crop and Pasture Science*, **56**, 1159–1168.

Bullard M (2001) 8. Economics of Miscanthus production. *Miscanthus for Energy and Fibre*, **155**, 155–171.

Byrt CS, Grof CPL, Furbank RT (2011) C4 Plants as biofuel feedstocks: optimising biomass production and feedstock quality from a lignocellulosic perspective. *Journal of Integrative Plant Biology*, **53**, 120–135.

Chaves MM, Maroco JP, Pereira JS (2003) Understanding plant responses to drought – from genes to the whole plant. *Functional Plant Biology*, **30**, 239–264.

Clifton-Brown JC, Lewandowski I, Bangerth F, Jones MB (2002) Comparative responses to water stress in stay-green, rapid- and slow senescing genotypes of the biomass crop, Miscanthus. *New Phytologist*, **154**, 335–345.

Clifton-Brown JC, Chiang YC, Hodkinson TR (2008) Miscanthus: genetic resources and breeding potential to enhance bioenergy production. In: *Genetic Improvement of Bioenergy Crops* (ed. Vermerris W), pp. 273–294. Springer Science + Business Media LLC, New York.

Da Costa RMF, Lee SJ, Allison GG, Hazen SP, Winters A, Bosch M (2014) Genotype, development and tissue-derived variation of cell-wall properties in the lignocellulosic energy crop Miscanthus. *Annals of Botany*, **114**, 1265–1277.

Dai A (2013) Increasing drought under global warming in observations and models. *Nature Climate Change*, **3**, 52–58.

Dien BS (2010) Mass balances and analytical methods for biomass pretreatment experiments. In: *Biomass to Biofuels: Strategies for Global Industries* (eds Vertès AA, Qureshi N, Blaschek HP, Yukawa H), pp. 213–232. Blackwell, Oxford, UK.

Doblin MS, Pettolino F, Bacic A (2010) Plant cell walls: the skeleton of the plant world. *Functional Plant Biology*, **37**, 357–381.

Emerson R, Hoover A, Ray A et al. (2014) Drought effects on composition and yield for corn stover, mixed grasses, and Miscanthus as bioenergy feedstocks. *Biofuels*, **5**, 275–291.

Farooq M, Wahid A, Kobayashi N, Fujita D, Basra S (2009) Plant drought stress: effects, mechanisms and management. In: *Sustainable Agriculture*, pp. 153–188. Springer, Dordrecht.

Frei M (2013) Lignin: characterization of a multifaceted crop component. *The Scientific World Journal*, 2013, Article ID 436517.

Goering HK, Van Soest PJ (1970) Forage fiber analyses (apparatus, reagents, procedures, and some applications). In: *Agricultural Handbook No. 379*, pp. 1–20. USDA-ARS, Washington, DC, USA.

Guenni O, Marín D, Baruch Z (2002) Responses to drought of five *Brachiaria* species. I. Biomass production, leaf growth, root distribution, water use and forage quality. *Plant and Soil*, **243**, 229–241.

Hatfield RD, Ralph J, Grabber JH (1999) Cell wall cross-linking by ferulates and diferulates in grasses[1]. *Journal of the Science of Food and Agriculture*, **79**, 403–407.

Heaton EA, Dohleman FG, Miguez AF et al. (2010) Miscanthus: a promising biomass crop. *Advances in Botanical Research*, **56**, 76–137.

Himmel ME, Picataggio SK (2008) Our challenge is to acquire deeper understanding of biomass recalcitrance and conversion. In: *Biomass Recalcitrance* (ed. Himmel ME), pp. 1–6. Blackwell Publishing Ltd, Oxford, UK.

Hu Y, Li WC, Xu Y, Li G, Liao Y, Fu F-L (2009) Differential expression of candidate genes for lignin biosynthesis under drought stress in maize leaves. *Journal of Applied Genetics*, **50**, 213–223.

Iraki NM, Bressan RA, Hasegawa P, Carpita NC (1989a) Alteration of the physical and chemical structure of the primary cell wall of growth-limited plant cells adapted to osmotic stress. *Plant Physiology*, **91**, 39–47.

Iraki NM, Singh N, Bressan RA, Carpita NC (1989b) Cell walls of tobacco cells and changes in composition associated with reduced growth upon adaptation to water and saline stress. *Plant Physiology*, **91**, 48–53.

Jiang Y, Yao Y, Wang Y (2012) Physiological response, cell wall components, and gene expression of switchgrass under short-term drought stress and recovery. *Crop Science*, **52**, 2718–2727.

Jones MB, Walsh M (2001) *Miscanthus for Energy and Fibre*, James & James, London.

Jung H, Casler M (2006) Maize stem tissues: impact of development on cell wall degradability. *Crop Science*, **46**, 1801–1809.

Lam MSJWQ, Martinez Y, Barbier O, Jauneau A, Pichon M (2013) Maize cell wall degradability, from whole plant to tissue level: different scales of complexity. *Maydica*, **58**, 103–110.

Le Gall H, Philippe F, Domon J-M, Gillet F, Pelloux J, Rayon C (2015) Cell wall metabolism in response to abiotic stress. *Plants*, **4**, 112–166.

Meibaum B, Riede S, Schröder B, Manderscheid R, Weigel H-J, Breves G (2012) Elevated CO_2 and drought stress effects on the chemical composition of maize plants, their ruminal fermentation and microbial diversity in vitro. *Archives of Animal Nutrition*, **66**, 473–489.

Moore JP, Vicré-Gibouin M, Farrant JM, Driouich A (2008) Adaptations of higher plant cell walls to water loss: drought vs desiccation. *Physiologia Plantarum*, **134**, 237–245.

Moura JCMS, Bonine CAV, De Oliveira Fernandes Viana J, Dornelas MC, Mazzafera P (2010) Abiotic and biotic stresses and changes in the lignin content and composition in plants. *Journal of Integrative Plant Biology*, **52**, 360–376.

Pauly M, Keegstra K (2010) Plant cell wall polymers as precursors for biofuels. *Current Opinion in Plant Biology*, **13**, 304–311.

Perlack RD, Wright LL, Turhollow AF, Graham RL, Stokes BJ, Erbach DC (2005) *Biomass as Feedstock for a Bioenergy and Bioproducts Industry: The Technical Feasibility of a Billion-Ton Annual Supply.* Oak Ridge National Laboratory, Oak Ridge.

Piro G, Leucci MR, Waldron K, Dalessandro G (2003) Exposure to water stress causes changes in the biosynthesis of cell wall polysaccharides in roots of wheat cultivars varying in drought tolerance. *Plant Science*, **165**, 559–569.

Quinn LD, Straker KC, Guo J et al. (2015) Stress-tolerant feedstocks for sustainable bioenergy production on marginal land. *BioEnergy Research*, **8**, 1081–1100.

Rakszegi M, Lovegrove A, Balla K, Láng L, Bedő Z, Veisz O, Shewry PR (2014) Effect of heat and drought stress on the structure and composition of arabinoxylan and β-glucan in wheat grain. *Carbohydrate Polymers*, **102**, 557–565.

Selig M, Weiss N, Ji Y (2008) Enzymatic saccharification of lignocellulosic biomass. In: *Laboratory Analytical Procedure*, pp. 1–8. National Renewable Energy Laboratory, Golden, CO.

Sheffield J, Wood E (2008) Projected changes in drought occurrence under future global warming from multi-model, multi-scenario, IPCC AR4 simulations. *Climate Dynamics*, **31**, 79–105.

Shenk JS, Westerhaus MO (1991) Populations structuring of near infrared spectra and modified partial least squares regression. *Crop Science*, **31**, 1548–1555.

Tenhaken R (2015) Cell wall remodeling under abiotic stress. *Frontiers in Plant Science*, **5**, 771.

Torres A, Noordam-Boot CM, Dolstra O et al. (2014) Cell wall diversity in forage maize: genetic complexity and bioenergy potential. *BioEnergy Research*, **8**, 187–202.

Torres AF, Slegers PM, Noordam-Boot CMM et al. (2016) Maize feedstocks with improved digestibility reduce the costs and environmental impacts of biomass pretreatment and saccharification. *Biotechnology for Biofuels*, **9**, 1–15.

Van Der Weijde T, Alvim Kamei CL, Torres AF, Vermerris W, Dolstra O, Visser RGF, Trindade LM (2013) The potential of C4 grasses for cellulosic biofuel production. *Frontiers in Plant Science*, **4**, 1–18.

Van Der Weijde T, Torres AF, Dolstra O, Dechesne A, Visser RGF, Trindade LM (2016) Impact of different lignin fractions on saccharification efficiency in diverse species of the bioenergy crop Miscanthus. *BioEnergy Research*, **9**, 146–156.

Van Soest PJ (1967) Development of a comprehensive system of feed analyses and its application to forages. *Journal of Animal Science*, **26**, 119–128.

Vincent D, Lapierre C, Pollet B, Cornic G, Negroni L, Zivy M (2005) Water deficits affect caffeate O-methyltransferase, lignification, and related enzymes in maize leaves. A proteomic investigation. *Plant Physiology*, **137**, 949–960.

Wyman CE (2007) What is (and is not) vital to advancing cellulosic ethanol. *Trends in Biotechnology*, **25**, 153–157.

Xu N, Zhang W, Ren S et al. (2012) Hemicelluloses negatively affect lignocellulose crystallinity for high biomass digestibility under NaOH and H_2SO_4 pretreatments in Miscanthus. *Biotechnology for Biofuels*, **5**, 1–12.

Zhao X, Zhang L, Liu D (2012) Biomass recalcitrance. Part I: the chemical compositions and physical structures affecting the enzymatic hydrolysis of lignocellulose. *Biofuels, Bioproducts and Biorefining*, **6**, 465–482.

13

Effects of biochar application on root traits: a meta-analysis

YANGZHOU XIANG[1],[#], QI DENG[2],[3],[#] (iD), HONGLANG DUAN[4] and YING GUO[1]

[1]*Guizhou Institute of Forest Inventory and Planning, Guiyang 550003, China*, [2]*Key Laboratory of Aquatic Botany and Watershed Ecology, Wuhan Botanical Garden, Chinese Academy of Sciences, Wuhan 430074, China*, [3]*Department of Biological Sciences, Tennessee State University, Nashville, TN 37209, USA*, [4]*Jiangxi Provincial Key Laboratory for Restoration of Degraded Ecosystems & Watershed Ecohydrology, Nanchang Institute of Technology, Nanchang 330099, China*

Abstract

Roots are the interfaces between biochar particles and growing plants. Biochar application may alter root growth and traits and thereby affect plant performance. However, a comprehensive understanding of the effects of biochar on root traits is lacking. We conducted a meta-analysis with 2108 paired observations from 136 articles to evaluate the responses of root traits associated with 13 variables under biochar application. Overall, biochar application increased root biomass (+32%), root volume (+29%) and surface area (39%). The biochar-induced increases in root length (+52%) and number of root tips (+17%) were much larger than the increase in root diameter (+9.9%); this result suggests that biochar application benefits root morphological development to alleviate plant nutrient and water deficiency rather than to maximize biomass accumulation. Biochar application did not change root N concentration but significantly increased root P concentration (+22%), particularly when combined with N fertilization. Biochar application also affected root-associated microbes and significantly increased the number of root nodules (+25%). The responses of root traits to biochar application were generally greater in annual plants than in perennial plants and were affected by soil texture and pH values. Moreover, it appears that biochar production process (pyrolysis temperature and time) plays a more important role in regulating root growth than does biochar source. Together, findings obtained from this meta-analysis may have significant implications for the future sustainable development of biochar management to improve plant growth and functioning.

Keywords: biochar, fertilization, meta-analysis, plant growth, root functioning, root morphology

Introduction

Biochar is a predominantly stable, recalcitrant organic carbon (C) compound that is produced from biomass via pyrolysis (Lehmann, 2007; Laird *et al.*, 2009). Due to the potential to sequester C in the soil, biochar application is currently considered as a means to help mitigate greenhouse gas (GHG) emissions and climate change (Marris, 2006; Woolf *et al.*, 2010). Recent reviews have highlighted that biochar application can also stimulate plant growth and yield (Jeffery *et al.*, 2011; Biederman & Harpole, 2013; Liu *et al.*, 2013) and thereby enhance the sequestration of carbon dioxide (CO_2) from the atmosphere (Lehmann *et al.*, 2006). Plant root systems play an important role in plant growth and soil C sequestration (Matamala *et al.*, 2003;

Nie *et al.*, 2013) because they not only take up soil nutrients and water to support plant production but also transport photosynthetically fixed C to soil organic matter pools (Jackson *et al.*, 1997; Li *et al.*, 2015; Peng *et al.*, 2017). Biochar application may have significant effects on plant root morphology and functioning because biochar particles contact plant roots directly (Prendergast-Miller *et al.*, 2014). Therefore, it is critical to determine how root traits respond to biochar application for sustainable biochar management (Laird, 2008; Jeffery *et al.*, 2015).

Numerous case studies have been conducted to examine how plant roots respond to biochar application (Rillig *et al.*, 2010; Prendergast-Miller *et al.*, 2011; Brennan *et al.*, 2014; Vanek & Lehmann, 2015). Some root traits, including root biomass, morphology (Prendergast-Miller *et al.*, 2011; Brennan *et al.*, 2014; Keith *et al.*, 2015), nutrient concentration, and root-associated microbes (Rondon *et al.*, 2007; Rillig *et al.*, 2010; Vanek & Lehmann, 2015), can be significantly influenced by biochar application. However, due to differential

---

[#]These authors contributed equally to this work.

Correspondence: Qi Deng
e-mail: qdeng@tnstate.edu

research objectives, root traits in these case studies are often studied independently (Prendergast-Miller *et al.*, 2014; Jeffery *et al.*, 2015). In addition, root traits are usually specific and play different roles in plant growth (Jackson *et al.*, 1997; Matamala *et al.*, 2003; Nie *et al.*, 2013). For example, root length is usually assumed to be proportional to water or nutrient acquisition, while root diameter is thought to be beneficial for biomass accumulation (Eissenstat & Yanai, 1997). Thus, it remains unclear how these root traits respond differently to biochar application. To better understand the underlying mechanism of root growth in response to biochar application, quantitative synthesis across these root traits is required (Lehmann *et al.*, 2011; Jeffery *et al.*, 2015).

To date, the effects of biochar application on root traits remain controversial and highly variable. For example, root biomass may increase (Prendergast-Miller *et al.*, 2011; Varela Milla *et al.*, 2013), decrease (Aguilar-Chávez *et al.*, 2012; Van De Voorde *et al.*, 2014) or remain relatively stable (Macdonald *et al.*, 2014; Keith *et al.*, 2015) under biochar application. These highly diverse results are not surprising because multiple factors and processes are involved in root responses (Prendergast-Miller *et al.*, 2014). For instance, root establishment in the soils can be enhanced by biochar addition (Brennan *et al.*, 2014). The choice of biochar type is also important as biochar generated from different materials or pyrolysis conditions varies significantly in structure, nutrient content, pH, and phenolic content (Novak *et al.*, 2009; Lehmann *et al.*, 2011). The characteristics of biochar, as well as its application rate and cumulative amount, may affect the soil environment and thereby alter root traits (Taghizadeh-Toosi *et al.*, 2012; Reverchon *et al.*, 2014; Reibe *et al.*, 2015). Biochar application promotes plant growth mainly by improving characteristics of the soil environment, such as nutrient status, pH and cation-exchange capacity (CEC) (Lehmann *et al.*, 2011; Noguera *et al.*, 2012; Vanek & Lehmann, 2015). Consequently, the responses of root traits to biochar application may depend on soil conditions (Macdonald *et al.*, 2014; Olmo *et al.*, 2016). In addition, biochar is often applied with fertilizer to the soil; this combined application may interactively regulate root growth (Alburquerque *et al.*, 2015). However, to our knowledge, no synthesis has revealed any general patterns of responses of root traits to biochar application. The lack of a comprehensive synthesis significantly prevents biochar from being widely popularized as a highly efficient, sustainable soil management practice for food security under climate change (Jeffery *et al.*, 2015).

The objective of this study was to quantitatively evaluate biochar-induced changes in root traits and their potentially determining factors. We compiled a large global dataset of root traits including 13 variables totaling 2108 paired observations from 136 articles and conducted a comprehensive meta-analysis. The two major questions we aimed to answer here were as follows: (i) how do root traits change with biochar application and (ii) how do fertilization, experimental method and duration, plant species, soil conditions, biochar materials, and application rate affect the response of root traits? More specifically, we hypothesized that (i) biochar application would improve root growth and morphology with an overall greater increase in root length than root diameter in response to biochar-induced rapid plant growth and an increased demand for nutrients and water; (ii) the responses of root traits to biochar application would vary with fertilizer inputs; and (iii) experimental method and duration, plant species, soil conditions, biochar materials, and application rate would contribute significantly to variation in root trait responses to biochar application.

Materials and methods

Dataset assembly

We searched journal articles that reported root trait responses to biochar application using ISI Web of Science (Thomson Reuters, New York, NY, USA) and Google and Google Scholar (Google Inc., Mountain View, CA, USA). Searches included combinations of the terms 'root', 'belowground*', 'biochar', 'char', 'black carbon', 'charcoal', 'pyrogenic organic matter', and 'agchar'. We also screened previous reviews and meta-analyses that investigated the impact of biochar application on plant growth (Atkinson *et al.*, 2010; Jeffery *et al.*, 2011; Biederman & Harpole, 2013; Liu *et al.*, 2013) for additional sources.

To avoid bias in reference selection, the following four criteria were applied: (i) studies were selected in which at least one of the target variables (root traits) was measured under both biochar treatment and control; (ii) the initial environmental conditions, plant species, and soil properties of each experiment were the same, and experiments were performed at the same temporal and spatial scales in the control and treatment plots; (iii) to avoid possible interactive effects, studies that included soil contaminants, tested allelopathic interference, or contained other treatments (except for fertilization in both control and treatment plots because combined application of biochar plus fertilizer was common) were excluded; and (iv) studies containing multiple levels of biochar application, biochar types, soil conditions, plant species, or fertilization statuses were treated as multiple independent studies. Finally, our searches yielded a total of 2108 paired observations from 136 useful articles (Table S1, Appendix S1). The compiled dataset of root traits included 13 variables [root biomass, root : shoot ratio, root volume, root surface area, root diameter, root length, specific root length, number of root tips, number of nodules, fungal colonization, root length colonized, and root nitrogen (N) and phosphorus (P) concentrations] (Table S2).

Most of the data were either obtained from tables or extracted from figures using the GETDATA GRAPH DIGITIZER (version 2.24, http://www.getdata-graph-digitizer.com/). Additionally, we calculated some indices such as root : shoot ratio (root biomass/shoot biomass) and specific root length (root length/root biomass) when data were available. All of the data represented the entire root level, as only one study reported fine root data (George et al., 2016).

We also collected background information from the selected studies, including data source, location (country), experimental type (field and pot), experimental duration, fertilization status (fertilized or unfertilized), plant functional group (life form, life cycle duration, or legumes), soil conditions (soil type, pH, total C, total N, and C/N ratio), and biochar characteristics (biochar feedstock types, rate of biochar application, temperature and time of pyrolysis, pH, and C/N ratio) (Table S1). The plant life forms included crop, grass, and woody plants. The examined plants were also assigned as annual or perennial based on their life cycle duration. The biochar feedstock types included crop, grass, woody plants, manure, and waste. The rate of biochar application was transformed to mass per area (expressed as t ha^{-1}) according to the soil layer reported in each study and the soil bulk density (BD). If the BD was not directly provided in the studies, we assumed a soil BD of 1.5 g cm^{-3} (Biederman & Harpole, 2013). Similarly, we assumed a soil layer of 20 cm when the soil depth was not provided (Liu et al., 2016). Soil texture was divided into sand, loam, and clay. The variables listed as background information were used as factors (either categorical or continuous) explaining the variation in root traits in response to biochar application.

Statistical analysis

We quantified the effects of biochar on root traits by calculating the natural log of the response ratio (RR), which is a metric commonly used in meta-analyses (Hedges & Olkin, 1985; Luo et al., 2006; Deng et al., 2015):

$$RR = \ln\left(\frac{\overline{X}_t}{\overline{X}_c}\right) = \ln(\overline{X}_t) - \ln(\overline{X}_c) \tag{1}$$

where RR is the ratio of the mean value of the chosen variable in the biochar treatment group $\overline{X}_t$ to that in the control group $\overline{X}_c$ and is an index of the effect of the experimental treatment on the target variable. The variance (v) of each individual RR is then estimated as:

$$v = \frac{S_t^2}{n_t \times X_t^2} + \frac{S_c^2}{n_c \times X_c^2} \tag{2}$$

where n_t and n_c are the sample sizes of the variable in the treatment and control groups, and S_t and S_c are the standard deviations for the treatment and control groups, respectively. When standard error, standard deviation, or confidence interval (CI) was not provided, we assumed a standard deviation of one-tenth of the mean value (Luo et al., 2006; Deng et al., 2015). A weighted RR (RR$_{++}$) and a 95% CI were calculated from the

individual RR by giving greater weight (W) to study estimates of which have greater precision (lower v) (Hedges & Olkin, 1985; Luo et al., 2006; Deng et al., 2015).

$$W = \frac{1}{v} \tag{3}$$

$$RR_{++} = \frac{\sum_{i=1}^{m}\sum_{j=1}^{k} W_{ij}RR_{ij}}{\sum_{i=1}^{m}\sum_{j=1}^{k} W_{ij}} \tag{4}$$

$$S(RR_{++}) = \sqrt{\frac{1}{\sum_{i=1}^{m}\sum_{j=1}^{w} W_{ij}}} \tag{5}$$

$$95\% \text{ CI} = RR_{++} \pm 1.96S(RR_{++}) \tag{6}$$

where m is the number of groups and k is the number of comparisons in the ith group. The treatment effect of biochar was considered significant if the 95% CI of the mean RR did not overlap with zero (Luo et al., 2006; Deng et al., 2015). The mean RR$_{++}$ and 95% CIs were then transformed back (i.e., exponentially transformed) and presented as percentage change.

To test the reliability of the data used in the meta-analysis, the frequency distribution of the individual RR was tested by a Normal test and fitted by a Gaussian function (Fig. S1) using Eqn (7).

$$y = \alpha \exp\left(-\frac{(x - \mu)^2}{2\sigma}\right) \tag{7}$$

where x is the RR of a variable; y is the frequency (i.e., number of RR values); α is a coefficient showing the expected number of RR values at $x = \mu$; and μ and σ are the mean and variance of the frequency distributions of RR, respectively.

To examine whether the mean RR$_{++}$ of root traits differed among treatments, the total heterogeneity among groups (Q_{total}) was partitioned into within-group heterogeneity (Q_w) and between-group heterogeneity (Q_B). A significant Q_B suggested that the mean RR$_{++}$ differed among these categorical factors (Hedges et al., 1999). When Q_B was not significant, there was no statistical justification for the further subdivision of the data. If the 95% CI of one group did not overlap with another group within a categorical factor, there was a significant difference between these two groups (Luo et al., 2006; Deng et al., 2015; He et al., 2016). A random-effect model was used to explore the continuous factors that may explain the response of root traits to biochar application. We also conducted a meta-regression analysis to examine the relationships between the RRs and the continuous factors (Luo et al., 2006; Deng et al., 2015; He et al., 2016).

To test the publication bias, funnel plots were presented as scatter plots of the RRs against their standard errors (Egger et al., 1997). In the absence of publication bias, studies should be distributed symmetrically in a 'funnel' shape around the mean RR. The potential asymmetry of the funnel plot was assessed by Egger's regression (Borenstein, 2005). If the mean RR was significantly different from zero and if publication bias existed, we further calculated the fail-safe number using a weighted method to estimate whether our conclusion was affected by the nonpublished data (Rosenberg, 2005). If the fail-safe number was over $5n + 10$ (where n is the number of cases in the analysis), we concluded that our result strongly suggested against publication bias. Otherwise, significant publication bias for the analysis existed.

All data were analyzed using SAS software (SAS Institute Inc., Cary, NC, USA), and the results were considered significant at $\alpha = 0.05$. The graphs were drawn with SIGMAPLOT software (SigmaPlot 12.5 for windows; Systat Software Inc., San Jose, CA, USA).

Results

The results obtained when considering the entire dataset suggested that biochar application significantly increased root biomass, root volume, and surface area by an average of 32% (95% CI: +26%, +37%), 29% (95% CI: +13%, +48%), and 39% (95% CI: +25%, +54%), respectively (Fig. 1a). Biochar application also increased the root diameter, root length, and number of root tips by an average of +9.9% (95% CI: +1.8%, +19%), 52% (95% CI: +42%, +63%), and 17% (95% CI: +9.1%, +26%), respectively (Fig. 1a). However, significant publication bias on root diameter was suggested by Egger's regression and Rosenthal's fail-safe number (Table S3). The root : shoot ratio was not significantly altered by biochar application, while the specific root length was significantly enhanced by 17% (95% CI: +7.7%, +27%) (Fig. 1a). Biochar application did not significantly change the root N concentration (+5.5%; 95% CI: −5.1%, +17%) but significantly increased the root P concentration (+22%; 95% CI: +14%, +30%) (Fig. 1b). Additionally,

biochar application significantly increased the number of root nodules and root length colonized by 25% (95% CI: +15%, +36%) and 35% (95% CI: +31%, +61%), respectively, but had no significant effects on fungal colonization (−4.0%; 95% CI: −11%, +4.0%) (Fig. 1c).

Biochar-induced changes in root biomass, morphology, and root N concentration were not significantly different between unfertilized and fertilized soil (Fig. 2a, Table S4), while responses of root P concentration varied (Table S4). Plant P concentration increased by 33% (95% CI: +26%, +40%) in fertilized soil; this increase was greater than that in unfertilized soil (18%, 95% CI: +13%, +24%) (Fig. 2b). Nonetheless, due to insufficient sample sizes ($n < 5$), it remained unclear whether fertilization could affect the responses of root-associated microbes to biochar application. By contrast, experimental duration and type appear to have little effect on the responses of root traits to biochar application (Table S4).

The effects of biochar application on root biomass and length varied remarkably among plant functional groups (Table S4). For instance, the RRs of root biomass (41%; 95% CI: +34%, +49%), root length (24%; 95% CI: +17%, +32%), and specific root length (60%; 95% CI: +53%, +67%) in annual plants were higher than those in perennial plants (Fig. 3a–c). The biochar-induced increase in root biomass was significantly higher for legumes (45%, 95% CI: +36%, +56%) than for non-legumes (25%, 95% CI: +18%, +32%) (Fig. 3d). In addition, the root length of crops (62%; 95% CI: +55%, +71%) under biochar treatment increased more than did the root length of trees (Fig. 3e).

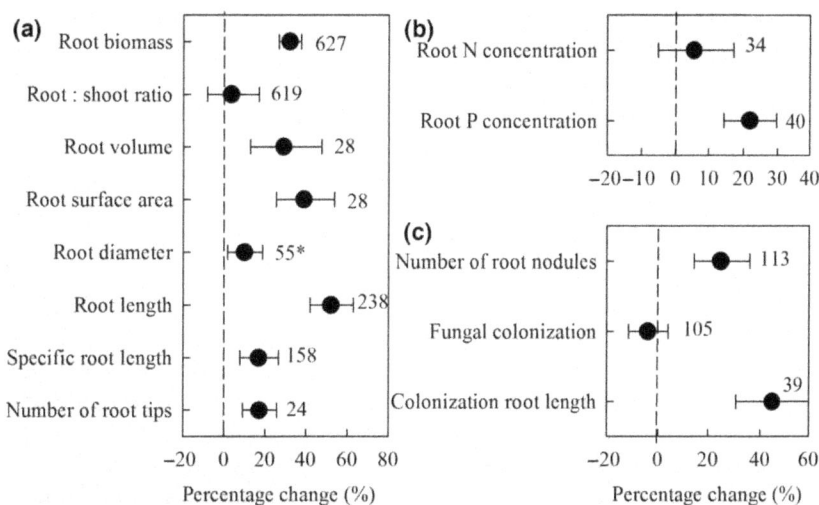

Fig. 1 Responses of root traits to biochar application. (a) Percentage change (%) of root biomass and morphology in response to biochar application. (b) Percentage change (%) of root nutrient concentrations in response to biochar application. (c) Percentage change (%) of root-associated microbes in response to biochar application. Error bars represent 95% bootstrap confidence intervals. The sample size for each variable is shown next to the error bar. *Significant publication bias for the analysis suggested by Egger's regression and Rosenthal's fail-safe number.

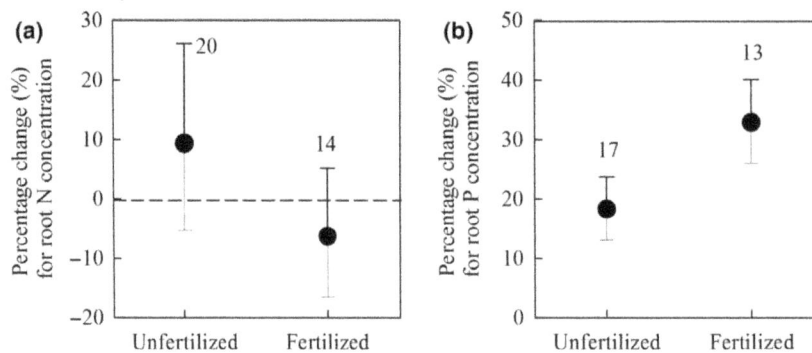

Fig. 2 Percentage change (%) in (a) root N concentration and (b) root P concentration in unfertilized and fertilized soils in response to biochar application. Error bars represent 95% confidence intervals. The sample size for each variable is shown next to the error bar. The statistical significance of the effects of fertilization is evaluated in Table S4.

The soil pH value significantly affected the responses of root P concentration, number of root nodules, and root length to biochar application (Table S4). As the soil pH value at the control sites increased, the log RR of the number of root nodules decreased (Fig. 4a). A negative linear relationship between the log RRs of root length and the soil pH value at the control sites was also observed (Fig. 4b). Only soil type had significant effects on the response of root biomass to biochar application (Table S4); root biomass increased more in sandy soils (42%, 95% CI: +34%, +51%) than in loamy soils (20%, 95% CI: +14%, +27%) (Fig. 4c).

There was no significant effect of biochar feedstock type, pH, or C : N ratio on the responses of root traits to biochar application (Table S4). Conversely, significantly positive linear correlations between pyrolysis temperature and the log RR of root length or specific root length were observed (Fig. 5a, b). Biochar application with faster pyrolysis generally increased root length and specific root length more strongly than did biochar application with slower pyrolysis (Fig. 5c, d).

Discussion

As hypothesized, our meta-analysis demonstrated that biochar application significantly enhanced root growth, which included greater root biomass, root volume, and surface area (Fig. 1a). The increased root biomass contradicts a recent meta-analysis reported by Biederman & Harpole (2013) that found that biochar application had no significant effect on root biomass. This inconsistency may be due to the different sample sizes between our study and their study (627 vs. 28, respectively). Our larger dataset may be more representative. The unchanged root : shoot ratio suggests that biomass allocation was not altered and thus that root growth is coordinated with shoot growth (Jeffery et al., 2011; Biederman & Harpole, 2013; Liu et al., 2013).

Despite the increased root biomass, our results suggested that plants under biochar application tended to invest their root biomass more efficiently to absorb soil water and nutrients rather than accumulate root biomass. This result is not surprising because biochar application stimulates plant growth and increases the demand for nutrients and water (Jeffery et al., 2011; Biederman & Harpole, 2013; Liu et al., 2013). In addition, biochar tends to absorb nutrient ions, particularly inorganic N, which may result in greater nutrient deficiency (Steiner et al., 2008; Clough et al., 2013). As a result, root length (+52%) under biochar application increased more than did the root diameter (+9.9%), which resulted in an overall enhancement in specific root length (Fig. 1a). The increased root length and specific root length suggest that biochar application is beneficial because it enlarges the plant rhizosphere to absorb water and nutrients that otherwise could not be reached by the roots (Eissenstat, 1992; Prendergast-Miller et al., 2014). Moreover, the increased number of root tips under biochar application could further extend root surface area and thus accelerate the exchange rates of resources at the plant–soil interface (Eissenstat & Yanai, 1997). The improvement in root morphological development would benefit plants grown under biochar application by alleviating nutrient and water deficiency.

Although biochar application improved root morphological development by alleviating nutrient deficiency, its effects on different nutrient elements appear to differ. Biochar application significantly enhanced root P concentration but had little effect on root N concentration (Fig. 1b). One possible reason is that biochar may contain an imbalance of nutrient elements and may release more available P than N via weathering (Yamato et al., 2006; Rajkovich et al., 2012). This effect, however, depends on the biochar feedstock type and pyrolysis processes (Hass et al., 2012). Another possible reason is that biochar-induced increases in soil alkalinity may

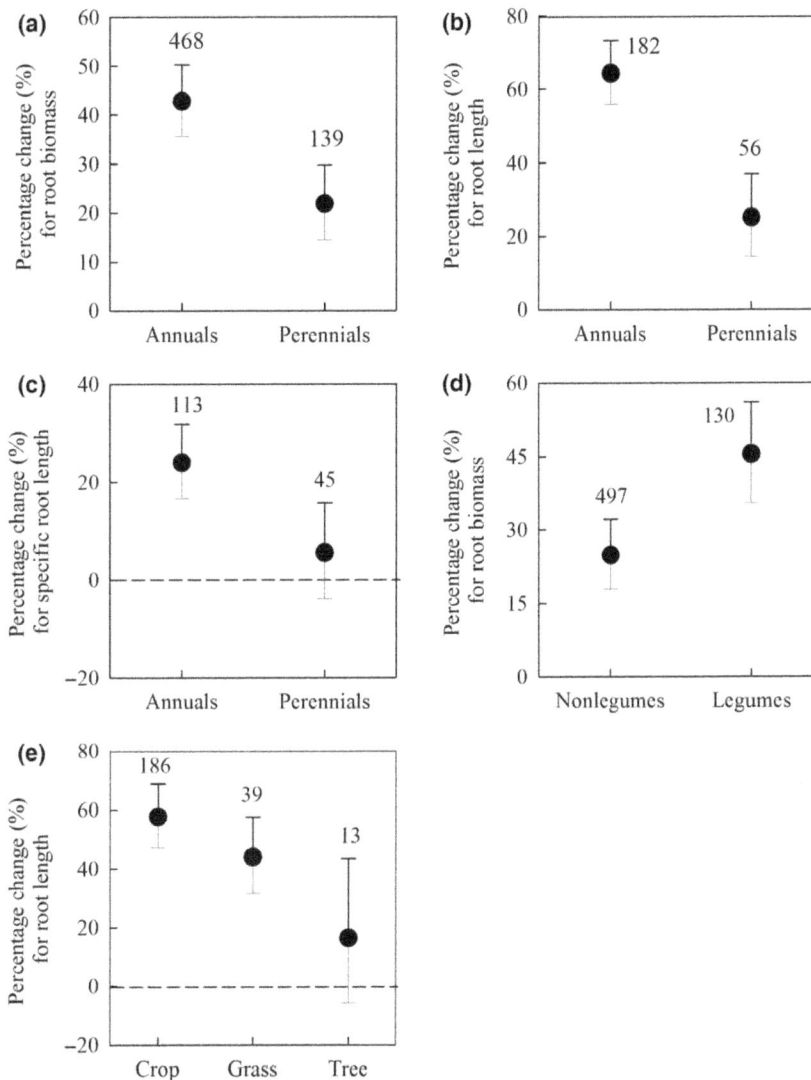

Fig. 3 Responses of root traits to biochar application for different plant functioning groups. (a–c) Percentage change (%) in root biomass (a), root length (b), and specific root length (c) in response to biochar application for annual and perennial plants. (d) Percentage change (%) in root biomass in response to biochar application for nonlegume and legume plants. (e) Percentage change (%) in root length in response to biochar application for crop, grass, and tree plants. Error bars represent 95% confidence intervals. The sample size for each variable is shown next to the error bar. The statistical significance of the effects of fertilization is evaluated in Table S4.

increase the soil pH value and thereby enhance soil available P (Cui *et al.*, 2011). Additionally, soil P is less mobile than N (Walker & Syers, 1976; Vitousek *et al.*, 2010). The increased root length under biochar application may benefit the plant absorption of relatively immobile nutrients such as P and K (Eissenstat, 1992).

Biochar application also inconsistently affected root-associated microbes. Our meta-analysis showed that root nodulation by rhizobia generally increased under biochar application (Fig. 1c). Root nodules are sensitive to P availability (Graham, 1981; Rondon *et al.*, 2007; Lehmann *et al.*, 2011). Biochar application often enhances P availability (Cui *et al.*, 2011) and thus promotes root nodulation. Nevertheless, biochar did not

significantly alter root colonization by mycorrhizal fungi but increased the root length colonized (Fig. 1c). The higher root length colonized under biochar application can help plants to explore more soil nutrients and thus may reduce plant dependence on mycorrhizae (Eissenstat, 1992; Raznikiewicz *et al.*, 1994).

Fertilization appeared to not alter the response magnitude of root biomass and morphology to biochar application (Table S4), which means that biochar-stimulated root growth was more profound under fertilizer input, as we previously expected. Moreover, the response magnitude of root P concentration to biochar application was significantly enhanced by fertilization (Fig. 2b). According to our dataset, N fertilizer was the

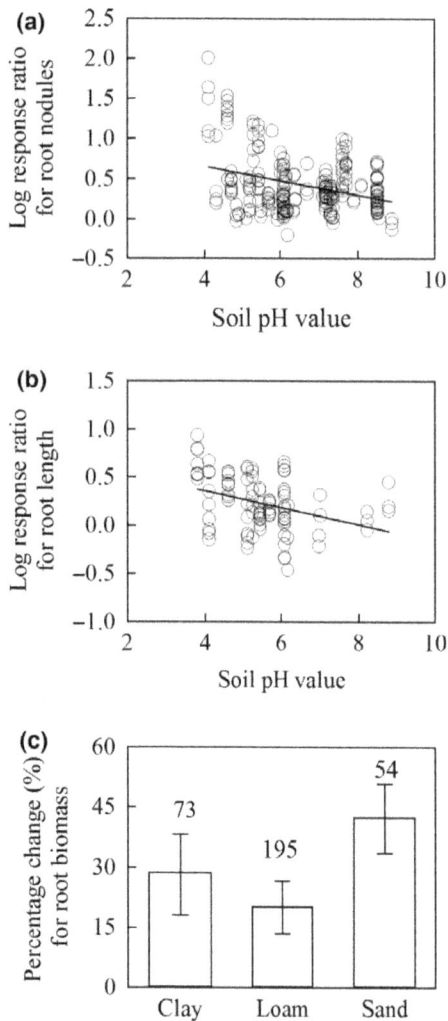

Fig. 4 Response of root traits to biochar application under different soil conditions. (a) Relationship between control soil pH values and biochar-induced response ratio for root biomass. (d) Relationship between control soil pH values and biochar-induced response ratio for root length. (c) Percentage change (%) in root biomass in response to biochar application in different soil types. Error bars represent 95% confidence intervals. The sample size for each variable is shown next to the error bar. The statistical significance of the effect of fertilization is evaluated in Table S4.

most applied fertilizer (Table S1). More available inorganic N sources under N fertilization could limit the root uptake of P due to nutrient imbalance and soil acidification (Vitousek *et al.*, 2010; Tian & Niu, 2015). Biochar tended to absorb NH_4^+ and NO_3^- after fertilizers were applied and hence could decrease the available N and soil N : P ratio (Steiner *et al.*, 2008; Clough *et al.*, 2013). Biochar's liming effect may also improve fertilization-induced soil acidification (Cui *et al.*, 2011). Therefore, the contrasting effects of fertilization and biochar application on soil P cycling might be the main cause

for the higher biochar-induced increase in root P concentration in the fertilized soils. Accordingly, the biochar immobilization of N might be responsive to the slight suppression of root N concentration in the fertilized soils (Steiner *et al.*, 2008; Clough *et al.*, 2013). These findings indicate that the combined applications of biochar and fertilizer may improve soil nutrient balances compared to biochar application or fertilization alone.

The duration of biochar application had no significant impact on root traits (Table S4); this result suggests that the root traits are altered quickly after biochar application. However, it remains unclear how long the biochar effects will last due to the lack of long-term experiments in our dataset (Table S1). Compared with the pot experiment, plants grown in a field plot may have more space for root extension in response to biochar application. However, such a short-term experimental dataset makes it difficult to detect a significant difference in root response between pot and field experiments. Thus, more long-term biochar field experiments are needed to generate a clear temporal trend of root responses.

The responses of root traits to biochar application differed among plant functional groups (Table S4). Our meta-analysis showed that biochar application generally induced a higher increase in root biomass, root length, and specific root length in annual plants than in perennial plants (Fig. 3a–c). These results may be associated with the different root strategies between the two plant life forms. In comparison with perennial plants that often develop root systems to maximize nutrient conservation, annual plants generally promote root growth to support greater nutrient acquisition (Chapin, 1980). Therefore, annual plants are likely to show more efficient root growth than perennial plants due to the improved soil nutrient environment under biochar application. The responses of root length to biochar application also varied with plant life form; the root length increased more in crops than in trees (Table S4). Nevertheless, because of the limited number of studies and sample size (201 for crops vs. 13 for trees; Fig. 3e), we were unable to draw a solid conclusion regarding how biochar will affect root traits among plant life forms. The greater root biomass responses of legumes than of nonlegumes (Fig. 3c) likely resulted from the increased root nodulation and more efficient N fixation for legumes treated by biochar application (Graham, 1981; Rondon *et al.*, 2007; Lehmann *et al.*, 2011).

The responses of root traits to biochar application also varied with soil pH value. For example, biochar application increases the root P concentration in acidic soils because increased soil alkalinity (liming effect) by biochar can directly increase available P by unbinding phosphate ions with Al and Fe (Busman *et al.*, 2002; Devau *et al.*, 2009). No significant

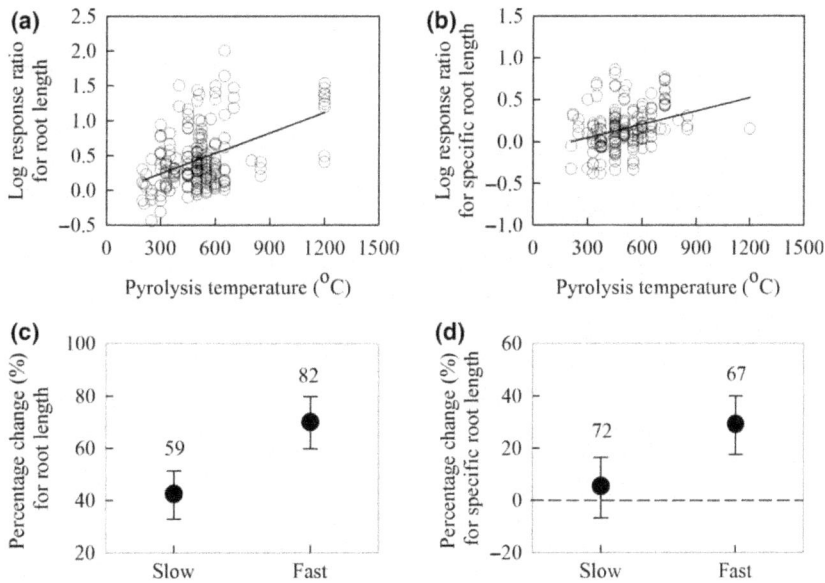

Fig. 5 Responses of root traits to biochar application under different pyrolysis temperatures and rates. (a,b) Relationships between pyrolysis temperature and the biochar-induced response ratios for root length (a) and specific root length (b). (c,d) Percentage change (%) in root length (c) and specific root length (d) in response to biochar application with different pyrolysis rates. Error bars represent 95% confidence intervals. The sample size for each variable is shown next to the error bar. The statistical significance of the effects of fertilization is evaluated in Table S4.

relationship was observed between the log RR of root nodulation and the soil pH values at the control sites probably due to insufficient data (*n* = 35 with only six studies). However, biochar-induced changes in soil P availability significantly affected root nodules (Fig. 4a); the log RR of the number of root nodules was positively correlated with the soil pH values (Graham, 1981; Rondon *et al.*, 2007; Lehmann *et al.*, 2011). Acidic, high-Al soils may also serve as a barrier to the growth of roots toward deeper soil horizons (Soethe *et al.*, 2006). Biochar application likely broke down these barriers by elevating soil alkalinity and soil porosity and hence causing a greater increase in root length (Fig. 4b). Several studies have suggested that biochar application could improve the soil nutrient and water environment in response to faster plant growth (Graber *et al.*, 2010; Major *et al.*, 2010). We did not detect any significant effect of soil nutrients, such as SOC, total N, and C : N ratio, on root traits in response to biochar application. However, biochar application increased root biomass more strongly in sandy soils than in the other soils (Fig. 4c) presumably because sandy soil is generally more acidic, dry, and nutrient-poor than are loam or clay soils.

The response of root traits to biochar application may be affected by the physicochemical properties and nutrient contents of biochar depending on the feedstock type and application rate. For instance, manure biochar generally contains large amounts of available nutrients,

while municipal waste biochar may produce toxic effects on plant growth (Sohi *et al.*, 2010). However, the nutrient or toxic contents of biochar and its application rate appeared to have little effect on root growth in our meta-analysis (Table S4). The effects of biochar pH on root traits were not significant (Table S4); this result suggests that the liming effect of biochar may not be directly induced by biochar pH. Alternatively, biochar produced at higher temperatures or under faster pyrolysis was generally more effective at promoting root length (Fig. 5). This result is not surprising because biochar characteristics partly depend on pyrolysis temperature (Sohi *et al.*, 2010). For example, high-temperature biochar tends to be alkaline (Bagreev *et al.*, 2001; Novak *et al.*, 2009). Processing temperature is also the main factor governing surface area; the surface area is greater under high temperature than under low temperature (Sohi *et al.*, 2010). In addition, biochar produced at low temperatures usually contains more nutrients and water and may limit root growth. It is noteworthy that pyrolysis temperature and rate are often negatively correlated with each other (Sohi *et al.*, 2010).

This study, to the best of our knowledge, is the first comprehensive synthesis to quantitatively evaluate the responses of root traits to biochar application. The findings in this meta-analysis may provide some insights into how biochar application would improve root traits for faster plant growth and may offer suggestions for the future sustainable development of biochar management. Plant growth is generally subjected to water and nutrient limitations. To address these deficiencies, plants often extend root length and other traits (Eissenstat & Yanai, 1997; Prendergast-Miller *et al.*, 2014). However, the responses of root traits to nutrient and water deficiencies are also limited by soil physical and chemical conditions (Soethe *et al.*, 2006). Biochar application likely improved soil environments and, in turn, significantly promoted root growth mainly by increasing root length, and hence eventually alleviating the deficiency of water and nutrients, particularly soil P. We also recommended that biochar can be applied with fertilizer to maximize plant growth because the combined application of biochar and fertilizer stimulated root growth and improved soil nutrient balance more effectively than did the application of biochar or fertilizer alone. Biochar application improved root growth and morphology more efficiently in annual plants than in perennial plants, particularly in acidic and sandy soils. Moreover, it appears that biochar production plays a more important role in regulating root growth than does biochar source. Thus, future biochar management should fully consider the factors associated with plant species, soil conditions, and biochar type to develop higher plant productivity.

Our results from this meta-analysis may also have significant implications for soil C cycling under biochar application. Biochar is characterized by stable and recalcitrant organic C with an extremely low decomposition rate and hence has been selected to increase soil C sequestration and climate change mitigation (Marris, 2006; Lehmann, 2007; Laird *et al.*, 2009; Woolf *et al.*, 2010). In this meta-analysis, we found that biochar application did not change the root : shoot ratio but significantly increased root length and the number of root tips more strongly than on root diameter. An increased number of thin roots may stimulate soil microbial activity due to the greater input of root exudates and may have priming effects on soil organic matter decomposition (Wang *et al.*, 2016). Indeed, several studies have reported that biochar application significantly altered soil microbial community composition and increased soil respiration (Lehmann *et al.*, 2011; He *et al.*, 2016; Liu *et al.*, 2016). Thus, the potential soil C sequestration of biochar may be re-estimated depending on the trade-off between biomass C inputs and soil C losses. Future land surface models must account for the shift in root morphology due to biochar application and its effects on soil C cycling.

Acknowledgements

We gratefully acknowledge financial support for this research from the Forestry Science and Research Program of Guizhou Province (GZFSTC[2015] NO. 6), the National Natural Science Foundation of China (31200372, 31428001, 31600483), and Jiangxi Provincial Department of Education (GJJ151097).

Conflict of interest

The authors declare that there is no conflict of interest.

References

Aguilar-Chávez A, Díaz-Rojas M, Del Rosario Cárdenas-Aquino M *et al.* (2012) Greenhouse gas emissions from a wastewater sludge-amended soil cultivated with wheat (*Triticum* spp. L.) as affected by different application rates of charcoal. *Soil Biology & Biochemistry*, **52**, 90–95.

Alburquerque JA, Cabello M, Avelino R *et al.* (2015) Plant growth responses to biochar amendment of Mediterranean soils deficient in iron and phosphorus. *Journal of Plant Nutrition and Soil Science*, **178**, 567–575.

Atkinson CJ, Fitzgerald JD, Hipps NA (2010) Potential mechanisms for achieving agricultural benefits from biochar application to temperate soils: a review. *Plant and Soil*, **337**, 1–18.

Bagreev A, Bandosz TJ, Locke DC (2001) Pore structure and surface chemistry of adsorbents obtained by pyrolysis of sewage sludge-derived fertilizer. *Carbon*, **39**, 1971–1979.

Biederman LA, Harpole WS (2013) Biochar and its effects on plant productivity and nutrient cycling: a meta-analysis. *Global Change Biology Bioenergy*, **5**, 202–214.

Borenstein M (2005) Software for publication bias. In: *Publication Bias in Meta-analysis: Prevention, Assessment and Adjustments* (eds Rothstein H, Sutton AJ, Borenstein M), pp. 391–403. John Wiley & Sons, Ltd, Chichester, UK.

Brennan A, Jiménez EM, Puschenreiter M *et al.* (2014) Effects of biochar amendment on root traits and contaminant availability of maize plants in a copper and arsenic impacted soil. *Plant and Soil*, **379**, 351–360.

Busman L, Lamb J, Randall G *et al.* (2002) The nature of phosphorus in soils. Available at: http://www.extension.umn.edu/agriculture/nutrient-management/phosphorus/the-nature-of-phosphorus/ (accessed 15 November 2016).

Chapin FS III (1980) The mineral nutrition of wild plants. *Ecology, Evolution, and Systematics*, **11**, 233–260.

Clough TJ, Condron LM, Kammann C *et al.* (2013) A review of biochar and soil nitrogen dynamics. *Agronomy*, **3**, 275–293.

Cui HJ, Wang MK, Fu ML *et al.* (2011) Enhancing phosphorus availability in phosphorus-fertilized zones by reducing phosphate adsorbed on ferrihydrite using rice straw-derived biochar. *Journal of Soils and Sediments*, **11**, 1135–1141.

Deng Q, Hui DF, Luo YQ *et al.* (2015) Down-regulation of tissue N: P ratios in terrestrial plants by elevated CO_2. *Ecology*, **96**, 3354–3363.

Devau N, Cadre EL, Hinsinger P *et al.* (2009) Soil pH controls the environmental availability of phosphorus: experimental and mechanistic modelling approaches. *Applied Geochemistry*, **24**, 2163–2174.

Egger M, Smith GD, Schneider M *et al.* (1997) Bias in meta-analysis detected by a simple, graphical test. *British Medical Journal*, **315**, 629–634.

Eissenstat DM (1992) Costs and benefits of constructing roots of small diameter. *Journal of Plant Nutrition*, **15**, 763–782.

Eissenstat DM, Yanai RD (1997) The ecology of root lifespan. *Advances in Ecological Research*, **27**, 1–60.

George C, Kohler J, Rillig MC (2016) Biochars reduce infection rates of the root-lesion nematode *Pratylenchus penetrans* and associated biomass loss in carrot. *Soil Biology & Biochemistry*, **95**, 11–18.

Graber ER, Harel YM, Kolton M *et al.* (2010) Biochar impact on development and productivity of pepper and tomato grown in fertigated soilless media. *Plant and Soil*, **337**, 481–496.

Graham PH (1981) Some problems of nodulation and symbiotic nitrogen fixation in *Phaseolus vulgaris* L.: a review. *Field Crops Research*, **4**, 93–112.

Hass A, Gonzalez JM, Lima IM *et al.* (2012) Chicken manure biochar as liming and nutrient source for acid Appalachian soil. *Journal of Environmental Quality*, **41**, 1096–1106.

He YH, Zhou XH, Jiang LL *et al.* (2016) Effects of biochar application on soil greenhouse gas fluxes: a meta-analysis. *Global Change Biology Bioenergy*, **9**, 743–755.

Hedges LV, Olkin I (1985) *Statistical Methods for Meta-Analysis*. Academic Press, Orlando, FL, USA.

Hedges LV, Gurevitch J, Curtis PS (1999) The meta-analysis of response ratios in experimental ecology. *Ecology*, **80**, 1150–1156.

Jackson RB, Mooney HA, Schulze ED (1997) A global budget for fine root biomass, surface area, and nutrient contents. *Proceedings of the National Academy of Sciences of the United States of America*, **94**, 7362–7366.

Jeffery S, Verheijen F, Van Der Velde M et al. (2011) A quantitative review of the effects of biochar application to soils on crop productivity using meta-analysis. *Agriculture, Ecosystems & Environment*, **144**, 175–187.

Jeffery S, Bezemer TM, Cornelissen G et al. (2015) The way forward in biochar research: targeting trade-offs between the potential wins. *Global Change Biology Bioenergy*, **7**, 1–13.

Keith A, Singh B, Dijkstra FA (2015) Biochar reduces the rhizosphere priming effect on soil organic carbon. *Soil Biology & Biochemistry*, **88**, 372–379.

Laird DA (2008) The charcoal vision: a win-win-win scenario for simultaneously producing bioenergy, permanently sequestering carbon, while improving soil and water quality. *Agronomy Journal*, **100**, 178–181.

Laird DA, Brown RC, Amonette JE et al. (2009) Review of the pyrolysis platform for coproducing bio-oil and biochar. *Biofuels, Bioproducts and Biorefining*, **3**, 547–562.

Lehmann J (2007) A handful of carbon. *Nature*, **447**, 143–144.

Lehmann J, Gaunt J, Rondon M (2006) Bio-char sequestration in terrestrial ecosystems – a review. *Mitigation and Adaptation Strategies for Global Change*, **11**, 395–419.

Lehmann J, Rillig MC, Thies J et al. (2011) Biochar effects on soil biota – a review. *Soil Biology & Biochemistry*, **43**, 1812–1836.

Li WB, Jin CJ, Guan DX et al. (2015) The effects of simulated nitrogen deposition on plant root traits: a meta-analysis. *Soil Biology & Biochemistry*, **82**, 112–118.

Liu X, Zhang A, Ji C et al. (2013) Biochar's effect on crop productivity and the dependence on experimental conditions – a meta-analysis of literature data. *Plant and Soil*, **373**, 583–594.

Liu S, Zhang Y, Zong Y et al. (2016) Response of soil carbon dioxide fluxes, soil organic carbon and microbial biomass carbon to biochar amendment: a meta-analysis. *Global Change Biology Bioenergy*, **8**, 392–406.

Luo YQ, Hui DF, Zhang DQ (2006) Elevated CO₂ stimulates net accumulations of carbon and nitrogen in land ecosystems: a meta-analysis. *Ecology*, **87**, 53–63.

Macdonald LM, Farrell M, Van Zwieten L et al. (2014) Plant growth responses to biochar addition: an Australian soils perspective. *Biology and Fertility of Soils*, **50**, 1035–1045.

Major J, Rondon M, Molina D et al. (2010) Maize yield and nutrition during 4 years after biochar application to a Colombian savanna oxisol. *Plant and Soil*, **333**, 117–128.

Marris E (2006) Putting the carbon back: black is the new green. *Nature*, **442**, 624–626.

Matamala R, Gonza'Lez-Meler MA, Jastrow JD et al. (2003) Impacts of fine root turnover on forest NPP and soil C sequestration potential. *Science*, **302**, 1385–1387.

Nie M, Lu M, Bell J et al. (2013) Altered root traits due to elevated CO₂: a meta-analysis. *Global Ecology and Biogeography*, **22**, 1095–1105.

Noguera D, Barot S, Laossi K-R et al. (2012) Biochar but not earthworms enhances rice growth through increased protein turnover. *Soil Biology & Biochemistry*, **52**, 13–20.

Novak JM, Lima I, Xing BS et al. (2009) Characterization of designer biochar produced at different temperatures and their effects on a loamy sand. *Annals of Environmental Science*, **3**, 195–206.

Olmo M, Villar R, Salazar P et al. (2016) Changes in soil nutrient availability explain biochar's impact on wheat root development. *Plant and Soil*, **399**, 333–343.

Peng YF, Guo DL, Yang YH (2017) Global patterns of root dynamics under nitrogen enrichment. *Global Ecology and Biogeography*, **26**, 102–114.

Prendergast-Miller MT, Duvall M, Sohi SP (2011) Localisation of nitrate in the rhizosphere of biochar-amended soils. *Soil Biology & Biochemistry*, **43**, 2243–2246.

Prendergast-Miller M, Duvall M, Sohi S (2014) Biochar-root interactions are mediated by biochar nutrient content and impacts on soil nutrient availability. *European Journal of Soil Science*, **65**, 173–185.

Rajkovich S, Enders A, Hanley K et al. (2012) Corn growth and nitrogen nutrition after additions of biochars with varying properties to a temperate soil. *Biology and Fertility of Soils*, **48**, 271–284.

Raznikiewicz H, Carlgren K, Maartensson A (1994) Impact of phosphorus fertilization and liming on the presence of arbuscular mycorrhizal spores in a Swedish long-term field experiment. *Swedish Journal of Agricultural Research*, **10**, 906–907.

Reibe K, Götz K-P, Roß C-L et al. (2015) Impact of quality and quantity of biochar and hydrochar on soil Collembola and growth of spring wheat. *Soil Biology & Biochemistry*, **83**, 84–87.

Reverchon F, Flicker RC, Yang H et al. (2014) Changes in δ¹⁵N in a soil–plant system under different biochar feedstocks and application rates. *Biology and Fertility of Soils*, **50**, 275–283.

Rillig MC, Wagner M, Salem M et al. (2010) Material derived from hydrothermal carbonization: effects on plant growth and arbuscular mycorrhiza. *Applied Soil Ecology*, **45**, 238–242.

Rondon MA, Lehmann J, Ramírez J et al. (2007) Biological nitrogen fixation by common beans (*Phaseolus vulgaris* L.) increases with bio-char additions. *Biology and Fertility of Soils*, **43**, 699–708.

Rosenberg MS (2005) The file-drawer problem revisited: a general weighted method for calculating fail-safe numbers in meta-analysis. *Evolution*, **59**, 464–468.

Soethe N, Lehmann J, Engels C (2006) The vertical pattern of rooting and nutrient uptake at different altitudes of a south Ecuadorian montane forest. *Plant and Soil*, **286**, 287–299.

Sohi SP, Krull E, Lopezcapel E et al. (2010) A review of biochar and its use and function in soil. *Advances in Agronomy*, **105**, 47–82.

Steiner C, Glaser B, Teixeira WG et al. (2008) Nitrogen retention and plant uptake on a highly weathered central Amazonian Ferralsol amended with compost and charcoal. *Journal of Plant Nutrition and Soil Science*, **171**, 893–899.

Taghizadeh-Toosi A, Clough TJ, Sherlock RR et al. (2012) Biochar adsorbed ammonia is bioavailable. *Plant and Soil*, **350**, 57–69.

Tian DS, Niu SL (2015) A global analysis of soil acidification caused by nitrogen addition. *Environmental Research Letters*, **10**, 024019.

Van De Voorde TFJ, Van Noppen F, Nachenius RW et al. (2014) Biochars produced from individual grassland species differ in their effect on plant growth. *Basic and Applied Ecology*, **15**, 18–25.

Vanek SJ, Lehmann J (2015) Phosphorus availability to beans via interactions between mycorrhizas and biochar. *Plant and Soil*, **395**, 105–123.

Varela Milla O, Rivera EB, Huang W et al. (2013) Agronomic properties and characterization of rice husk and wood biochars and their effect on the growth of water spinach in a field test. *Journal of Soil Science and Plant Nutrition*, **13**, 251–266.

Vitousek PM, Porder S, Houlton BZ et al. (2010) Terrestrial phosphorus limitation: mechanisms, implications, and nitrogen-phosphorus interactions. *Ecological Applications*, **20**, 5–15.

Walker TW, Syers JK (1976) The fate of phosphorus during pedogenesis. *Geoderma*, **15**, 1–19.

Wang J, Xiong Z, Kuzyakov Y (2016) Biochar stability in soil: meta-analysis of decomposition and priming effects. *Global Change Biology Bioenergy*, **8**, 512–523.

Woolf D, Amonette JE, Streetperrott FA et al. (2010) Sustainable biochar to mitigate global climate change. *Nature Communications*, **1**, 56. doi: 10.1038/ncomms1053.

Yamato M, Okimori Y, Wibowo IF et al. (2006) Effects of the application of charred bark of *Acacia mangium* on the yield of maize, cowpea and peanut, and soil chemical properties in South Sumatra, Indonesia. *Soil Science and Plant Nutrition*, **52**, 489–495.

14

Investigating the biochar effects on C-mineralization and sequestration of carbon in soil compared with conventional amendments using the stable isotope (δ^{13}C) approach

BALAL YOUSAF[1,2], GUIJIAN LIU[1,2], RUWEI WANG[1], QUMBER ABBAS[1], MUHAMMAD IMTIAZ[3] and RUIJIA LIU[1]

[1]CAS-Key Laboratory of Crust-Mantle Materials and the Environments, School of Earth and Space Sciences, University of Science and Technology of China, Hefei 230026, China, [2]State Key Laboratory of Loess and Quaternary Geology, Institute of Earth Environment, The Chinese Academy of Sciences, Xi'an, Shaanxi 710075, China, [3]Microelement Research Center, College of Resources and Environment, Huazhong Agricultural University, Wuhan 430070, China

Abstract

Biomass-derived black carbon (biochar) is considered to be an effective tool to mitigate global warming by long-term C-sequestration in soil and to influence C-mineralization via priming effects. However, the underlying mechanism of biochar (BC) priming relative to conventional biowaste (BW) amendments remains uncertain. Here, we used a stable carbon isotope (δ^{13}C) approach to estimate the possible biochar effects on native soil C-mineralization compared with various BW additions and potential carbon sequestration. The results show that immediately after application, BC suppresses and then increases C-mineralization, causing a loss of 0.14–7.17 mg-CO_2-C g^{-1}-C compared to the control (0.24–1.86 mg-CO_2-C g^{-1}-C) over 1–120 days. Negative priming was observed for BC compared to various BW amendments (-10.22 to -23.56 mg-CO_2-C g^{-1}-soil-C); however, it was trivially positive relative to that of the control (8.64 mg-CO_2-C g^{-1}-soil-C). Furthermore, according to the residual carbon and δ^{13}C signature of postexperimental soil carbon, BC-C significantly increased ($P < 0.05$) the soil carbon stock by carbon sequestration in soil compared with various biowaste amendments. The results of cumulative CO_2–C emissions, relative priming effects, and carbon storage indicate that BC reduces C-mineralization, resulting in greater C-sequestration compared with other BW amendments, and the magnitude of this effect initially increases and then decreases and stabilizes over time, possibly due to the presence of recalcitrant-C (4.92 mg-C g^{-1}-soil) in BC, the reduced microbial activity, and the sorption of labile organic carbon (OC) onto BC particles.

Keywords: biochar, biowaste, carbon mineralization, carbon stable isotope, C-sequestration, priming effects

Introduction

The rise of anthropogenic carbon dioxide (CO_2) emissions in the atmosphere, particularly through the combustion of fossil fuels and changes in land use, is the most noteworthy driver of radiative energy imbalance by positive radiative-forcing of the earth's climate system, which implies the warming of the global atmosphere since the mid-20th century (Meehl *et al.*, 2005; Wigley, 2005; Liepert, 2010; Crombie & Masek, 2015; Houghton *et al.*, 2015; Rogelj *et al.*, 2015; Scott *et al.*, 2015; IPCC, 2016; Tokarska *et al.*, 2016). It may sound strange, but this consequential increase in the mean global temperature has a significant impact on the rise of

sea levels due to thermal expansion of ocean water and is also responsible for the shrinking of land ice by the loss of mass from glaciers and polar ice (Oppenheimer, 1998; Raper & Braithwaite, 2006; Hay *et al.*, 2015; Slangen & Church, 2016). Keeping in view the potential deleterious impact of man-made GHG emissions on the global climate system, the first move toward integrated policies to overcome or stabilize this issue was made at the Earth Summit in 1992, under the platform of the UN Framework Convention on Climate Change (UNFCCC, 1992; Liepert, 2010; IPCC, 2016). Later on, this work was extended to the Kyoto Protocol in 1997, which formulated target policies in terms of 'equivalent carbon dioxide (CO_2eq)' and committed its signatories to mitigating GHG emissions by setting legally binding emission reduction targets (UNFCCC, 1998; Liepert, 2010). As the Protocol was signed, the sustainable strategy of long-

Correspondence: Guijian Liu
e-mail: lgj@ustc.edu.cn

term sequestration of atmospheric-C into soil to increase the carbon sink in the terrestrial environment is receiving increasing attention (Zavalloni et al., 2011).

Although the global carbon cycle is an immeasurably complex phenomenon, it has been widely accepted that both soil and atmospheric-C pools strongly interact with each other and that the natural two-way flux between them is one of the most influential processes (Fontaine et al., 2004). Indeed, the soil organic-C pool in the active atmosphere–ecosystem exchange comprises roughly two-thirds of the total carbon in terrestrial ecosystems (Post et al., 1982), which is approximately three times greater (~2400 Pg C to 2 m depth) than the current atmospheric-C (~830 Pg C) and 240 times higher compared with the current annual fossil fuel emission (~10 Pg C) (Schlesinger, 1995; Batjes, 1996; Fontaine et al., 2004; Paustian et al., 2016). Moreover, the long residence times of the relatively large C reservoirs in terrestrial ecosystems make this pool a potentially important sink for atmospheric carbon emitted by fossil fuel combustion, which could act as a buffer against increasing atmospheric CO_2 concentration (Post et al., 1982). Therefore, an increase of even a few percent in the net soil C-stock can considerably offset the increase of atmospheric CO_2.

Recent research suggests that the proximal controls on the balance of soil-C incorporate the rate of organic carbon input (agricultural wastes, sludge, and compost), subtracting C that is released through decomposition (Paustian et al., 2016). Therefore, the soil-C pool can be increased by adding exogenous sources of organic carbon or by minimizing the rate of decay/decomposition (for example, by minimum or no tillage) or by a combination of both, leading to enhanced sequestration of atmospheric-C (Karlen & Cambardella, 1996; Paustian et al., 1997; Novak et al., 2009; Zavalloni et al., 2011). However, a few long-term studies explain that the soil C-stock does not necessarily increase even if fresh organic matter in the form of manures and agricultural wastes is applied to soil in large quantities (Gill et al., 2002; Fontaine et al., 2004). This could be due to the decreasing rate of soil carbon buildup with time as the stockpile approaches the new equilibrium (West & Six, 2007).

Another novel strategy to increase the soil C-stock for the long term and to mitigate global warming by offsetting the atmospheric-C (up to 9.5 Pg C annually) is to add pyrogenic organic carbon (biochar) (Lehmann et al., 2006; Wardle et al., 2008; Woolf et al., 2010; Zavalloni et al., 2011; Ventura et al., 2015). Biochar (BC) is a carbonaceous material (a coproduct of pyrolysis) that is obtained from biomass pyrolysis and that is generally considered to be chemically and biologically inert due to its recalcitrant-C composition (Singh & Cowie, 2014;

Sagrilo et al., 2015). In the context of carbon sequestration, BC can stay in soil for a long time, with mean retention times of several centuries–millennia due to the high stability of the BC (Forbes et al., 2006). The archaic existence of the C-rich dark earth (Terra Preta) in the Amazon basin gives authentic proof of long-term C storage through a slash-and-char system (Lehmann et al., 2006; Singh et al., 2012).

Supplied organic carbon (fresh organic matter or BC) is reported to accelerate or suppress native SOC decomposition by so-called negative or positive priming effects of organic matter, respectively (Wardle et al., 2008; Keith et al., 2011; Luo et al., 2011; Zimmerman et al., 2011). Although many previous studies have reported priming effects (negative or positive) in terrestrial ecosystems, uncertainties still remain about the interactive influence of biochar and various biowaste amendments, such as the use of $\delta^{13}C$ unlabeled organic matter or nonuniform labeling of inputs on native SOC carbon mineralization (Zimmerman et al., 2011; Singh & Cowie, 2014; Ventura et al., 2015). To some extent, these experimental uncertainties could represent contradictions between the obtained results, that is, positive and negative priming effects of applied C sources on native SOC mineralization (Cardon et al., 2001).

To test the above-mentioned issue of uncertainty about the intensity and direction (positive or negative) of priming by BC and various BWs on the native SOC, we performed a two-phase experiment to compare the C-mineralization and C-storage potential of BC and various BWs (PrM, FM, CP, SS, and PM) in a laboratory incubation under controlled conditions, supported by a greenhouse experiment over a 120-day time period. The addition of BC and BW amendments with notably different $\delta^{13}C$ signatures compared to soil permitted the differentiation of C-mineralization between applied and native stocks. Our study was based on the following hypotheses: (i) at a very early stage, BC will negatively prime native soil C-mineralization/CO_2 emission, and then be triggered toward the peak (as positive priming); (ii) BC-induced positive priming effects will be minimized by gradual stabilization of the soil organic carbon pool, possibly due to the depletion of labile carbon content in soil; and (iii) interactive to various BW amendments, BC will show minimum C-mineralization but higher carbon sequestration potential for the long term.

Materials and methods

Soil, BC and BW amendments

The surface soil sample (0–15 cm) was collected from a research station near the peri-urban area of the city in a region that has been cultivated with fruits and vegetables over the past few decades. The collected soil sample was dried in air-shade conditions, ground, passed through a 10 mesh sieve (2 mm), and stored below 4 °C to restrict further biochemical changes prior to analysis. Comprehensive information about the soil characteristics (texture, saturation percentage, cation exchange capacity, pH, moisture content, organic carbon, and electrical conductivity) and nutrient contents (macronutrients: N, P, K) is summarized in Table 1.

The biochar (BC) was produced from wood sawdust (C 46.3, N 0.03%) as a feedstock via slow pyrolysis using an Isotemp muffle furnace (550 series; Fisher Scientific, Pittsburgh, PA, USA). The complete process and pyrolysis conditions used for the production of BC are described elsewhere (Yousaf et al., 2016). In brief, the wood sawdust BC was prepared at 450 °C under a continuous flow of nitrogen (50 sccm), with the temperature rising at a rate of 10 °C min^{-1} for a retention time of 60 min. The produced BC was then ground and sieved through a 2 mm sieve (10 mesh size sieve). Various BW materials (pressmud, farm manure, compost, sewage sludge, and poultry manure) were obtained from a local nursery, air-dried at 65 °C overnight, and passed through a 2 mm sieve. The physico-chemical characteristics of the BC and various BW amendments are given in Table 1.

Incubation experiment

Air-dried soil was uniformly mixed with BC at a 2% OC basis (BC: 6.34 g, 200 g-soil^{-1} DW basis) and various BW amendments, that is, PrM@2%OC (pressmud: 6.06 g, 200 g-soil^{-1} DW basis), FM@2%OC (farm manure: 7.40 g, 200 g-soil^{-1} DW basis), CP@2%OC (compost: 6.52 g, 200 g-soil^{-1} DW basis), SS@2%OC (sewage sludge: 18.17 g, 200 g-soil^{-1} DW basis), and PM@2%OC (poultry manure: 7.96, 200 g-soil^{-1} DW basis) in 500 mL plastic jars. Soil treatment mixtures were adjusted to a 65% water-holding capacity using a nutrient solution with ca. 400 mg N-kg^{-1} soil, 200 mg P-kg^{-1} soil, 100 mg-K kg^{-1} soil, and trace elements (Chapman, 1997; Singh & Cowie, 2014). All treatments, including control (without any amendment) and amended soils, were replicated three times. The soils were placed in 500 mL plastic containers in separate sealed experimental units containing subunits: (i) a 50 mL beaker containing 30 mL of double-distilled water (DDW) to maintain constant humidity and (ii) a 50 mL flask containing 30 mL of 2 M NaOH solution to trap CO$_2$–C evolved during C-mineralization. Each experimental unit was placed in dark conditions at a constant temperature of 25 ± 1 °C, and the incubation experiment was carried out over a time period of 120 days. The NaOH (2 M)-captured CO$_2$ was removed and replaced nine (9) times during incubation after 1, 5, 10, 20, 30, 50, 70, 90, and 120 days and was analyzed for total CO$_2$–C and CO$_2$–δ^{13}C. The blanks (triplicate) without soil were also set up to take into account the presence of atmospheric CO$_2$ in sealed units.

Measurement of carbon mineralization and priming effects

The total C mineralized (CO$_2$–C) in various treatments was estimated by precipitating 1 mL of 2 M NaOH with 5 mL of 0.4 M barium chloride (BaCl$_2$) followed by titration against 0.1 M HCl in the presence of phenolphthalein indicator (Singh et al., 2012). To determine the trapped CO$_2$ (as SrCO$_3$) in NaOH (C mineralized) from BC and various BW amendments, a 10 mL aliquot of NaOH (2 M) was precipitated with 10 mL of 1 M SrCl$_2$ (strontium chloride hexahydrate), and then the strontium carbonate (SrCO$_3$) precipitate was washed with double-distilled water (DDW) 7–10 times to obtain a neutral pH value. Finally, the precipitate was dried at 65 °C for 24 h and ground, followed by δ^{13}C analysis (1.5 mg SrCO$_3$ and 3 mg V$_2$O$_5$) by EA-IRMS (Finnigan MAT Delta Plus, Bremen, Germany) (Harris et al., 1997; Ventura et al., 2015).

The mineralizable C content (C$_M$) from BC and various BW amendments C$_M$ was calculated by following Eqn (1).

$$C_M = \frac{C_T(\delta_A^{13}CO_2 - \delta_C^{13}CO_2)}{\delta_I^{13}CO_2 - \delta_C^{13}CO_2}, \quad (1)$$

where C$_T$ is the total C mineralized from BC and various BW-amended soils; $\delta_A^{13}CO_2$ represents the δ^{13}C value of the CO$_2$–C released from BC and various BW-amended soils; $\delta_C^{13}CO_2$ indicates the δ^{13}C value of the CO$_2$–C released from the control (unamended soil); and $\delta_I^{13}CO_2$ is the initial δ^{13}C value of the BC and various BW amendments (Singh & Cowie, 2014; Ventura et al., 2015).

Priming effects (±) of BC and various BW amendments on native soil organic carbon (SOC) mineralization were calculated as described below in Eqn (2).

$$PE_C^A = C_{C,A} - C_C, \quad (2)$$

where C$_{C,A}$ is the value of the CO$_2$–C released from BC and various BW-amended soils extracted from Eqn (1), and C$_C$ is the value of the CO$_2$–C released from the control (unamended soil) (Zimmerman et al., 2011).

Interactive priming effects (IPE) of BC with respect to various BW amendments on C-mineralization were calculated as:

$$IPE_{BA}^B = C_{BA,B} - C_{BA}, \quad (3)$$

where C$_{BA,B}$ is the value of the CO$_2$–C released from BC-amended soils extracted from Eqn (1), and C$_{BA}$ is the value of the CO$_2$–C released from various BW amendments (Zimmerman et al., 2011).

Pot experiment

A pot experiment was conducted under greenhouse conditions to evaluate the labile organic carbon contents, total carbon sequestration potential, and nutrient behavior of BC and various BW amendments. Six treatments, biochar (BC), pressmud (PrM), farm manure (FM), compost (CP), sewage sludge (SS), and poultry manure (PM), were applied on a 2% organic carbon (OC) basis, with a similar rate as that used in incubation experiments. All the treatments were replicated three times, including an unamended soil as a control, and a total of 28 polyvinylchloride pots were filled with the soil (7 kg soil pot^{-1}, 30 cm height, and 20 cm diameter). The treatments were applied as C = control (unamended soil), BC@2% OC (biochar: 222.17 g pot^{-1}), PrM@2%OC (pressmud: 212.03 g pot^{-1}), FM@2%OC (farm manure: 259.16 g pot^{-1}),

Table 1 Physicochemical characteristics of soil and various amendments used in this study

Characteristic	Unit	Soil	Biochar (BC)	Biowaste amendments (BWs)				
				PrM	FM	CP	SS	PM
Texture	–	Loam*	–	–	–	–	–	–
Sand	%	33.48 ± 0.72	–	–	–	–	–	–
Silt	%	43.72 ± 0.39	–	–	–	–	–	–
Clay	%	22.80 ± 0.44	–	–	–	–	–	–
SP†	%	34 ± 1.49	–	–	–	–	–	–
CEC‡	cmol$_c$ kg^{-1}	17.06 ± 0.14	–	–	–	–	–	–
pH§	–	6.82 ± 0.03	10.83 ± 0.15	7.55 ± 0.29	7.69 ± 0.23	7.94 ± 0.24	7.64 ± 0.23	7.1 ± 0.18
MC (105 °C)¶	%	11.34 ± 0.98	4.23 ± 0.27	33.48 ± 2.45	25.22 ± 2.25	17.35 ± 1.86	34.35 ± 3.50	32.84 ± 2.75
Ash content¶	%	–	27.65 ± 0.28	12.35 ± 0.75	11.78 ± 0.91	13.16 ± 0.95	31.58 ± 0.65	10.85 ± 1.06
OC**	%	0.40 ± 0.06	63.03 ± 1.35	66.03 ± 3.31	54.02 ± 2.29	61.34 ± 2.85	22.01 ± 1.48	50.25 ± 3.86
δ^{13}C	‰	−16.18 ± 0.05	−28.16 ± 0.11††	−13.01 ± 0.08	−22.96 ± 0.07	−23.35 ± 0.07	−21.59 ± 0.08	−22.38 ± 0.06
EC‡‡	dS m^{-1}	3.37 ± 0.02	0.84 ± 0.21	4.37 ± 0.66	7.85 ± 1.25	6.82 ± 0.42	8.62 ± 0.74	7.56 ± 0.29
Macronutrients								
Total N	%	0.19 ± 0.07	1.52 ± 0.22	3.88 ± 0.31	2.24 ± 0.33	3.88 ± 0.31	1.56 ± 0.35	4.46 ± 0.81
P	ppm	8.45 ± 0.24§§	308.44 ± 5.15	426.50 ± 10.45	455.35 ± 13.60	622.25 ± 25.37	386.12 ± 8.45	437.41 ± 14.47
K	%	0.49 ± 0.08§§	0.87 ± 0.04	1.05 ± 0.09	0.93 ± 0.05	1.24 ± 0.12	0.96 ± 0.08	0.82 ± 0.06

PrM, pressmud; FM, farm manure; CP, compost; SS, sewage sludge; PM, poultry manure.

*USDA soil classification system.

†Saturation percentage.

‡Cation exchange capacity.

§pH of soil saturated paste.

¶Heating at 750 °C.

**Soil organic carbon contents.

††δ^{13}C ‰ value for biochar feedstock was −27.63 ± 0.14 ¶ moisture content ($n = 3$).

‡‡Electrical conductivity of soil saturated paste extract.

§§Available.

CP@2%OC (compost: 228.24 g pot^{-1}), SS@2%OC (sewage sludge: 636.07 g pot^{-1}), and PM@2%OC (poultry manure: 278.61 g pot^{-1}). The soil was fertilized with the recommended rates of nitrogen, phosphorus, and potassium. Half of the nitrogen (0.3 g N pot^{-1}) was applied with a full dose of phosphorus (0.5 g P pot^{-1}) and potassium (0.24 g K pot^{-1}) as the basal dose. The remaining half of the nitrogen was applied in two parts (1st with the first irrigation and 2nd at the milking stage) (Yousaf *et al.*, 2016). Wheat seeds were sterilized for 10 min using 30% H_2O_2 solution, followed by washing with deionized water (Miché & Balandreau, 2001). The sterilized seeds were soaked in deionized water overnight, and ten wheat seeds were spread in each pot. Then, after sprouting, the seedlings were thinned to four plants pot^{-1}. All necessary agronomic practices were followed, and pots were also randomized in alternate weeks to minimize positional effects.

DOC and MBC analysis

Dissolved organic carbon was determined by UV/persulfate oxidation method: combines the sample with an acid, lowering the sample pH to 2.0. In this process, inorganic carbon (IC) is converted to dissolve CO_2–C and eliminate from the sample. Potassium persulfate reagent was added to the sample followed by oxidization of remaining carbon by ultraviolet (UV) radiation, which was detected by the nondispersive infrared detector (NDIR) sensor (Eykelbosh *et al.*, 2015).

The chloroform-fumigated extraction (CFE) method was used to determine the microbial biomass carbon at 1, 30, 60, 90, and 120 days using a recovery factor (K_E) of 0.67 (for 10 days fumigation) (Sparling *et al.*, 1993), instead of 0.45 (for 1 day fumigation) (Vance *et al.*, 1987). A ten (10) g of soil (fumigated and nonfumigated) was extracted (for 1 h) using 0.5 M K_2SO_4 reagent. The microbial biomass carbon was calculated by following Eqn (7) (Fontaine *et al.*, 2004; Benbi *et al.*, 2015).

$$MBC = \frac{(C_{fumigated} - C_{nonfumigated})}{K_E}, \quad (4)$$

where $C_{fumigated}$ is the organic carbon extracted from fumigated soil sample, $C_{nonfumigated}$ is organic carbon extracted from nonfumigated soil sample; K_E represents the recovery factor (0.67 for 10 days fumigation) to adjust the greater release of microbial cytoplasm (Sparling *et al.*, 1993).

TOC, oxidizable carbon and recalcitrant-C contents

Total carbon content (TC) in soil samples was determined by dry-combustion technique using a combustion analyzer (G4 ICARUS HF; Bruker Elemental GmbH, Kalkar, Germany) with high frequency furnace and infrared detector. IC was also analyzed by dilute acid titration method using 0.01 N HCl solution (Jackson, 1967). Total organic carbon (TOC) was calculated by difference method as described in Eqn (4).

$$TOC = TC - IC, \quad (5)$$

where TC is total carbon content (TC) in soil samples determined by dry combustion, and IC represents the inorganic carbon contents analyzed by dilute HCl.

The oxidizable carbon was measured by a method described by Walkley–Black (Jackson, 1967). From the prepared soil sample, 1 g of soil was placed in a 500 mL conical flask followed by 10 mL of 1 N $K_2Cr_2O_7$ solution and 20 mL of concentrated H_2SO_4. After 30 min, 200 mL of double-distilled water (DDW), 0.2 g of NaF, and 10 mL of H_3PO_4 were added to the material and titrated against $FeSO_4·7H_2O$ to a dull green endpoint in the presence of 30 drops of diphenylamine indicator. A blank solution was also run without soil for calculation of the corrected oxidizable carbon. The oxidizable-C was calculated by the following formula:

$$\% \text{ Oxidizable-C} = \frac{V_{blank} - V_{sample}}{Wt} \times 0.3 \times M, \quad (6)$$

where V_{blank} is volume of $FeSO_4·7H_2O$ solution used to titrate the blank (mL), V_{sample} is volume of $FeSO_4·7H_2O$ solution used to titrate the soil sample (mL), Wt is the weight of air-dried soil used for analysis, 0.3 is the $3 \times 10^{-3} \times 100$ (3 is equivalent wait of carbon), and M is the molarity of $FeSO_4·7H_2O$ solution.

The carbon content not oxidized by potassium dichromate ($K_2Cr_2O_7$) referred as recalcitrant carbon (Benbi *et al.*, 2015). Recalcitrant-C was estimated by the difference between total organic carbon and oxidizable carbon as represented in Eqn (6).

$$\text{Recalcitrant-C} = \text{Total organic carbon (TOC)} - \text{Oxidizable-C} \quad (7)$$

Statistical analysis and quality control

To ensure accuracy in data with reference to quality control, all the samples were analyzed repeatedly for CO_2 emission, C-mineralization priming effects, DOC, MBC, and TOC. All the descriptive data were statistically analyzed by two-way ANOVA using SPSS 16.0 (SPSS Inc., Chicago, IL, USA) at various incubation times. The BC and BW amendments were used as interactive sources of variance. The predicted mean values were compared using LSD (at 0.05 probability), where F-tests were significant. SIGMAPLOT 11.0 (Systat Software Inc., San Jose, CA, USA) was employed for graph plotting.

Results

Influence of BC and various BW amendments on soil–crop characteristics and nutrient concentrations

The BC and various BW amendments significantly affected many physicochemical characteristics of the soil, including the organic carbon (OC), electrical conductivity (EC), pH, and cation exchange capacity (CEC), and their mean values are given in Table 2. During the course of the study, BC and various BW amendments had a statistically significant ($P \leq 0.05$) influence on the soil organic carbon (OC) content, with a highest value of $1.2 \pm 0.07\%$ when BC was applied to a 2% organic carbon basis followed by PrM, CP, FM, PM, and SS. However, the minimum OC content ($0.39 \pm 0.08\%$) was observed with the nonamended soil (Table 2). Similarly,

Table 2 Postexperiment organic carbon (OC) contents, electrical conductivity (EC), pH, and cation exchange capacity (CEC) of unamended and amended soils (biochar and various biowaste amendments)

Soil characteristics	Control	Biochar (BC)	Biowaste amendments (BWs)				
			PrM	FM	CP	SS	PM
Organic carbon (%)	0.39 ± 0.08	1.20 ± 0.07	0.95 ± 0.07	0.81 ± 0.14	0.90 ± 0.08	0.62 ± 0.16	0.73 ± 0.20
EC_e (dS m^{-1})	3.36 ± 0.17	5.59 ± 0.41	3.96 ± 0.30	6.02 ± 0.38	4.49 ± 0.44	4.50 ± 0.27	8.62 ± 0.34
pH*	7.67 ± 0.33	8.20 ± 0.05	7.45 ± 0.26	8.19 ± 0.05	7.53 ± 0.34	7.83 ± 0.32	7.35 ± 0.41
CEC (cmol$_c$ kg^{-1})	17.27 ± 0.68	23.82 ± 1.63	20.68 ± 0.82	19.79 ± 0.22	20.82 ± 0.41	17.40 ± 0.98	19.79 ± 0.80

PrM, pressmud; FM, farm manure; CP, compost; SS, sewage sludge; PM, poultry manure ($n = 3$).
*pH of soil saturated paste.

the highest values of the pH (8.20 ± 0.05) and cation exchange capacity (23.82 ± 2.63 cmol$_c$ kg^{-1}) were recorded in the case of biochar (BC) amendment, regardless of other BW amendments, while the electrical conductivity (EC) was extremely high in the PM-amended soil (8.62 ± 0.34 dS m^{-1}), followed by FM (6.02 ± 0.38 dS m^{-1}) and BC (5.59 ± 0.41 dS m^{-1}) (Table 2).

Fresh shoot biomass, dry shoot biomass, and grain weight increased significantly ($P \leq 0.01$) with BC and most of the BW amendments compared with that of the control, except for with SS and PM. On the other hand, the fresh root biomass and dry root biomass were nonsignificant in the BC and all BW-amended soils (Table 3). Moreover, the plant physiological characteristics, that is, photosynthetic rate (Pn), transpiration rate (Tr), and stomatal conductance (gs), were statistically significant ($P \leq 0.05$) with the BC and all BW amendments. However, there was no significant difference in the calculated leaf area (cm^2) with all amendments, except for with BC and CP. The mean values of FSB (46.15 ± 3.82 g pot^{-1}) and DSB (30.18 ± 3.09 g pot^{-1}) in the BC-amended soil were 23.7% and 25.2% greater than the control (37.32 ± 3.58 g pot^{-1}), respectively. Similarly, the average harvested grain weight with the BC-amended treatment (15.35 g pot^{-1}) was greater (44.1%) than that observed with the control (10.65 g pot^{-1}). Furthermore, the photosynthetic rate (44.4%), the transpiration rate (36.8%), the stomatal conductance (109%), and the leaf area (18.1%) for BC-amended soil were also greater than that for control soil (Table 3). Soil and plant (root, shoot, and grain) macronutrient (NPK) contents were also significantly ($P \leq 0.05$) affected by the application of BC and various BW amendments compared with that of the control (Fig. 1).

Total carbon mineralization/CO$_2$ emission and interactive priming effects

The δ^{13}C signature of CO$_2$ released form BC and various BW-amended soils are presented in Table S1.

Similarly, C-mineralization in soils amended with BC and various BW amendments compared with the control are shown in Fig. 2. The δ^{13}C signature of BC ($-28.16 \pm 0.09‰$) and other BW amendments (PrM, FM, CP, SS and PM) are summarized in Table 1. The δ^{13}C ‰ value of pine wood ($-27.23 \pm 0.11‰$), which was used as the feedstock for BC production and produced BC ($-28.16 \pm 0.09‰$), showed a remarkable difference between the raw material and final product, indicating the significant ($P \leq 0.05$) effect of pyrolysis on the δ^{13}C signature (Table 1). On the other hand, the δ^{13}C signature of CO$_2$–C evolved during the incubation period from all amendments on every sampling day and was highly significant ($P \leq 0.01$) compared to that of the control. The only exception was observed in the case of BC-amended soil on the first day (Table S1).

The addition of BW amendments increased the soil respiration rate and cumulative CO$_2$ emissions up to tenfold compared to the unamended soil. However, the total BC-C-mineralization/CO$_2$ emission rate decreased up to 0.14 ± 0.07 mg CO$_2$–C g^{-1}-C day^{-1} after immediate application (1st day) compared to that of the control (0.25 ± 0.06 mg CO$_2$–C g^{-1}-C day^{-1}), and then increased, causing a loss of 2.57 mg CO$_2$–C g^{-1}-C compared to the control (0.43 mg CO$_2$–C g^{-1}-C) over 1–5 days. After this, the BC-C-mineralization decreased again from 2.43 ± 0.09 to 0.52 ± 0.05 mg CO$_2$–C g^{-1}-C day^{-1} over 30 days and then stabilized over time (over 120 days). On the other hand, the C-mineralization rates of all BW amendments were higher on day 1 (5.42 ± 0.12, 6.51 ± 0.14, 5.73 ± 0.12, 7.13 ± 0.14, and 6.89 ± 0.13 mg CO$_2$–C g^{-1}-C day^{-1} for PrM, FM, CP, SS, and PM, respectively) and then decreased in an exponential manner during the experimental time period. Throughout the experiment, the highest C-mineralization rate was observed in the case of SS followed by PM and FM. However, in the case of cumulative C-mineralization, 10.50 ± 0.39 mg of CO$_2$–C g^{-1}-C was released from the BC-amended soil over 120 days ($P \leq 0.05$). All the BW amendments significantly increased ($P \leq 0.05$) the cumulative C-mineralization

Table 3 Biomass production and plant physiological characteristics affected by biochar and various biowaste amendments

Characteristics	Control	Biochar (BC)	Biowaste amendments (BWs)				
			PrM	FM	CP	SS	PM
FSB/pot (g)	37.32 ± 3.58	46.15 ± 3.82	52.95 ± 6.50	52.15 ± 5.13	51.75 ± 2.95	40.75 ± 4.91	48.32 ± 1.05
DSB/pot (g)	24.11 ± 1.72	30.18 ± 3.09	30.70 ± 2.57	31.44 ± 2.84	32.20 ± 5.03	26.41 ± 3.45	28.57 ± 2.74
FRB/pot (g)	12.20 ± 1.48	13.75 ± 2.96	16.50 ± 2.68	15.85 ± 4.00	12.60 ± 1.33	14.80 ± 1.03	13.75 ± 1.15
DRB/pot (g)	6.09 ± 1.18	6.82 ± 2.46	8.69 ± 2.03	7.762 ± 3.21	8.15 ± 0.40	7.30 ± 2.36	6.63 ± 0.75
GW/pot (g)	10.65 ± 1.33	15.35 ± 2.57	15.26 ± 1.53	15.10 ± 2.23	16.25 ± 1.66	10.85 ± 1.27	12.30 ± 1.20
Photosynthetic rate (Pn)	26.46 ± 0.95	38.21 ± 1.18	35.35 ± 0.93	38.85 ± 2.55	43.18 ± 2.25	35.34 ± 1.88	36.85 ± 3.54
Transpiration rate (Tr)	3.18 ± 0.08	4.35 ± 0.35	3.99 ± 0.45	4.16 ± 0.62	4.95 ± 0.55	3.81 ± 0.21	4.05 ± 0.25
Stomatal conductance (gs)	0.11 ± 0.01	0.23 ± 0.03	0.20 ± 0.03	0.23 ± 0.02	0.31 ± 0.05	0.19 ± 0.03	0.21 ± 0.05
Leaf Area (cm^2)	28.62 ± 2.27	33.80 ± 1.54	30.88 ± 3.41	32.72 ± 2.25	32.32 ± 1.14	29.22 ± 0.53	31.30 ± 1.80

FSB, fresh shoot biomass; DSB, dry shoot biomass; FRB, fresh root biomass; DRB, dry root biomass; GW, grain weight; PrM, pressmud; FM, farm manure; CP, compost; SS, sewage sludge; PM, poultry manure ($n = 3$).

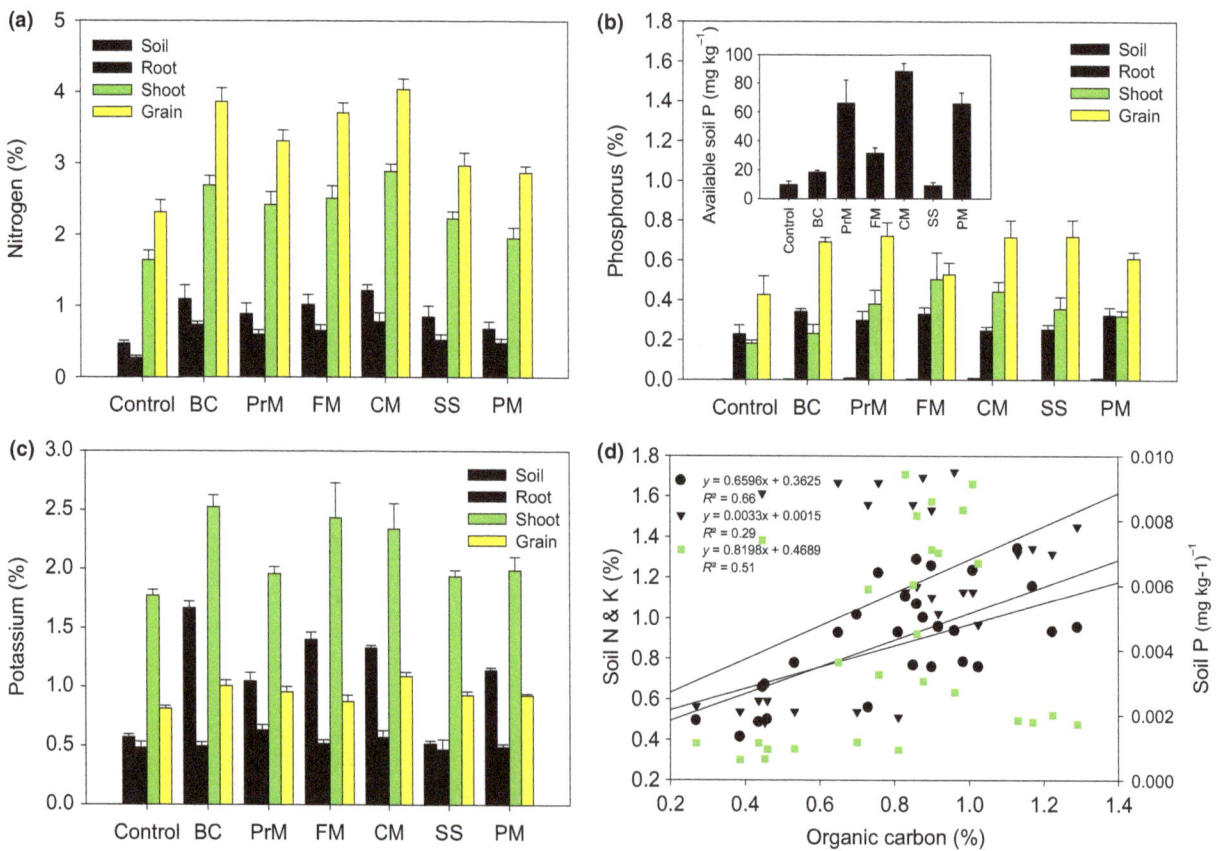

Fig. 1 Nutrient contents of unamended and amended soils (biochar, pressmud, farm manure, compost, sewage sludge, and poultry manure), and wheat plant (root, shoot, and grain): (a) total nitrogen contents in soil and plant parts; (b) phosphorus contents in soil (available phosphorus) and plant parts; (c) potassium contents in soil (extractable potassium) and plant parts; (d) relationship between soil organic carbon and macronutrients (NPK).

(PrM: 20.72 ± 0.73 mg CO_2–C g^{-1}-C, FM: 28.67 ± 0.93 mg CO_2–C g^{-1}-C, CP: 23.57 ± 1.07 mg CO_2–C g^{-1}-C, SS: 34.06 ± 1.11 mg CO_2–C g^{-1}-C, and PM: 31.34 ± 0.15 mg CO_2–C g^{-1}-C) compared with that of the control (1.86 ± 0.24 mg CO_2–C g^{-1}-C) as well as with BC.

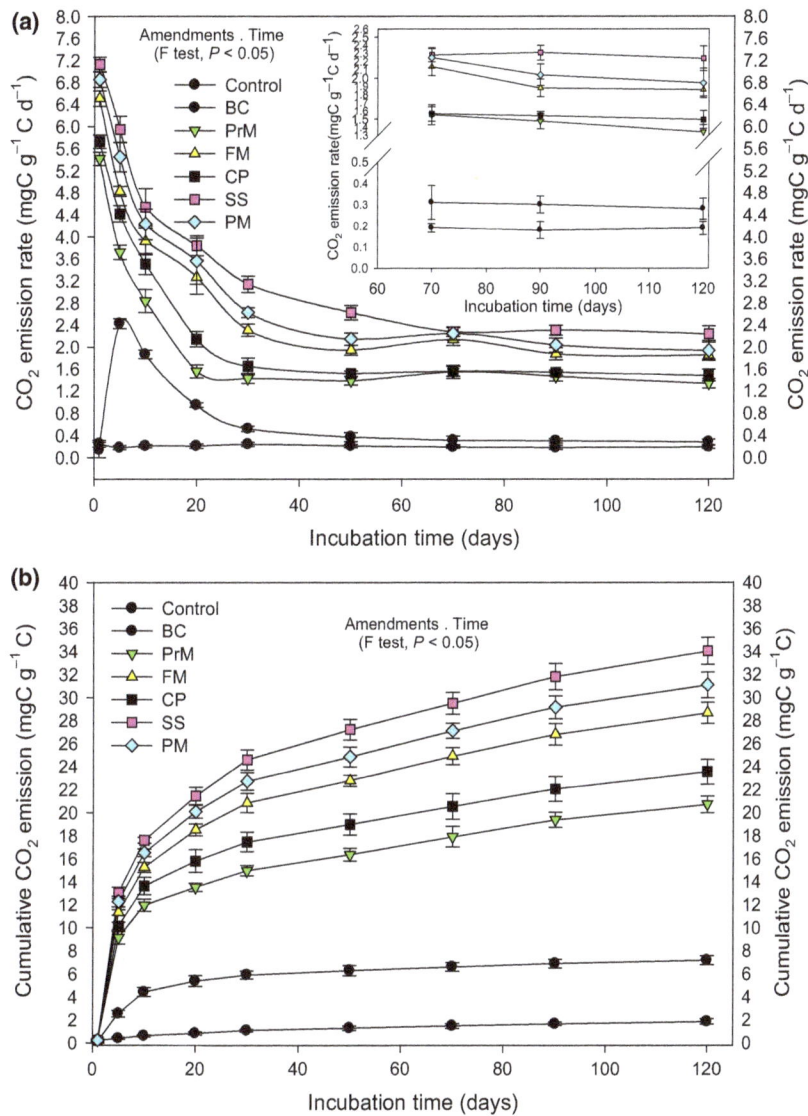

Fig. 2 Carbon mineralization and CO_2 emission under aerobic conditions: (a) carbon mineralization/CO_2 emission rate in una-mended and amended soils (biochar, pressmud, farm manure, compost, poultry manure, and sewage sludge) during 120 days experimental time period. (b) Cumulative CO_2 emission in unamended and amended soils (biochar, pressmud, farm manure, compost, sewage sludge, and poultry manure) during 120 days experimental time period.

Immediately after the application, BC suppressed the native soil C-mineralization (-0.11 mg CO_2–C g^{-1}soil-C day^{-1}), indicating a negative priming effect ($P \leq 0.05$) (Fig. 3). Afterward, BC triggered the increase of the rate of soil C-mineralization toward the peak within 5–20 days, and this positive priming effect remained over the 120 days of the experimental time period. However, large positive priming effects were observed with all BW amendments (PrM, FM, CP, SS, and PM) relative to the control. Detailed information on a cumulative basis shows that C-mineralization of native soil due to the

positive priming effect was higher for SS, followed by FM and CP = PM (Fig. 3).

Dissolved organic carbon and microbial organic carbon

At the beginning, the DOC slightly decreased (2.65 ± 0.42 μg-C g^{-1} soil) in the BC-amended soil compared to the control (3.18 ± 0.35 μg-C g^{-1} soil), then gradually increased between 5 and 30 days (up to 16.35 ± 0.85 μg-C g^{-1} soil) and remained uncertain (12.34 ± 1.19 μg-C g^{-1} soil to 13.87 ± 2.18 μg-C g^{-1}

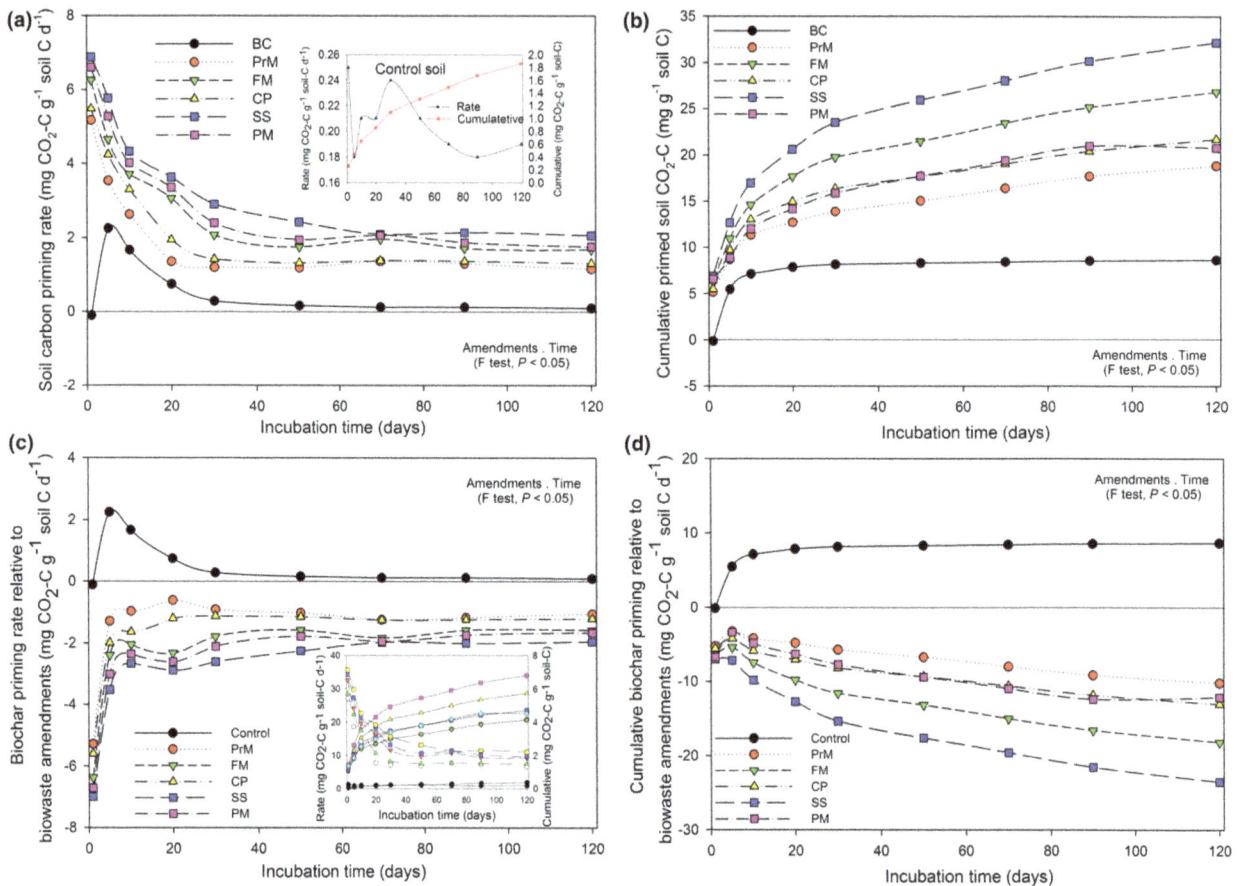

Fig. 3 Interactive priming effect of biochar and biowaste amendments: (a) soil carbon priming rate (mg CO_2–C g^{-1} soil C day^{-1}); (b) soil priming effect on cumulative CO_2 per gram soil carbon [CO_2–C (mg g^{-1} soil C)]; (c) effect of various biowaste amendments on biochar priming rate (mg CO_2–C g^{-1} soil C day^{-1}); (d) effect of various biowaste amendments on biochar cumulative priming CO_2–C (mg g^{-1} soil C) in unamended and amended soils (biochar, pressmud, farm manure, compost, sewage sludge, and poultry manure) during 120 days experimental time period.

soil) from 30 to 120 days (Fig. 4a). However, various BW amendments significantly increased the DOC content ($P \leq 0.05$), showing a maximum for SS within the range of 19.75 ± 1.23 µg-C g^{-1} soil to 35.85 ± 2.16 µg-C g^{-1} soil, followed by PM, FM, PrM, and CP, compared to that of the un-amended soil as well as to BC during the experimental time period of 120 days. These large DOC contents in various BW-amended soils indicate higher microbial activity, which is responsible for more C-mineralization.

Similar to DOC, BC addition to soil did not influence the MBC on day 1 (32.45 ± 1.90 µg-C g^{-1} soil) compared to that of the control (31.45 ± 1.54 µg-C g^{-1} soil). On this other hand, this influence resulted in an increase in BC compared to the control on day 30 and continued to decrease slowly till the end of the experiment (Fig. 4b). However, analogous to the inceptive sequence of the C-priming and the comparative DOC in various amendments, the MBC was significantly higher for the SS (at 1, 30, 60, 90, and 120 days by

51.15 ± 1.28 µg-C g^{-1} soil, 131.82 ± 4.34 µg-C g^{-1} soil, 122.35 ± 3.64 µg-C g^{-1} soil, 95.21 ± 3.87 µg-C g^{-1} soil, and 91.34 ± 2.01 µg-C g^{-1} soil, respectively), with a decreasing order of SS < PM < FM < CP < PrM < BC < control. Although MBC was significantly higher in all BW-amended soils, the maximum MBC level was sustained in the SS-amended soil comparative to the control, over 120 days.

C-sequestration potential

The change in total organic carbon content and the post-experiment soil δ^{13}C value with the addition of BC and various BW amendments, compared to that of the control over an experimental time period of 120 days, are presented in Fig 5a and Table 4. The results obtained from the present study indicate that BC had a clear influence ($P \leq 0.05$) on TOC compared with those of various BW amendments, except for PrM, which was not significant on the 30th, 60th, and 90th days. However, on day 120,

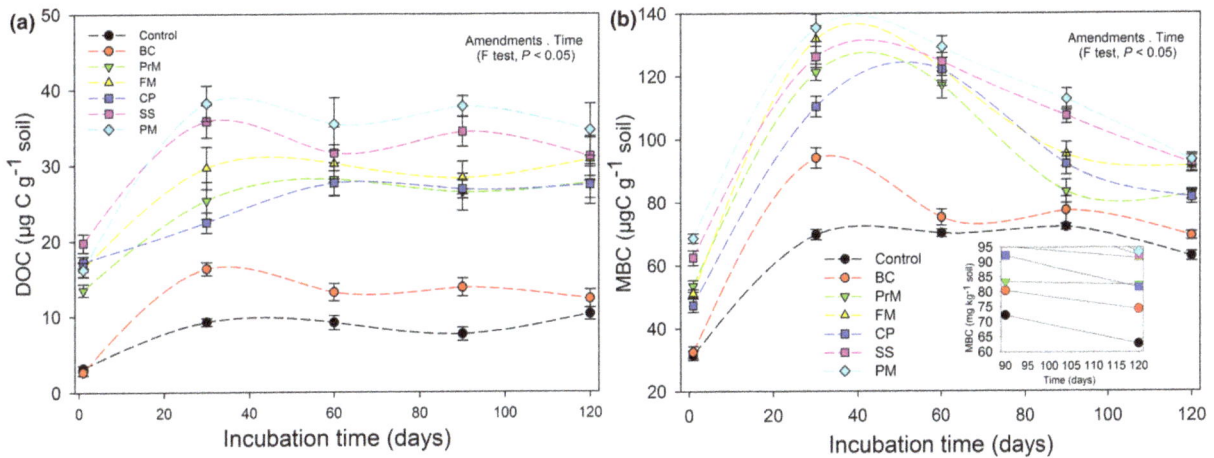

Fig. 4 Microbial biomass carbon (MOC) and dissolved organic carbon (DOC) contents in soil: (a) Estimated MOC in unamended and amended soils (biochar, pressmud, farm manure, compost, sewage sludge, and poultry manure) during 120 days experimental time period (b) DOC contents in unamended and amended soils (biochar, pressmud, farm manure, compost, sewage sludge, and poultry manure) during 120 days experimental time period.

all the BW amendments showed a significant ($P \leq 0.05$) decrease in TOC compared to that of the BC-amended soil. On the other hand, no significant changes were observed in the TOC of the control (un-amended soil) throughout the experimental time period (1–120 days). All the amendments (BC and various BW amendments) had a positive and significant ($P \leq 0.05$) effect on TOC, OC, and RC contents, which are presented in Fig 5b. Similarly, a strong, positive relationship ($P \leq 0.05$) was observed between TOC and other forms of carbons (OC and RC). TOC, OC, and RC contents were significantly increased with the addition of all amendments with maximum contents for BC (TOC: 262.24%, OC: 214% and RC: 405.98%) compared to that for the control (TOC: 3.88 ± 0.81 mg-C g^{-1} soil, OC: 2.90 ± 0.15 mg-C g^{-1} soil and RC: 0.97 ± 0.07 mg-C g^{-1} soil). It was also observed that BC had a highly significant ($P \leq 0.05$) effect on the TOC, OC, and RC contents compared to all BW amendments, except in the case of OC for PrM at the termination of the experiment. The relationships between the incubation study and pot experiment, with respect to total organic carbon and recalcitrant carbon, are presented in Fig. S1. Although both total and recalcitrant carbon contents were higher in the incubation study, the increasing pattern was almost similar to the pot experiment, indicating that the rate of decomposition was fast in the pot experiment compared with incubation.

Discussion

In the case of BC, the higher values of pH and CEC may be attributed to the highly stable organic carbon (OC) content of BC, which provides more surface functional groups and a large surface area due to micropores and aromatic ring structures, which are the main causes of the high pH and CEC values, respectively (Jien & Wang, 2013; Li et al., 2013; Chintala et al., 2014; Gai et al., 2014; Obia et al., 2015). On the other hand, the high pH values of FM- and SS-amended soils could be attributed to the dissolved organic compounds (DOCs) with huge amounts of organic acids, due to the high rate of decomposition of organic matter (OM) (Riffaldi et al., 1998; Stumpe & Marschner, 2010; Liu et al., 2013). These results indicate that BC is composed of a highly persistent form of organic-C having strong carbon–carbon bonds (Singh et al., 2012). In addition, BC has a remarkably lower decomposition rate compared with various BW amendments having the same inceptive organic carbon (OC) contents (Lu et al., 2014; Jiang et al., 2016). Zhang et al. (2012) studied the change in the soil C-stock through the addition of BW amendments. Various BW amendments, such as manures, sewage sludge, and pressmud, contained large amounts of DOCs that were easily mineralized and converted into CO_2 and other inorganic forms of carbon, resulting in a reduction of the soil-OC content (Walker et al., 2003; Nett et al., 2010; Romanyà et al., 2012).

The results obtained from this study indicate that addition of BC to soil significantly improved the wheat yield, plant biomass, and various physiological characteristics. Previous studies have shown an increase in the growth and yield of rapeseed, wheat, rice, canola, sweet potato, and maize with the addition of BC (Zhang et al., 2012; Liu et al., 2014; Martinsen et al., 2014; Tammeorg et al., 2014; Ahmed & Schoenau, 2015). In the present study, soils that were amended with carbonized wheat straw and olive tree pruning had a low observed plant

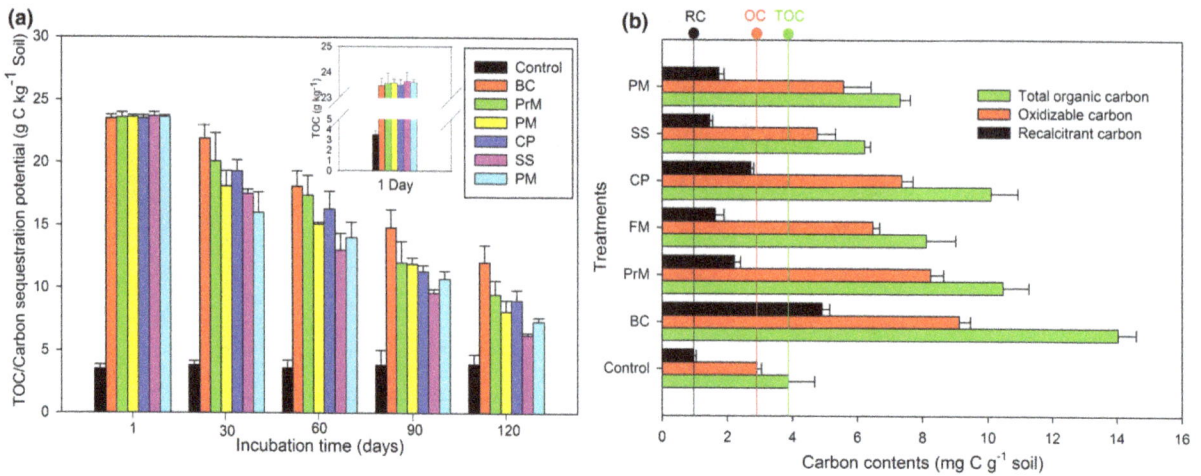

Fig. 5 (a) Potential carbon sequestration of biochar and various biowaste amendments (pressmud, farm manure, compost, sewage sludge, and poultry manure) compared with unamended soil during 120 days experimental time period. (b) Total, oxidizable, and recalcitrant carbon contents of biochar and various biowaste-amended soils at the termination of experiment period (120 days). The black, red, and green lines are representing the controlled values for recalcitrant carbon, oxidizable carbon, and total organic carbon, respectively.

Table 4 Mean $\delta^{13}C$ signatures of total soil carbon in both incubation and pot experiments

Treatments	Incubation experiment ($\delta^{13}C$ ‰)	Pot experiment
Control	−16.67 ± 0.11	−23.41 ± 0.13
BC	−27.49 ± 0.09	−25.82 ± 0.17
PrM	−14.32 ± 0.16	−16.69 ± 0.11
FM	−21.74 ± 0.12	−20.98 ± 0.19
CP	−22.62 ± 0.15	−22.08 ± 0.17
SS	−20.22 ± 0.14	−19.34 ± 0.14
PM	−21.53 ± 0.11	−20.74 ± 0.18

Control (unamended) and amended soil (biochar and various biowaste amendments); ($n = 3$).

biomass and grain yield compared with those that were amended with the application of BC (Antonio *et al.*, 2013). The high value of biomass and greater grain yield of BC-amended soils could be attributed to the various mechanisms that are responsible for the change in soil physicochemical and biological properties (Lehmann *et al.*, 2011; Herath *et al.*, 2013; Jien & Wang, 2013; Mukherjee *et al.*, 2014) and the phyto-availability of numerous macronutrients, that is, N, P, and K.

Soil and plant (root, shoot and grain) N contents increased significantly ($P \leq 0.05$) with the application of BC and various BW amendments compared with that of the control (Fig. 1a). Similar results were observed by Saarnio *et al.* (2013) on the availability and uptake of N when BC spruce chips were applied to soil cultivated with Phleum pretense plants. Khan *et al.* (2014)

described the constructive relationship between the plant uptake of N and the higher available content of N (NO₃–N and NH₄–N) in BC-amended soil compared with the control. In addition, Saarnio *et al.* (2013) also argued that the higher uptake of N in BC-amended soil could be attributed to the low availability of N to microbes due to the retention of N in BC-amended soil, which includes numerous mechanisms such as the adsorption of NH₃–N and NO₃–N onto the BC surface and ion-exchange reactions.

Unlike N, available and plant-uptaken P contents were nonsignificant in BC-amended soil (Fig. 1b). However, the maximum available and plant-uptaken P contents found with CP, PrM, and PM were significantly higher ($P \leq 0.05$) than that of the control. This decreased P content in plant parts could be attributed to the adsorption of PO₄–P onto BC partials. A few previous studies have shown alternative results, that is, in the BC-amended soils, the release of PO₄–P was higher than that of the control, and this could be available to the plants (Hale *et al.*, 2013; Khan *et al.*, 2014).

Among the carbon sources, the maximum concentration of available K was recorded with BC, which was significantly higher compared to that with the control, followed by FM and CP (Fig. 1c). Similarly, the K content of shoots was greater compared with the control, but the K content of roots and grains was observed as being nonsignificant. Overall, BC, FM, and CP significantly ($P > 0.05$) affected the availability of K as well as its accumulation and uptake by wheat. The relationship between organic carbon and soil macronutrients (NPK) is presented in Fig. 1d. The BC differed significantly in its macronutrient availability and uptakes (Khan *et al.*, 2014). The release of macronutrients with BC could also probably be attributed to various factors, that is, feedstock, pyrolytic conditions, and biochar pH (Nett *et al.*, 2010; Kim *et al.*, 2013; Mukherjee *et al.*, 2014; Ahmed & Schoenau, 2015).

The findings of this study highlight that BC results in manifold stabilization of soil-C and mitigation of CO_2 emissions, compared with various BW amendments, while maintaining the C-mineralization rate approximately equal to that of the control. In this study, the results of positive priming effects of BC relative to the control over the short–medium term are entirely consistent with previous investigations of Luo *et al.* (2011) and others (Farrell *et al.*, 2013; Singh & Cowie, 2014). The priming effects (positive and negative) indicate ancillary mineralization of native soil-C or stabilization with the addition of BC and various BW amendments (Singh & Cowie, 2014; Hernandez-soriano *et al.*, 2016). However, in the case of BC, large negative priming effects were observed relative to various BW amendments [PrM: -10.22 CO_2–C (mg g^{-1} soil-C), FM: -18.17 CO_2–C (mg g^{-1} soil-C), CP: -13.07 CO_2–C (mg g^{-1} soil-C), SS: -23.56 CO_2–C (mg g^{-1} soil-C), and PM: -12.14 CO_2–C (mg g^{-1} soil-C)]. This negative priming effect of BC was large during days 1–5 [-3.24 to -7.18 CO_2–C (mg g^{-1} soil-C)], then slowly reduced over time (Fig. 3). These negative priming effects indicate that BC is highly stable and persistent in soil systems having long-term carbon sequestration potential compared to the other BW amendments (Zimmerman *et al.*, 2011; Tarquis *et al.*, 2014). Based on the C-mineralization/CO_2 emission rate, labile-C, and stability, treatments can be ranked in the following order: BC < PrM < CP < SS < FM < PM.

Although the BC used in the present study maintained or slightly increased cumulative CO_2 emission compared to the control soil, it was still significantly lower than those of various biowaste amendments. Hernandez-soriano *et al.* (2016) described that this slight increase in CO_2 emission in BC-amended soil may be due to the unexceptional existence of mineralizable carbon content in BC. Furthermore, the initially high respiration rate in BC-amended soil may be attributable to the release of surface-bound CO_2 and in precipitated inorganic carbon on BC particles in the form of carbonate during the production of BC (Méndez *et al.*, 2013). Some previous studies also suggest that BC may also stabilize native soil organic carbon by forming organo-mineral interactions as well as through the sorption of DOC on to the surface of BC and within the pore spaces due to the exceptionally high surface area and void fraction (Keith *et al.*, 2011; Lehmann *et al.*, 2011; Zimmerman *et al.*, 2011; Singh & Cowie, 2014).

In contrast to the various BW amendments, DOC was significantly decreased by the addition of BC. The change in aboriginal microbial processes and the decrease in bioavailability were due to the high adsorption affinity of BC for organic compounds and may influence the DOC content of soil (Song *et al.*, 2012; Suresh *et al.*, 2012; Zhou *et al.*, 2013; Shan *et al.*, 2015). However, less MBC in the presence of BC indicates that microbial growth was restricted as a result of low organic carbon availability, which contributed to the decrease in DOC content (Singh & Cowie, 2014). However, previous studies have explained that the soil microbial activities might be an indicator of the influential role of BC in C-metabolism (Lehmann *et al.*, 2011; Farrell *et al.*, 2013). On the other hand, some studies suggest that the addition of BC to soil increases the aromaticity in the form of humic-like substances in DOC (Smebye *et al.*, 2016)

Similarly, many previous studies are consistent with our findings describing that BC could inhibit many enzymatic activities and reduce MBC in soil (Singh & Cowie, 2014; Shan *et al.*, 2015). This inhibition of microbial activity with the addition of BC may be attributable to the variability of soil physicochemical characteristics that are induced by BC (organo-mineral interactions), that is, adsorption of soluble compounds (organic and inorganic) onto the BC surface and change in water retention and microaggregates (Jin *et al.*, 2013; Hernandez-soriano *et al.*, 2016).

Our results highlight the potential increase in recalcitrant-C as a result of the addition of BC to soil rather than various BW amendments. Similarly, Lehmann *et al.* (2015) have also described that BC has higher recalcitrant-C content than fresh biowaste amendments and uncharred biomass, which are highly stable and not easily decomposable (Singh & Cowie, 2014). Furthermore, the complex structure (aromaticity) of biochar plays an important role in its long-term stay in the soil system. However, the stability of biochar may also be attributed to its low accessibility (Czimczik & Masiello, 2007) due to the existence of microaggregates, which are the indicators of organo-mineral interactions, in soil (Hernandez-soriano *et al.*, 2016). Similarly, Ventura *et al.* (2015) used isotopic ($\delta^{13}CO_2$) measurements to assess the stability and carbon sequestration potential of biochar in two European soils (Italy and UK). They observed less mineralization and a significant increase in soil organic carbon with the addition of biochar in both sites.

In a terrestrial ecosystem, soils act as a major sink for atmospheric carbon and can be helpful for storing atmospheric carbon for an indefinite period (Paustian *et al.*, 2016). Exogenous addition of stable carbon sources, that is, biochar and compost, can significantly contribute to the enhancement of soil C-sequestration and may reduce net CO_2 emission/C-mineralization (Lehmann et al., 2015; Ryals et al., 2015). Recent investigations have described that the rate of biochar decomposition is extremely low compared with various conventional organic amendments (fresh manures, sewage sludge, plant residues, etc.). Generally, biochar-C mineralizes several times slower than uncarbonized biomass, with mean retention times (MRT) 10- to 100-fold greater than those of unamended soil (Lehmann et al., 2015). Consequently, a large proportion of additive C from biomass-derived black carbon (biochar) can stay in soil systems for the long term, up to hundreds of years, due to more stable and recalcitrant-C. However, the retention time for sequestration of biochar-C varies depending on various biochar-soil factors such as feedstocks, pyrolytic conditions, and soil physicochemical characteristics (Schmidt et al., 2011).

Our findings describe that although exogenous addition of different sorts of organic amendments may produce varying effects on C-mineralization, microbial activity, and soil C-stocks (labile-C or recalcitrant-C etc.), biochar has a significantly higher influence on sequestration of carbon in soil as compared with various conventional amendments. Recently, the addition of biochar to soil has progressively increased as a potential tool to mitigate global warming (Shan et al., 2015).

In conclusion, the current study presents, for the first time, the priming effect of biochar-C-mineralization with respect to various BW amendments using a stable isotope ($\delta^{13}C$) approach and the possible impact of biochar-C-mineralization on C-sequestration and the labile organic carbon pool. During incubation, BC showed a negative priming effect, indicating a potentially higher carbon sequestration ability compared with those of various other BW amendments. However, recalcitrant-C and incubation time can also significantly affect the route and intensity of the priming effect. Overall, the results of this study suggest that the application of BC reduces C-mineralization compared with various BW amendments through soil carbon stabilization due to the presence of recalcitrant biochar-C and sorption of labile organic carbon onto BC particles, which results in less decomposition and significant increase in sequestration of carbon in soil over the long term (Singh & Cowie, 2014; Hernandez-soriano et al., 2016). In this circumstance, our results have important implications for an understanding of the priming effects of biochar-C-mineralization relative to various BW amendments. However, further studies are needed to evaluate the exact priming mechanism of biochar-C-mineralization in multisource mineralization systems using C-isotopic fractionation and a compound-specific stable isotope approach with a combination of spectro-microscopic techniques.

Acknowledgements

The authors greatly acknowledged the National Basic Research Program of China (973 Program, 2014CB238903) and the National Natural Science Foundation of China (No. 41373110) for financial support for this study. The Chinese Academy of Science, China (CAS), and The World Academy of Science, Italy (TWAS), are also greatly acknowledged for providing the CAS-TWAS President's fellowship (CAS-TWAS No. 2014-179). We also greatly appreciate the thoughtful comments and valuable suggestions from anonymous reviewers for the improvement of this manuscript.

References

Ahmed HP, Schoenau JJ (2015) Effects of biochar on yield, nutrient recovery, and soil properties in a canola (Brassica napus L)-wheat (Triticum aestivum L) rotation grown under controlled environmental conditions. BioEnergy Research, 8, 1183–1196.

Antonio J, Salazar P, Barron V et al. (2013) Enhanced wheat yield by biochar addition under different mineral fertilization levels. Agronomy for Sustainable Development, 33, 475–484.

Batjes NH (1996) Total carbon and nitrogen in the soils of the world. European Journal of Soil Science, 47, 151–163.

Benbi DK, Brar K, Toor AS et al. (2015) Total and labile pools of soil organic carbon in cultivated and undisturbed soils in northern India. Geoderma, 237, 149–158.

Cardon ZG, Hungate BA, Cambardella CA et al. (2001) Contrasting effects of elevated CO_2 on old and new soil carbon pools. Soil Biology and Biochemistry, 33, 365–373.

Chapman SJ (1997) Carbon substrate mineralization and sulphur limitation. Soil Biology and Biochemistry, 29, 115–122.

Chintala R, Mollinedo J, Schumacher TE et al. (2014) Effect of biochar on chemical properties of acidic soil. Archives of Agronomy and Soil Science, 60, 393–404.

Crombie K, Masek O (2015) Pyrolysis biochar systems, balance between bioenergy and carbon sequestration. GCB Bioenergy, 7, 349–361.

Czimczik CI, Masiello CA (2007) Controls on black carbon storage in soils. Global Biogeochemical Cycles, 21, 1–8.

Eykelbosh AJ, Johnson MS, Couto EG (2015) Biochar decreases dissolved organic carbon but not nitrate leaching in relation to vinasse application in a Brazilian sugarcane soil. Journal of Environmental Management, 149, 9–16.

Farrell M, Kuhn TK, Macdonald LM et al. (2013) Microbial utilisation of biochar-derived carbon. Science of the Total Environment, 465, 288–297.

Fontaine S, Bardoux G, Abbadie L et al. (2004) Carbon input to soil may decrease soil carbon content. Ecology Letters, 7, 314–320.

Forbes MS, Raison RJ, Skjemstad JO (2006) Formation, transformation and transport of black carbon (charcoal) in terrestrial and aquatic ecosystems. Science of the Total Environment, 370, 190–206.

Gai X, Wang H, Liu J et al. (2014) Effects of feedstock and pyrolysis temperature on biochar adsorption of ammonium and nitrate. PLoS One, 9, 1–19.

Gill RA, Polley HW, Johnson HB et al. (2002) Nonlinear grassland responses to past and future atmospheric CO_2. Nature, 417, 279–282.

Hale SE, Alling V, Martinsen V et al. (2013) The sorption and desorption of phosphate-P, ammonium-N and nitrate-N in cacao shell and corn cob biochars. Chemosphere, 9, 1612–1619.

Harris D, Porter LK, Paul EA (1997) Continuous flow isotope ratio mass spectrometry of carbon dioxide trapped as strontium carbonate. Communications in Soil Science & Plant Analysis, 28, 747–757.

Hay CC, Morrow E, Kopp RE et al. (2015) Probabilistic reanalysis of twentieth-century sea-level rise. Nature, 517, 481–484.

Herath HMSK, Camps-Arbestain M, Hedley M (2013) Effect of biochar on soil physical properties in two contrasting soils: an Alfisol and an Andisol. Geoderma, 209–210, 188–197.

Hernandez-soriano MC, Kerré B, Kopittke PM et al. (2016) Biochar affects carbon composition and stability in soil: a combined spectroscopy-microscopy study. Scientific Reports, 6, 1–13.

Houghton RA, Byers B, Nassikas AA (2015) A role for tropical forests in stabilizing atmospheric CO_2. Nature Climate Change, 5, 1022–1023.

IPCC (2016) Assessment Reports I–IV. Available at: (https://www.ipcc.ch/publications_and_data/publications_and_data_reports.shtml#1) FAR, 1990, SAR 1995, TAR 2001, AR4 2007 (accessed 19 June 2016).

Jackson ML (1967) Soil Chemical Analysis. Prentice Hall International Inc., London, UK.

Jiang X, Denef K, Stewart CE et al. (2016) Controls and dynamics of biochar decomposition and soil microbial abundance, composition, and carbon use efficiency during long-term biochar-amended soil incubations. Biology and Fertility of Soils, 52, 1–14.

Jien SH, Wang CS (2013) Effects of biochar on soil properties and erosion potential in a highly weathered soil. Catena, 110, 225–233.

Jin L, Son Y, Yoon TK et al. (2013) High concentrations of single-walled carbon nanotubes lower soil enzyme activity and microbial biomass. Ecotoxicology and Environmental Safety, 88, 9–15.

Karlen DL, Cambardella CA (1996) Conservation strategies for improving soil quality and organic matter storage. In: Structure and Organic Matter Storage in Agricultural Soils (eds Carter R, Stewart BA), pp. 395–420. CRC Press, Boca Raton, FL.

Keith A, Singh B, Singh BP (2011) Interactive priming of biochar and labile organic matter mineralization in a smectite-rich soil. Environmental Science & Technology, 45, 9611–9618.

Khan S, Reid BJ, Li G et al. (2014) Application of biochar to soil reduces cancer risk via rice consumption: a case study in Miaoqian village, Longyan, China. Environment International, 68, 154–161.

Kim P, Johnson AM, Essington ME et al. (2013) Effect of pH on surface characteristics of switchgrass-derived biochars produced by fast pyrolysis. Chemosphere, 90, 2623–2630.

Lehmann J, Gaunt J, Rondon M (2006) Bio-char sequestration in terrestrial ecosystems – a review. Mitigation and Adaptation Strategies for Global Change, 11, 403–427.

Lehmann J, Rillig MC, Thies J et al. (2011) Biochar effects on soil biota – a review. Soil Biology and Biochemistry, 43, 1812–1836.

Lehmann J, Joseph S et al. (2015) Stability of biochar in soil. In: Biochar for Environmental Management: Science, Technology and Implementation (eds Lehmann J, Joseph S), pp. 235–282. Taylor and Francis, London, UK.

Li X, Shen Q, Zhang D et al. (2013) Functional groups determine biochar properties (pH and EC) as studied by two-dimensional ^{13}C NMR correlation spectroscopy. PLoS One, 8, 1–7.

Liepert BG (2010) The physical concept of climate forcing. Wiley Interdisciplinary Reviews: Climate Change, 1, 786–802.

Liu E, Yan C, Mei X et al. (2013) Long-term effect of manure and fertilizer on soil organic carbon pools in dryland farming in northwest China. PLoS One, 8, 1–9.

Liu Z, Chen X, Jing Y et al. (2014) Effects of biochar amendment on rapeseed and sweet potato yields and water stable aggregate in upland red soil. Catena, 123, 45–51.

Lu W, Ding W, Zhang J et al. (2014) Biochar suppressed the decomposition of organic carbon in a cultivated sandy loam soil: a negative priming effect. Soil Biology and Biochemistry, 76, 12–21.

Luo Y, Durenkamp M, Nobili MD et al. (2011) Short term soil priming effects and the mineralisation of biochar following its incorporation to soils of different pH. Soil Biology and Biochemistry, 43, 2304–2314.

Martinsen V, Mulder J, Shitumbanuma V et al. (2014) Farmer-led maize biochar trials: effect on crop yield and soil nutrients under conservation farming. Journal of Plant Nutrition and Soil Science, 177, 681–695.

Meehl GA, Washington WM, Collins WD et al. (2005) How much more global warming and sea level rise? Science, 307, 1769–1772.

Méndez A, Tarquis AM, Saa-Requejo A et al. (2013) Influence of pyrolysis temperature on composted sewage sludge biochar priming effect in a loamy soil. Chemosphere, 93, 668–676.

Miché L, Balandreau J (2001) Effects of rice seed surface sterilization with hypochlorite on inoculated Burkholderia vietnamiensis. Applied and Environmental Microbiology, 67, 3046–3052.

Mukherjee A, Lal R, Zimmerman AR (2014) Effects of biochar and other amendments on the physical properties and greenhouse gas emissions of an artificially degraded soil. Science of the Total Environment, 487, 26–36.

Nett L, Averesch S, Ruppel S et al. (2010) Does long-term farmyard manure fertilization affect short-term nitrogen mineralization from farmyard manure? Biology and Fertility of Soils, 46, 159–167.

Novak JM, Frederick JR, Bauer PJ et al. (2009) Rebuilding soil organic matter contents in sandy Coastal Plain soils using conservation tillage management systems. Soil Science Society of American Journal, 73, 622–629.

Obia A, Cornelissen G, Mulder J et al. (2015) Effect of soil pH Increase by biochar on NO, N_2O and N_2 production during denitrification in acid soils. PLoS One, 10, 1–19.

Oppenheimer M (1998) Global warming and the stability of the West Antarctic Ice Sheet. Nature, 393, 325–332.

Paustian K, Andrén O, Janzen H et al. (1997) Agricultural soil as a C sink to offset CO_2 emissions. Soil Use Manage, 13, 230–244.

Paustian K, Lehmann J, Ogle S et al. (2016) Climate-smart soils. Nature, 532, 49–57.

Post WM, Emanuel WR, Zinke PJ et al. (1982) Soil carbon pool and life zones. Nature, 298, 156–159.

Raper SCB, Braithwaite RJ (2006) Low sea level rise projections from mountain glaciers and icecaps under global warming. Nature, 439, 311–313.

Riffaldi R, Levi-Minzi R, Saviozzi A et al. (1998) Adsorption on soil of dissolved organic carbon from farmyard manure. Agriculture, Ecosystems & Environment, 69, 113–119.

Rogelj J, Luderer G, Pietzcker RC et al. (2015) Energy system transformations for limiting end-of-century warming to below 1.5 °C. Nature Climate Change, 5, 519–538.

Romanyà J, Arco N, Solà-Morales I et al. (2012) Carbon and nitrogen stocks and nitrogen mineralization in organically managed soils amended with composted manures. Journal of Environmental Quality, 41, 1337–1347.

Ryals R, Hartman MD, Parton WJ et al. (2015) Long-term climate change mitigation potential with organic matter management on grasslands. Ecological Applications, 25, 531–545.

Saarnio S, Heimonen K, Kettunen R (2013) Biochar addition indirectly affects N_2O emissions via soil moisture and plant N uptake. Soil Biology and Biochemistry, 58, 99–106.

Sagrilo E, Jeffery S, Hoffland E et al. (2015) Emission of CO_2 from biochar-amended soils and implications for soil organic carbon. GCB Bioenergy, 7, 1294–1304.

Schlesinger WH (1995) An overview of the carbon cycle. In: Soils and Global Change (eds Lal R, Kimble J, Levine E, Stewart BA), pp. 9–27. CRC Press, Boca Raton, FL.

Schmidt MWI, Torn MS, Abiven S et al. (2011) Persistence of soil organic matter as an ecosystem property. Nature, 478, 49–56.

Scott V, Haszeldine RS, Tett SFB et al. (2015) Fossil fuels in a trillion tonne world. Nature Climate Change, 5, 419–423.

Shan J, Ji R, Yu Y et al. (2015) Biochar, activated carbon, and carbon nanotubes have different effects on fate of 14 C-catechol and microbial community in soil. Scientific Reports, 5, 1–11.

Singh BP, Cowie AL (2014) Long-term influence of biochar on native organic carbon mineralisation in a low-carbon clayey soil. Scientific Reports, 4, 1–9.

Singh BP, Cowie AL, Smernik RJ (2012) Biochar carbon stability in a clayey soil as a function of feedstock and pyrolysis temperature. Environmental Science & Technology, 46, 11770–11778.

Slangen A, Church J (2016) Burning fossil fuels is responsible for most sea-level rise since 1970. Available at: http://theconversation.com/burning-fossil-fuels-is-responsible-for-most-sea-level-rise-since-1970-57286 (accessed 19 June 2016).

Smebye A, Alling V, Vogt RD et al. (2016) Biochar amendment to soil changes dissolved organic matter content and composition. Chemosphere, 142, 100–105.

Song Y, Wang F, Bian Y et al. (2012) Bioavailability assessment of hexachlorobenzene in soil as affected by wheat straw biochar. Journal of Hazardous Materials, 217–218, 391–397.

Sparling GP, Gupta VVSR, Zhu C (1993) Release of ninhydrin-reactive compounds during fumigation of soil to estimate microbial C and N. Soil Biology and Biochemistry, 25, 1803–1805.

Stumpe B, Marschner B (2010) Dissolved organic carbon from sewage sludge and manure can affect estrogen sorption and mineralization in soils. Environmental Pollution, 158, 148–154.

Suresh S, Srivastava VC, Mishra IM (2012) Adsorption of catechol, resorcinol, hydroquinone, and their derivatives: a review. International Journal of Energy and Environmental Engineering, 3, 32.

Tammeorg P, Simojoki A, Mäkelä P et al. (2014) Short-term effects of biochar on soil properties and wheat yield formation with meat bone meal and inorganic fertiliser on a boreal loamy sand. Agriculture, Ecosystems & Environment, 191, 108–116.

Tarquis M, Platonov A, Matulka A et al. (2014) Application of multifractal analysis to the study of SAR features and oil spills on the ocean surface. Nonlinear Processes in Geophysics, 21, 439–450.

Tokarska KB, Gillett NP, Weaver AJ et al. (2016) The climate response to five trillion tonnes of carbon. Nature Climate Change, 6, 851–855.

UNFCCC (1992) First steps to a safer future: Introducing the United Nations Framework Convention on Climate Change. Available at: http://unfccc.int/essential_background/convention/items/6036.php (accessed 19 March 2016).

UNFCCC (1998) Kyoto Protocol to the United Nations Framework Convention on Climate Change. Available at: http://unfccc.int/kyoto_protocol/items/2830.php (accessed 19 March 2016).

Vance ED, Brookes PC, Jenkinson DS (1987) An extraction method for measuring microbial biomass C. *Soil Biology and Biochemistry*, **19**, 703–707.

Ventura M, Alberti G, Viger M *et al.* (2015) Biochar mineralization and priming effect on SOM decomposition in two European short rotation coppices. *GCB Bioenergy*, **7**, 1150–1160.

Walker DJ, Clemente R, Roig A *et al.* (2003) The effects of soil amendments on heavy metal bioavailability in two contaminated Mediterranean soils. *Environmental Pollution*, **122**, 303–312.

Wardle DA, Nilsson M, Zackrisson O (2008) Fire-derived charcoal causes loss of forest humus. *Science*, **320**, 629.

West TO, Six J (2007) Considering the influence of sequestration duration and carbon saturation on estimates of soil carbon capacity. *Climatic Change*, **80**, 25–41.

Wigley TM (2005) The climate change commitment. *Science*, **307**, 1766–1769.

Woolf D, Amonette JE, Street-Perrott FA *et al.* (2010) Sustainable biochar to mitigate global climate change. *Nature Communications*, **1**, 56.

Yousaf B, Liu G, Wang R *et al.* (2016) Investigating the potential influence of biochar and traditional organic amendments on the bioavailability and transfer of Cd in the soil–plant system. *Environmental Earth Sciences*, **75**, 1–10.

Zavalloni C, Alberti G, Biasiol S *et al.* (2011) Microbial mineralization of biochar and wheat straw mixture in soil: a short-term study. *Applied Soil Ecology*, **50**, 45–51.

Zhang W, Xu M, Wang X *et al.* (2012) Effects of organic amendments on soil carbon sequestration in paddy fields of subtropical China. *Journal of Soils and Sediments*, **12**, 457–470.

Zhou J, Wang J-J, Baudon A *et al.* (2013) Improved florescence excitation-emission matrix regional integration to quantify spectra for florescent dissolved organic matter. *Journal of Environmental Quality*, **42**, 925–930.

Zimmerman AR, Gao B, Ahn M-Y (2011) Positive and negative carbon mineralization priming effects among a variety of biochar-amended soils. *Soil Biology and Biochemistry*, **43**, 1169–1179.

Permissions

All chapters in this book were first published in GCB BIOENERGY, by John Wiley & Sons Ltd.; hereby published with permission under the Creative Commons Attribution License or equivalent. Every chapter published in this book has been scrutinized by our experts. Their significance has been extensively debated. The topics covered herein carry significant findings which will fuel the growth of the discipline. They may even be implemented as practical applications or may be referred to as a beginning point for another development.

The contributors of this book come from diverse backgrounds, making this book a truly international effort. This book will bring forth new frontiers with its revolutionizing research information and detailed analysis of the nascent developments around the world.

We would like to thank all the contributing authors for lending their expertise to make the book truly unique. They have played a crucial role in the development of this book. Without their invaluable contributions this book wouldn't have been possible. They have made vital efforts to compile up to date information on the varied aspects of this subject to make this book a valuable addition to the collection of many professionals and students.

This book was conceptualized with the vision of imparting up-to-date information and advanced data in this field. To ensure the same, a matchless editorial board was set up. Every individual on the board went through rigorous rounds of assessment to prove their worth. After which they invested a large part of their time researching and compiling the most relevant data for our readers.

The editorial board has been involved in producing this book since its inception. They have spent rigorous hours researching and exploring the diverse topics which have resulted in the successful publishing of this book. They have passed on their knowledge of decades through this book. To expedite this challenging task, the publisher supported the team at every step. A small team of assistant editors was also appointed to further simplify the editing procedure and attain best results for the readers.

Apart from the editorial board, the designing team has also invested a significant amount of their time in understanding the subject and creating the most relevant covers. They scrutinized every image to scout for the most suitable representation of the subject and create an appropriate cover for the book.

The publishing team has been an ardent support to the editorial, designing and production team. Their endless efforts to recruit the best for this project, has resulted in the accomplishment of this book. They are a veteran in the field of academics and their pool of knowledge is as vast as their experience in printing. Their expertise and guidance has proved useful at every step. Their uncompromising quality standards have made this book an exceptional effort. Their encouragement from time to time has been an inspiration for everyone.

The publisher and the editorial board hope that this book will prove to be a valuable piece of knowledge for researchers, students, practitioners and scholars across the globe.

List of Contributors

Erik O. Ahlgren, Martin Börjesson Hagberg and Maria Grahn
Department of Energy and Environment, Chalmers University of Technology, G€oteborg, Sweden

Sylvestre Njakou Djomo, Marie Trydeman Knudsen, Ranjan Parajuli, Mikael Skou Andersen and John Erik Hermansen
Department of Agroecology, Aarhus University, Blichers Alle 20, DK-8830 Tjele, Denmark

Morten Ambye-Jensen
Department of Engineering, Aarhus University, Hangovej 2, DK-8200 Aarhus Denmark

Gerfried Jungmeier
Joanneum Research Forschungsgesellschaft mbH, Elisabethstraße 18/II, 8010 Graz, Austria

Benoît Gabrielle
EcoSys Research Unit, AgroParisTech, INRA, F-78850 Thiverval-Grignon, France

Jonathan J. Ojeda
National Research Council (CONICET), Oro Verde, Argentina
Facultad de Ciencias Agropecuarias, Universidad Nacional de Entre Rıos, Ruta 11, km 10.5 (3101), Oro Verde

Jeffrey J. Volenec and Sylvie M. Brouder
Department of Agronomy, Purdue University, West Lafayette, IN, USA

Octavio P. Caviglia
National Research Council (CONICET), Oro Verde, Argentina
Facultad de Ciencias Agropecuarias, Universidad Nacional de Entre Rıos, Ruta 11, km 10.5 (3101), Oro Verde, Entre Rıos, Argentina
Instituto Nacional de Tecnologıa Agropecuaria (INTA), Estacion Experimental Agropecuaria Parana, Ruta 11, km 12.5 (3101), Oro Verde, Entre Rıos, Argentina

Mónica G. Agnusdei
Instituto Nacional de Tecnologıa Agropecuaria (INTA), Estacion Experimental Agropecuaria Balcarce, Ruta 226, km 73.5 (7620), Balcarce, Buenos Aires, Argentina

Dener M. S. Oliveira
Natural Resource Ecology Laboratory, Colorado State University, Fort Collins, CO, 80523-1499, USA
Department of Soil Science, Luiz de Queiroz College of Agriculture, University of São Paulo, Piracicaba, 13418-900, Brazil

Stephen Williams
Natural Resource Ecology Laboratory, Colorado State University, Fort Collins, CO, 80523-1499, USA

Carlos E. P. Cerri
Department of Soil Science, Luiz de Queiroz College of Agriculture, University of São Paulo, Piracicaba, 13418-900, Brazil

Keith Paustian
Natural Resource Ecology Laboratory, Colorado State University, Fort Collins, CO, 80523-1499, USA
Department of Soil and Crop Sciences, Colorado State University, Fort Collins, CO, 80523-1170, USA

Kaile I Tang
Centre for Crop Systems Analysis, Department of Plant Sciences, Wageningen University & Research, Wageningen, 6700 AK, The Netherlands
Department of Sustainable Crop Production, Universita Cattolica del Sacro Cuore, via Emilia Parmense 84, Piacenza, 29122, Italy

Paul C. Struik, Tjeerd-Jan Stomph and Xinyou Yin
Centre for Crop Systems Analysis, Department of Plant Sciences, Wageningen University & Research, Wageningen, 6700 AK, The Netherlands

Stefano Amaducci
Department of Sustainable Crop Production, Universita Cattolica del Sacro Cuore, via Emilia Parmense 84, Piacenza, 29122, Italy

Kari Väätäinen, Robert Prinz, Juha Laitila and Lauri Sikanen
Natural Resources Institute Finland (LUKE), Joensuu, Finland

Jukka Malinen
University of Eastern Finland (UEF), Joensuu, Finland

Niclas Ericsson, Åke Nordberg, Serina Ahlgren, Torun Hammar and Per-Anders Hansson
Department of Energy and Technology, Swedish University of Agricultural Sciences (SLU), SE 750 07, Uppsala, Sweden

Cecilia Sundberg
Department of Energy and Technology, Swedish University of Agricultural Sciences (SLU), SE 750 07, Uppsala, Sweden
KTH Royal Institute of Technology, School of Architecture and the Built Environment, Department of Sustainable Development, Environmental Science and Engineering, Unit of Industrial Ecology, Stockholm, Sweden
Division of Industrial Ecology, Department of Sustainable Development, Environmental Science and Engineering, KTH Royal Institute of Technology, SE-100 44, Stockholm, Sweden

Mengmeng Hao, Dong Jiang, Jingying Fu and Yaohuan Huang
Institute of Geographical Sciences and Natural Resources Research, Chinese Academy of Sciences, 11A Datun Road, Beijing 100101, China
College of Resources and Environment, University of Chinese Academy of Sciences, Beijing 100049, China

Jianhua Wang
State Key Laboratory of Simulation and Regulation of Water Cycle in River Basin, Department of Water Resources, China Institute of Hydropower & Water Resources Research, Beijing 100038, China

Anne L. Maddison, Anyela Camargo-Rodriguez, Ian M. Scott, Charlotte M. Jones, Sarah Hawkins, Alice Massey, John Clifton-Brown, Iain S. Donnison and Sarah J. Purdy
Institute of Biological, Environmental and Rural Sciences, Aberystwyth University, Plas Gogerddan SY23 3EB, UK

Dafydd M. O. Elias and Niall P. McNamara
Centre for Ecology and Hydrology, Lancaster Environment Centre, Library Avenue, Bailrigg, Lancaster LA1 4AP, UK

Melissa Wagner
School of Geographical Sciences and Urban Planning, Arizona State University, Tempe, AZ 85287-5302, USA

Meng Wang
School of Mathematical and Statistical Sciences, Arizona State University, Tempe, AZ, USA

Gonzalo Miguez-Macho
Universidade de Santiago de Compostela, Galicia, Spain

Jesse Miller
Department of Plant Biology, University of Illinois, Urbana, IL, USA

Andy Vanloocke
Department of Agronomy, Iowa State University, Ames, IA, USA

Justin E. Bagley
Climate and Ecosystems Science Division, Lawrence Berkeley National Laboratory, Berkeley, CA, USA

Carl J. Bernacchi
Department of Plant Biology, University of Illinois, Urbana, IL, USA
Global Change and Photosynthesis Research Unit, United States Department of Agriculture Agricultural Research Service, Urbana, IL 61801, USA

Matei Georgescu
School of Geographical Sciences and Urban Planning, Arizona State University, Tempe, AZ 85287-5302, USA, School of Mathematical and Statistical Sciences, Arizona State University, Tempe, AZ, USA

Tim Van Der Weijde
Wageningen UR Plant Breeding, Wageningen University and Research, 6700 AJ Wageningen, The Netherlands
Graduate School Experimental Plant Sciences, Wageningen University, Droevendaalsesteeg 1, 6708 PB Wageningen, The Netherlands

Laurie M. Huxley, Sarah Hawkins and Kerrie Farrar
Institute of Biological, Environmental & Rural Sciences (IBERS), Aberystwyth University, Plas Gogerddan, Aberystwyth, Ceredigion SY23 3EE, UK

Eben Haeser Sembiring, Oene Dolstra, Richard G. F. Visser and Luisa M. Trindade
Wageningen UR Plant Breeding, Wageningen University and Research, 6700 AJ Wageningen, The Netherlands

Yangzhou Xiang and Ying Guo
Guizhou Institute of Forest Inventory and Planning,
Guiyang 550003, China

Qi Deng
Key Laboratory of Aquatic Botany and Watershed
Ecology, Wuhan Botanical Garden, Chinese
Academy of Sciences, Wuhan 430074, China
Department of Biological Sciences, Tennessee State
University, Nashville, TN 37209, USA

Honglang Duan
Jiangxi Provincial Key Laboratory for Restoration of
Degraded Ecosystems & Watershed Ecohydrology,
Nanchang Institute of Technology, Nanchang
330099, China

Balal Yousaf and Guijian Liu
CAS-Key Laboratory of Crust-Mantle Materials
and the Environments, School of Earth and Space
Sciences, University of Science and Technology of
China, Hefei 230026, China
State Key Laboratory of Loess and Quaternary
Geology, Institute of Earth Environment, The Chinese
Academy of Sciences, Xi'an, Shaanxi 710075, China

Ruwei Wang, Qumber Abbas and Ruijia Liu
CAS-Key Laboratory of Crust-Mantle Materials
and the Environments, School of Earth and Space
Sciences, University of Science and Technology of
China, Hefei 230026, China

Muhammad Imtiaz
Microelement Research Center, College of Resources
and Environment, Huazhong Agricultural University,
Wuhan 430070, China

Index

www.ingramcontent.com/pod-product-compliance
Lightning Source LLC
Chambersburg PA
CBHW082028190326
41458CB00010B/3307